1st International Telecommunication Conference on Advanced Micro- and Nanoelectronic Systems and Technologies 2015

IOP Conference Series: Materials Science and Engineering
Volume 151

Moscow, Russia
22 – 23 December 2015

Editors:

V.S. Pershenkov
G.I. Zebrev

ISBN: 978-1-5108-3272-5
ISSN: 1757-8981

Printed from e-media with permission by:

Curran Associates, Inc.
57 Morehouse Lane
Red Hook, NY 12571

Some format issues inherent in the e-media version may also appear in this print version.

Printed by Curran Associates, Inc. (2016)

For permission requests, please contact the Institute of Physics at the address below.

Institute of Physics
Dirac House, Temple Back
Bristol BS1 6BE UK

Phone: 44 1 17 929 7481
Fax: 44 1 17 920 0979

techtracking@iop.org

Additional copies of this publication are available from:

Curran Associates, Inc.
57 Morehouse Lane
Red Hook, NY 12571 USA
Phone: 845-758-0400
Fax: 845-758-2633
Email: curran@proceedings.com
Web: www.proceedings.com

Table of contents

Volume 151

1st International Telecommunication Conference "Advanced Micro- and Nanoelectronic Systems and Technologies"22–23 December 2015, Moscow, Russia

Accepted papers received: 5 October 2016Published online: 17 November 2016

Preface

Papers

IOP Conf. Series: Materials Science and Engineering **151** (2016) 011001 doi:10.1088/1757-899X/151/1/011001

First International Telecommunication Conference on Advanced Micro- and Nanoelectronic Systems and Technologies

V S Pershenkov and G I Zebrev

National Research Nuclear University MEPhI (Moscow Engineering Physics Institute), Kashirskoe shosse 31, 115409 Moscow, Russian Federation

E-mail: VSPershenkov@mephi.ru, GIZebrev@mephi.ru

First International Telecommunication Conference on Advanced Micro- and Nanoelectronic Systems and Technologies took place from December 22nd to December 23rd 2015 at National Research Nuclear University MEPhI (Moscow Engineering Physics Institute). It was organized by the Micro- and Nanoelectronic Department of National Research Nuclear University MEPhI (the oldest Microelectronic department in Russia (USSR).

The Conference consists of one plenary and three oral and poster sessions. The plenary session presented the most significant results of Micro- and Nanoelectronic Department MEPHI and other participants. This Conference provides to participants an opportunity to discuss their scientific and technical results in a wide audience of leading industry experts.

The main topics of the Conference were:

- Semiconductor physics;
- Design of micro- and nanosystems;
- Manufacturing technology of micro- and nanosystems.

More than 150 specialists on Micro- and Nanoelectronics took part in the Conference. They represented leading Russian scientific research centres and universities (such as National Research Center "Kurchatov Institute", All-Russia Research Institute of Automatics (VNIIA), Bauman Moscow State Technical University, Moscow Aviation Institute (National Research University), Moscow Power Engineering Institute, Moscow State University of Mechanical Engineering (MAMI), D. Mendeleev University of Chemical Technology of Russia and many others) and universities from Armenia, Belarus, Ukraine, Germany, Colombia, Spain, South Korea, Republic of Uzbekistan.

We would like to thank all of the speakers, participants and organizing committee members for their contribution to the Conference.

IOP Conf. Series: Materials Science and Engineering **151** (2016) 011002 doi:10.1088/1757-899X/151/1/011002

Peer review statement

All papers published in this volume of *IOP Conference Series: Materials Science and Engineering* have been peer reviewed through processes administered by the proceedings Editors. Reviews were conducted by expert referees to the professional and scientific standards expected of a proceedings journal published by IOP Publishing.

IOP Conf. Series: Materials Science and Engineering **151** (2016) 012001 doi:10.1088/1757-899X/151/1/012001

Conversion model of radiation-induced interface-trap buildup and the some examples of its application

V S Pershenkov, A V Sogoyan and V A Telets

National Research Nuclear University MEPhI, (Moscow Engineering Physics Institute), Kashirskoe shosse 31, 115409 Moscow, Russian Federation

Corresponding author e-mail address: VSPershenkov@mephi.ru

Abstract. It is supposed that the rechargeable radiation-induced positive centers in the oxide must have energy levels within the Si forbidden gap. These assumptions served as the basis for the physical model of the interface-trap build-up and enhanced low dose rate sensitivity (ELDRS) in bipolar devices. As examples of using the proposed conversion model is considered for bipolar devices space application.

1. Introduction

In this paper, we describe the conversion model of interface-trap buildup which integrates conventional hydrogen and hydrogen-electron approaches. The experimental data confirming the validity of this type of combined model are presented. The energy level location of positive centers and its recharge mechanism using the reversibility of annealing mechanism is described. An account of the interaction of trapped positive charges and electrons from the substrate allows the development of the numerical model for the enhancement low dose rate sensitivity (ELDRS) effect in bipolar devices. The demonstration of using the conversion model for the numerical estimation of bipolar device behavior during a hypothetical space mission is presented.

2. Mechanism of interface-trap buildup

The radiation-induced interface-trap buildup is one of the major problems in the radiation physics of metal-oxide-semiconductor (MOS) structures. In the previous 30 years, two physical models with a description of the process for the radiation-induced interface-trap buildup were proposed. The first model was the hydrogen model (H-model) [1]. The second model was a conversion model where trapped holes are converted to interface traps by electron capture [2]. The last model is not well regarded today.

The hydrogen model proposed an active participation of various hydrogen species for an interface-trap generation. According to [1], the protons produced in bulk oxide by radiation move to the Si interface. When the protons reach the interface, they react, breaking the Si-H bonds, forming H_2 and a trivalent Si defect according to the reaction.

The hydrogen model successfully explains the field dependence of interface-trap buildup at constant and switching electric fields in the oxide. However, this concept, unfortunately, does not explain all known experimental data. The series of experimental results cannot be unambiguously interpreted by the traditional hydrogen model.

The second group of models is based on experimental data, and interface traps are formed by positive trapped charge annealing [2], when trapped holes are converted to interface traps. In [2], an injection of electrons into the oxide and a compensation of trapped holes led to interface-trap buildup. It allowed us to make an empirical assumption [2] that the process of recombining the positively charged defect with the electron can lead to the interface-trap formation ("conversion" model). The effect of the hydrogen species on interface-trap buildup was outside of the conversion mechanism. Meanwhile, even in early experiments [2,3], it was shown that the conversion process had a strong dependence on hydrogen concentration in the oxide.

3. Experiment

To check the hypothesis that interface-trap buildup is connected with the interaction between the positively charged hydrogen species in the oxide and electrons tunneling from the substrate, a special experiment was performed [4]. The result of this experiment is presented below.

The n-channel MOS transistors with a 30-nm gate oxide were irradiated by Cu-target X-rays sourced with a 1 krad(SiO2)/s dose rate to 3 Mrad (SiO2) total dose. For one hour after the end of irradiation, the devices were annealed in a molecular hydrogen atmosphere for 24 hours. The interface-trap buildup was registered during this annealing time. Four different tests were investigated. The different combinations of hydrogenous species and electrons present near the interface were varied in 4 tests. A maximum change of ΔV_{it} is observed in test 1, when both electrons and hydrogenous species are presented near the surface. In other cases, when there are no hydrogen species (test 3) or no electrons (test 2) or both near the interface (test 4), shift ΔV_{it} is essentially reduced.

The presented experimental data confirms the hypothesis that the presence of hydrogen is sufficient for effective interface-trap buildup. The interaction of hydrogenous species and electrons from the substrate is an important component of this process. In hydrogen-electron concept, both components - hydrogenous species and electrons - are factors that can restrict the rate of interface-trap buildup.

The presented approach corresponds to experimental data for the "conversion" concept and a large experimental basis for hydrogen models. It allows us to discuss the common hydrogen-electron (H-e) model of radiation-induced interface-trap buildup, which consolidates hydrogen and conversion concepts.

4. Thermoactivated tunneling mechanism

The post-irradiation relaxation of a positive oxide charge ("annealing") is traditionally considered as the superposition of two independent processes: a tunneling of electrons from the substrate and a thermal excitation of electrons from the oxide valence band. In [5], the effect of the reversibility of annealing was described, which could not be explained by previous models. The model [6] supposed that an interaction of the oxide defects and electrons from the substrate include the thermal excitation of the defect and tunneling of the electron. Simplistically speaking, the thermal atom fluctuation leads to the oscillation of energy levels. When the energy level of the defect in the oxide corresponds to the allowed electron level in Si, it is possible that the tunneling transitions the electron from the substrate to the defect.

The thermoactivated tunneling model describes the main features of the oxide charge relaxation, and it is the most important part of the interface-trap buildup mechanism presented above. In the framework of the hydrogen-electron model, the interface-trap formation at a long time interval is determined by electron tunneling from the substrate. On the other hand, the oxide defect can exhibit some characteristics of interface-traps if it is located close enough to the interface Si/SiO2 (less than 1 nm). Therefore, there is some contradiction between a long time electron transition to a defect during interface-trap formation and a short recharge time for the interface-trap after the buildup. The thermoactivated tunneling model eliminates this contradiction. The center located close to interface Si/SiO2 but with energy levels within the silicon forbidden gap (with large activation energy) is recharged during a long time interval. However, after electron capture and defect reconstruction, its energy level shifts and the defect acts as an interface-trap. For example, the typical transition time of the electron to the defect located at the depth 1 nm from the silicon interface and with an activation energy of 0.65 eV is greater than 10^6 s. However, the defect located at the same distance but with an activation energy of 0.1 eV has a recharge time less than 1 ms and acts as an interface-trap.

5. Conversion model

The low dose rate effect in bipolar transistors or the enhanced low dose rate sensitivity (ELDRS) effect is a more severe degradation of the bipolar structure current gain for a given total dose following a low dose rate. All aspects concerning this effect (mechanisms, modeling, and testing recent developments in ELDRS) are analyzed comprehensively in [7]. The ELDRS model in given the work is based on the hydrogen-electron conversion model. The motivation of this development is the creation of a model that is allowed to obtain a quantitative numerical estimation of radiation degradation of bipolar transistor current gain for arbitrary dose rate and temperature. Because the H-e model is based on the conversion of a radiation-induced positive trapped charge to interface traps, the model described below is called the ELDRS conversion model.

The key concept of the ELDRS conversion model is as follows. We suppose that there are two types of traps that are located in the oxide interface opposite the silicon forbidden gap: shallow traps with a short time of conversion responsible for the degradation at high dose rates, and deep traps that determine the excess base current at greater times of irradiation

As shown in [8], the degradation of the base current as a function of the dose rate (for irradiation time substantially more than 1 s) can be written as:

$$\Delta I_B = (K_D + K_S) \cdot D + \gamma \cdot K_D \cdot \tau_D \left(e^{-D/\gamma \cdot \tau_D} - 1\right), \tag{1}$$

where K_S is the excess base current per unit dose at a high dose rate; K_D is the excess base current per unit dose at a low dose rate; γ is dose rate; τ_D is conversion time of the deep traps and D is a total dose.

The ELDRS conversion model includes an experimental procedure for fitting parameters extraction. This feature is a distinguishing characteristic of the model that allows the reception of a quantitative numerical estimation for radiation degradation of a bipolar transistor base current for arbitrary dose rate and temperature.

A conversion of oxide charge to interface traps is a thermally stimulated process. To consider the temperature effect on the base current degradation, dependence on the deep trap conversion time from temperature is introduced. The temperature dependence of time constant τ_D can be described by the Arrhenius law:

$$\tau_D(T) = \tau_{D_0} \cdot exp\left(\frac{E_A}{kT}\right), \tag{2}$$

where T is the temperature; E_A is the activation energy of deep oxide trap thermal excitation; k is Boltzmann's constant and τ_{D_0} is pre-exponential coefficient.

6. Examples of using the conversion model for numerical estimation of bipolar device radiation degradation in space application

The conversion model can be used for a description practical and interesting applications when the dose rate and temperature are not constant. This takes place during real operating conditions for the on-board space apparatus when the dose rate cyclically varies when crossing the radiation belts of Earth and temperature changes according to the sun apparatus position. It is probable that short high-intensity radiation receives impacts from a solar flare. Below, the conversion model for the quantitative numerical hypothetical analysis of the radiation degradation LM111 comparator (as an example) for a possible on-board space apparatus operation is described. The technical details of this approach were described in [9].

6.1. Dose rate variation

Usually, the ELDRS analysis [7] is performed using a constant dose rate. In real operations, a periodical short increasing dose rate connects by crossing the radiation belts of Earth with the space apparatus.

The selection of averaged dose rates for estimating radiation degradation bipolar devices that are on-board is not obvious.

In the analyzed example, the dose rate varies from 10 mrad(SiO2)/s during 1 hour to less than 10^{-6} rad(SiO2)/s during 11 hours [10]. The averaged dose rate was 0.8 mrad(SiO2)/s. A simulation of the dose dependence at 25°C for the input current of the LM111 comparator, is presented in Fig. 1. For comparison, in Fig. 2, the excess input current versus the total dose is shown for the constant averaged dose rate. Fig. 1 shows changes in the input current on a larger scale. The results are practically the same for both cases. This means that for the case of periodic variation in the dose rate and short periods of dose rate variations compared to the total irradiation time, it is correct to use the averaged dose rate for the estimation of radiation degradation.

The obtained data shows, it is possible to use the averaged value of the dose rate during a space apparatus flight on orbit as the in-flight dose rate after switching the dose rate from high to low.

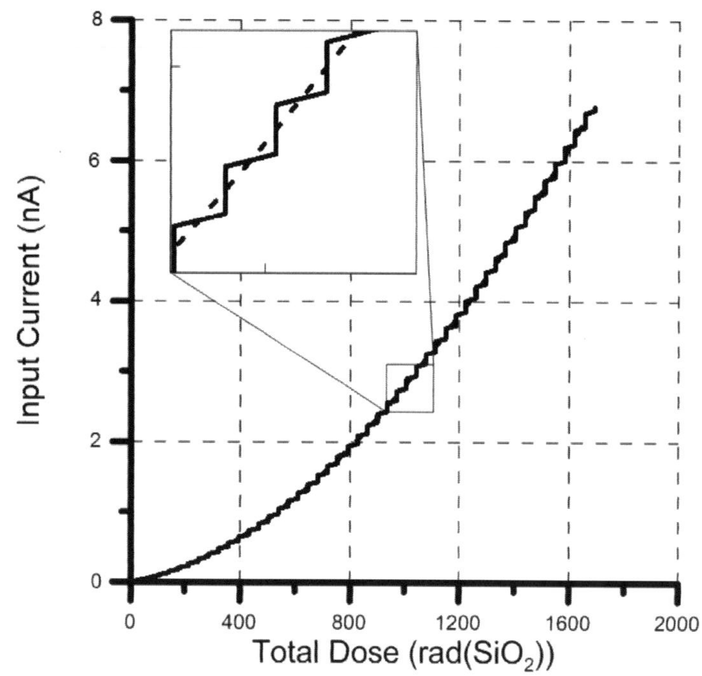

Figure 1. Excess input current of LM111 comparator versus total dose for 12-hour orbit dose rate variation (solid) and constant averaged dose rate (dashed).

6.2. Cyclic temperature variation

The effect of environment temperature on bipolar device operation during a long space mission was investigated, for example, with a sinusoidal temperature variation from 268 K (-5 °C) to 328 K (+55 °C) with the 12-hour period of variation. The average temperature during the period is 298K (25 °C).

As illustrated, the results of the simulation when using the ELDRS conversion model are presented in Fig. 2 for variable and constant averaged temperatures. The obtained results mean that using the averaged temperature can lead to a large error. From the illustration in Fig. 2, the total dose level of a 1.6 krad(SiO2) estimation error may be more than 300% when the averaged temperature level is used instead of the real dependence temperature on time. If the parametric failure level for the comparator input current is defined as 100 mA, the failure total dose level could be overestimated more than three times when using the averaged temperature value.

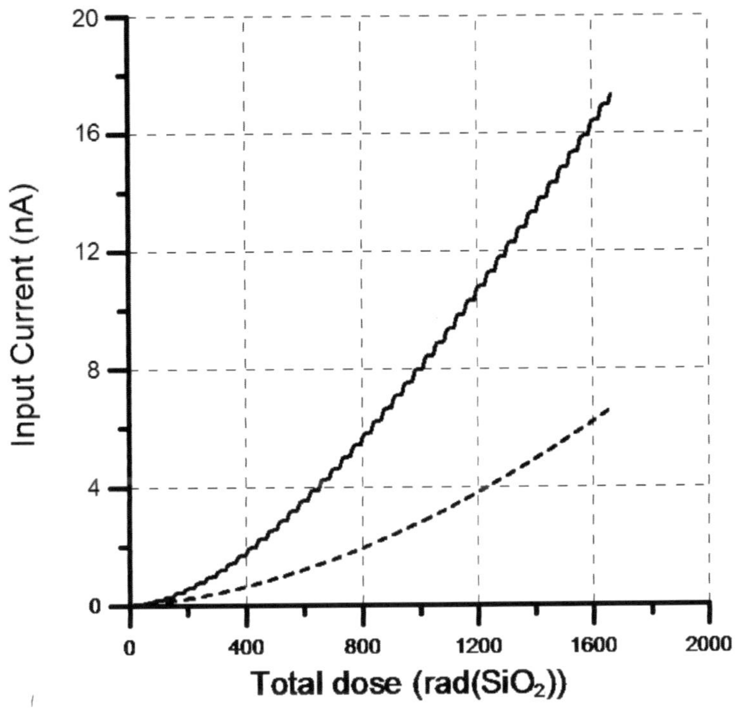

Figure 2. Radiation degradation of LM111 comparator input current with a total dose for variable (solid) and constant averaged (dashed) device temperature operation.

Consider the application of the result obtained above for the technique of a thermal annealing approach for regenerated electrical characteristic devices [11]. The technique of [11] is efficient on a thick oxide when electrical characteristic degradation is mainly caused by an oxide-trapped charge. Radiation-induced bipolar transistor gain degradation connects with processes in a thick oxide above the base region and is dominated by the formation of interface traps at the Si/SiO2 interface. For these devices, cycling elevated temperature annealing can lead to an increase in interface trap concentration and an increase in degradation. The following conclusion can be made: the technique for a thermal cycling annealing approach that extends electronic device lifetimes exposed to very high total dose [11] can be used for devices in which a role of radiation-induced interface trap buildup is not significant. It concerns the bipolar devices in which the ELDRS effect takes place, and the role of the interface-trap buildup is important. Using this technique [11] for these types of devices should be performed with caution.

7. Conclusion

The conversion model for modeling the radiation-induced degradation of bipolar device parameters for the impact of low dose rate irradiation is described. The model is based on the concept that the radiation-induced interface-trap buildup connects with the hydrogen-electron mechanism, where both hydrogenous species and electrons are responsible for radiation-induced interface-trap formation. The interaction of trapped positive charges (hydrogenous species) and electrons from the substrate leads to the formation of interface traps. The experimental data confirming the validity of this type of combined model is presented. To explain the reversibility effect of annealing, a thermoactivated tunneling mechanism of the defect recharge was proposed. It is supposed that the rechargeable positive centers in the oxide must have energy levels within the Si forbidden gap.

The applicability of the proposed conversion model for the estimation of the radiation degradation of bipolar devices on the impact of low dose rate irradiation is demonstrated. For one type of IC (LM111 comparator) for which the fitting parameters of the ELDRS conversion model were extracted, we try to demonstrate a numerical estimation of this circuit behavior during a space mission. The numerical analysis of the comparator input current radiation degradation during a long operation in space is simulated during a dose rate variation corresponding to the 12-hour orbit and cyclic device temperature variation. All of these quantitative numerical results may have practical application.

It was shown that in the case of a variable dose rate during a space mission, the averaged value of the dose rate could be used. However, using the averaged temperature can lead to essential error. The correct estimation of radiation-induced degradation should be accurate when considering the temperature time dependence during device operation.

Acknowledgements
This work was supported by the Competitiveness Program of NRNU MEPhI.

References

[1] McLean F B 1980 *IEEE T. Nucl. Sci.* **27** 1651
[2] Lai S K 1983 *J. Appl. Phys.* **54** 2540
[3] Sabnis A G 1983 *IEEE T. Nucl. Sci.* **30** 4094
[4] Sogoyan A V, Cherepko S V, Pershcnkov V S, Rogov V I, Ulinov V N and Emelianov V V 1997 *Proc. Eur. Conf. on Radiation and Its Effects on Components and Systems (Cannes)*(IEEE) p 69
[5] Lelis A J, Boesch H E, Oldham T R and McLean F B 1988 *IEEE T. Nucl. Sci.* **35** 1186
[6] Emelianov V V, Zebrev G I, Ulimov V N, Useinov R G, Belyakov V V and Pershenkov V S 1996 *IEEE T. Nucl. Sci.* **43** 805
[7] Pease R L, Schrimpf R D and Fleetwood D M 2009 *IEEE T. Nucl. Sci.* **56** 1894
[8] Pershenkov V S, Savchenkov D V, Bakerenkov A S, Ulimov V N, Nikiforov A Y and Chumakov A I 2009 *Proc. Eur. Conf. on Radiation and Its Effects on Components and Systems (Bruges)*(IEEE) p 286
[9] Pershenkov V S, Bakerenkov A S, Solomatin A V, Belyakov V V and Shurenkov V V 2013 *Proc. Eur. Conf. on Radiation and Its Effects on Components and Systems (Oxford)*(IEEE), p 1
[10] Turflinger T L, Schemichel W M, Krieg J F, Titus J L, Campbell A B, Reeves M, Walters R J, Marshall P W and Pease R L 2003 *IEEE T. Nucl. Sci.* **50** 2328
[11] Roig F, Dusseau L, Boch J, Saigne F, Vaille J-R, Touboul A, Adell P and Lorfevre E 2012 *Proc. Eur. Conf. on Radiation and Its Effects on Components and Systems (Biarritz)*(IEEE)

IOP Conf. Series: Materials Science and Engineering **151** (2016) 012002 doi:10.1088/1757-899X/151/1/012002

Taking into account the space environments variability for prediction of dose response in bipolar devices

I V Elushov and G I Zebrev

National Research Nuclear University MEPhI (Moscow Engineering Physics Institute), Kashirskoe shosse 31, 115409 Moscow, Russian Federation

Corresponding author e-mail address: ElushovIV@gmail.com

Abstract. The aim of this paper is modelling and simulation of impact of space variability on dose rates and total ionizing dose (TID) effects in space-borne electronics. It has been shown that the simultaneous thermal annealing may lead to non-stationary relaxation after dose-rate peaks and steady state degradation saturation depending on thermal and electrical operating modes.

1. Introduction

Most testing experiments on the total ionizing dose (TID) have been typically performed with a constant dose rate. However, the space environment can be highly variable. Dose rates are commonly orders of magnitude larger during solar bursts [1,2]. The influence of the space radiation environment variability on spacecraft can affect total dose effects in spaceborne microelectronic parts.

Onboard measurements of dose-rate variations can be conducted through different kinds of monitoring systems. NASA Space Weather Prediction Center contains various sets of data regarding solar activity and particle fluxes measurements. Most of space weather missions require complex equipment to obtain copious data sets, although simple dose detectors could be located on almost any spacecraft. For example, some data presented in [3] was obtained via space-borne segment, containing a set of small-size n-MOSFET based dosimeters with telemetric sub-systems for real-time transfer to Earth. One should note that it may be useful to collect onboard temperature data available considering RADFET temperature sensitivity drift.

Space weather variability makes dose response predictions for onboard devices difficult, in particular, because simultaneous time-dependent anneal effects, which could significantly modify radiation response in IC, especially for bipolar technology, under the solar flares against the long-term low dose-rate radiation background. The main objective of this report is to improve knowledge of space radiation variability impact on the circuit radiation response via considering approaches for degradation prediction modelling.

2. Rate equation approach in modelling

The radiation-induced defect build-up in the insulation layers and at the Si-SiO_2 interfaces is the common cause of parameter degradation in bipolar and CMOS technology components. The time-dependent thermal annealing of these defects has found to be a general effect, and the devices of all technologies to some degree are prone to it, although the annealing parameters (activation energies and the time constants) vary for different technologies, and even for different lots of devices.

The rate equation of the criterion parameter change can be written in the linear response approximation as:

$$\frac{d\Delta\Pi(t)}{dt}=A\eta_{eff}\left[P(t),T_{irr}(T),E_{ox}(t)\right]P(t)-\frac{\Delta\Pi(t)}{\tau_a\left[T_{irr}(t)\right]} \quad (1)$$

where A is a technology and lot dependent normalized factor, η_{eff} is the effective charge yield function, generally dependent on the electric field in the oxides E_{ox}, irradiation temperature T_{irr}, and a dose rate P. However, radiation-induced modification of electrical fields, which is considerable at high (~Mrad) doses, was described elsewhere [4]. The simultaneous thermal annealing model is elaborated in [5] in different cases of one or few types of defects and corresponding amount of temporal annealing constants.

Equation (1) has a general solution for arbitrary profile of dose rate $P(t)$:

$$\Delta\Pi(t)=A\int_0^t\eta_{eff}\left[P(t'),T_{irr}(t'),E_{ox}(t')\right]e^{-\frac{t'-t}{\tau_a\left[T_{irr}(t')\right]}}P(t')dt' \quad (2)$$

This solution expressly shows that the radiation degradation may be represented as a functional (a convolution) of the previous dose-rate, electric and thermal history of the object exposed. This means that parameters degraded under irradiation are affected by any non-stationary variation in dose-rate environments, thermal and electric conditions via transient effects.

3. Space Dose Rate Variations And Time-dependent effects

3.1. ELDRS problem

An increase in charge yield (η_{eff}) with decreasing dose rate in the bipolar transistors and ICs is known as the ELDRS effect [6]. The explicit form of this relation was found in [5,6]. According to [5], the increased η_{eff} at low dose rate is possible due to a decrease in electron-hole recombination (Langevin mechanism) through steady-state and dose-rate dependent density of localized holes. This recombination is not observed in the gate oxides of the MOSFETs due to suppression by strong electric fields in the insulating oxides. However, considering ELDRS as a property of thick layers of an amorphous insulator with low electric fields, it may be observed in the thick oxides of p-MOSFET based dosimeters [7].

One should note that even large-scaled dose-rate temporal fluctuations in the cosmic range of 10^{-6}-10^{-3} rad/s do not change the charge yield significantly. The charge yield evaluates the magnitudes near the maximum at a given temperature and electric field in the oxide.This means that ELDRS effect is a problem of the ground test adequacy (choice of fitting *P(t)* for irradiation), rather than a problem of space radiation environment.

3.2. Thermal anneal effects

In bipolar circuits, elevation of irradiation temperatures to a moderate amount (up to 100 °C) results in enhancement of degradation due to increase in charge yield η_{eff} [5]. This is true for short-term exposure, when the time-dependent anneal did not have time to become noticeable. However, under the long-term irradiation, the degradation is limited by simulteneous anneal effects for all types of devices. Result of competition between enhancement of the charge yield and the simultaneous thermal annealing appear to show the remarkable degradation maximum as a function of irradiation temperature in bipolar transistors [8]. Thereby a noticeable reduction in the conservatism of the hardness assessment may be achieved by taking into account the thermal annealing during the prediction procedure.

Particularly, the simultaneous thermal annealing shows the time-dependent phenomena of the accumulated degradation relaxation after the dose-rate peaks. As an example, Fig. 1 shows the simulation results for the bipolar IC degradation calculated for the actual dose-rate profile derived from the various satellite data.

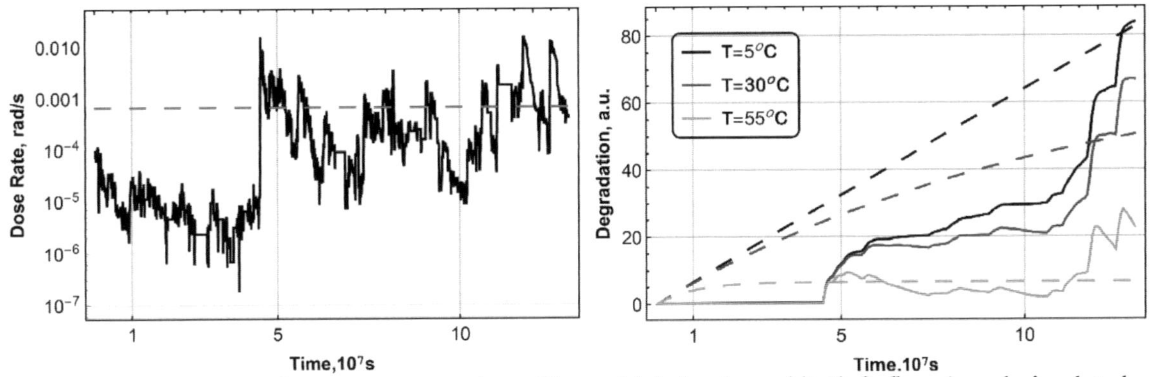

Figure. 1: A common dose rate temporal profile on highcircular orbit (left figure) and simulated degradation curve (right figure), calculated in arbitrary units for different irradiation temperatures. Temporal annealing constants used in simulation were obtained in [6] based on experimental data at different operation temperatures. Dashed lines show simulation result for average dose rate values.

It is also important to notice that elevated operation temperatures may result in occurrence of quasi-steady-state saturation of degradation due to dynamic compensation of the radiation-induced buildup and the simultaneous annealing processes. Particulary, the level of degradation saturation generally depends on dose rate, irradiation temperature and electric mode, implying that non-stationary and non-monotonic effects could be observed after variation in conditions. Quasi-steady-state saturation in degradation may also point out at sensitivity to operation temperature variations. An example of anti-correlation between operating temperature and RADFET readings was demonstrated in [9].

4. Conclusion

For estimation of degradation space-borne devices using data obtained from ground testing, one should consider variability of the space environment. By modeling dose rate variations and time-annealing effects, it has been shown that the real degradation in onboard devices, is a functional of its radiation, electric and thermal history. Simultaneous anneal effect is also should be taken into account as it is shown to relatedevice operation temperature and degradation magnitude level.

Acknowledgements
This work was supported by the Competitiveness Program of NRNU MEPhI.

References

[1] Barth J L, Dyer C S and Stassinopoulos E G 2003 *IEEE T. Nucl. Sci.* **50** 466-482
[2] Turflinger T 2004 *Solar Variability and the near-Earth environment - mining ELDRS data from the microelectronics and photonics test bed space experiment* NASA/TP - 2004-213339
[3] Anashin V S, Protopopov G A, Balashov S V, Gaidash S P and Sergeecheva N V 2011 *8 Eur. Space Weather Week (Namur).*
[4] Zebrev G I, Petrov A S, Useinov R G, Ikhsanov R S, Ulimov V N, Anashin V S, Elushov I V, Drosdetsky M G and Galimov A M 2014 *IEEE T. Nucl. Sci.* **61** 1785-1790
[5] Zebrev G I, Pavlov D Y, Pershenkov V S, Nikiforov A Y, Sogoyan A V, Boychenko D V, Ulimov V N and Emelyanov V V 2006 *IEEE T. Nucl. Sci.* **53** 1981-1987
[6] Zebrev G I and Gorbunov M S 2009 *IEEE T. Nucl. Sci.* **56** 2230-2236
[7] Kim S J, Seon J H, Min K W, Shin Y H and Choe W 2003 *Proc. Eur. Conf. on Radiation and Its Effects on Components and Systems (Noordwijk)*(IEEE) p 669
[8] Witczak S C, Schrimpf R D, Galloway K F, Fleetwood D M, Pease R L, Puhl J M, Schmidt D M, Combs W E and Suehle J S 1996 *IEEE T. Nucl. Sci.* **43** 3151-3160
[9] Harboe-Sorensen R, Poivey C, Zadeh A, Keating A, Fleurinck N, Puimege K, Guerre F X, Lochon F, Kaddour M, Li L and Walter D 2012 *IEEE T. Nucl. Sci.* **59** 1086-1091

The development of ion mobility technology

V V Beliakov, A V Golovin, V K Vasilev, E K Malkin, E A Gromov, I A Ivanov, M A Matusko and D Y Lipatov

National Research Nuclear University MEPhI (Moscow Engineering Physics Institute), Kashirskoe shosse 31, 115409 Moscow, Russian Federation

Corresponding author e-mail address: vvbelyakov@mail.ru

Abstract. Increased terrorist threat in recent years makes it especially important to improve the custom equipment including the development of ion mobility spectrometers for reliable, real-time and sensitive identification of illicit substances. The paper summarizes different approaches used in recent developments in the last years: statistical analysis, sampling automation and construction materials study.

1. Introduction

The border service interdicts illegal substances to be crossed the border [1-3]. The main problem in typical inspection area and border service is time-consuming inspection procedure of people, luggage, and parcels. The essential search purpose is to recognize explosive components that are the most harmful from illicit substances. Although the inspection is selective, even one incident of explosives on board can lead to a fatal crash. One of the most widely distributed technic for the detailed security inspection of passengers and cargo is the Ion Mobility Spectrometry (IMS). This method is capable of detecting volatile and semi-volatile illicit substances in a wide range of concentrations from ppq to ppm for custom service needs. The customs inspector assembles the illicit components from the surface of passenger's hands and luggage using a napkin, then he puts the napkin to the IMS system. The procedure may last for tens of seconds that is why selective inspection is used. However, two problems of inspection method do exist. Firstly, it is a long duration of probing. Secondly, it is expensive consumables because of high requirements to clearance. Solutions of the problems can be sampling, transport, and analysis of samples an automation, fast cleaning and multiple reuses of the collecting particles surface. In a case of close contact with explosive compounds, trace amounts of explosives stay on the skin surface for several hundred touchings even after washing. The suggested method includes following procedures:

- investigation of push mechanics;
- statistical analysis of fingerprint characteristics (figure 1);
- study of fingerprints traces allocation.

The system is used for analysis is fingerprint and documents which is described in previous papers [4,5]. The gas flow chart of fingerprint and documents system is presented in figure. 2-3. The ion mobility spectrometer [6-8] operates with substance in 3 seconds. The ideal probe preparations should include fast heating up to specified temperature and fast freezing of one object within the period of time range from ms to s.

The peculiarity of operation with fingerprint substances consists of sweat and fatty traces heating regime. The regime causes an increase of its evaporation time and limits heat temperature. The fingerprint should be heated in a mode that allows avoiding formation of soot smoke. In addition, the regime of heating could be divided into stages for separation of substances by their boiling temperature.

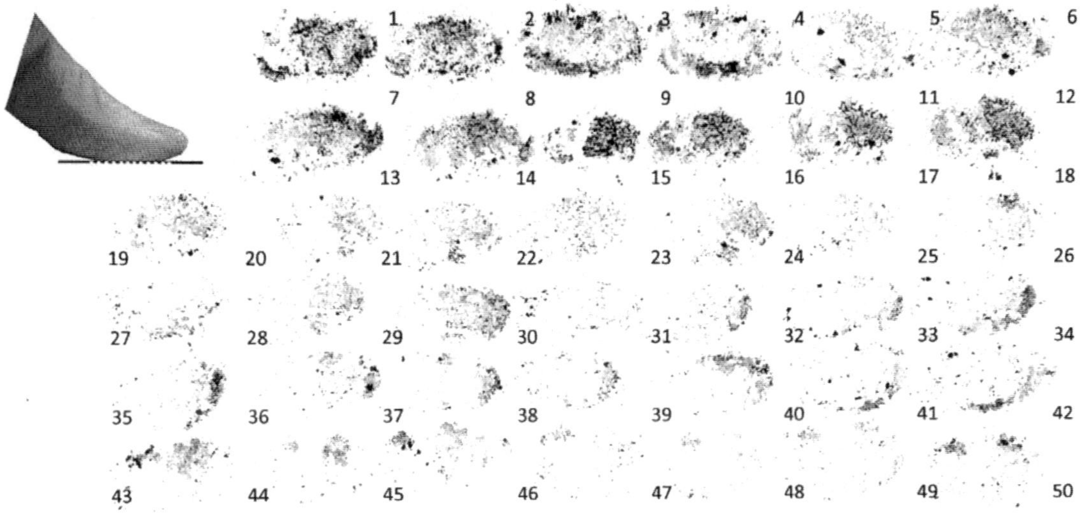

Figure 1. Statistical analysis of fingerprint characteristics.

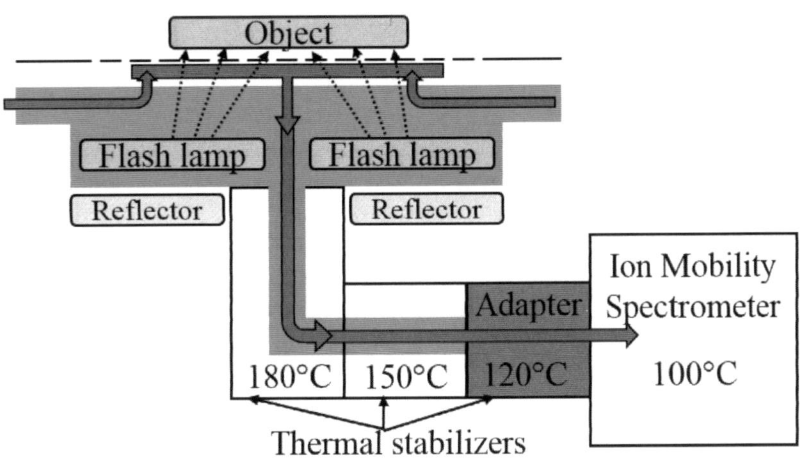

Figure 2. The gas flows chart of fingerprint and documents system.

The development of analytical instrument should involve material engineering and cleaning technology. The gas channels is transporting probe are made of tubes from different materials such as Teflon F-4, 2. Glass and silanized borosilicate glass, 3. Stainless steel 12X18H9T / SilcoSteel, 4. Quartz glass.

The transport channel should be cleaned for next probe after probe's shift. The chemical composition of analyzed substances are acids. As a result metal and other materials could react with the probe. In that case, passivation is used for the avoidance of probe contact with metal. One of the passivation methods is a silanization.

Every detail in the analytical system is cleaned and annealed for a long time up to hours and days. Then the detail is set to the device. A new experience is received after the start of mass instruments' production. Statistics results start to show a worsening of quality baseline spectrum after a long storage of devices up to 2 weeks or more (Fig. 4). The reasons are the lack of initial treatment and the processes of accumulation of "pollutant" elements in the spectrometry cell.

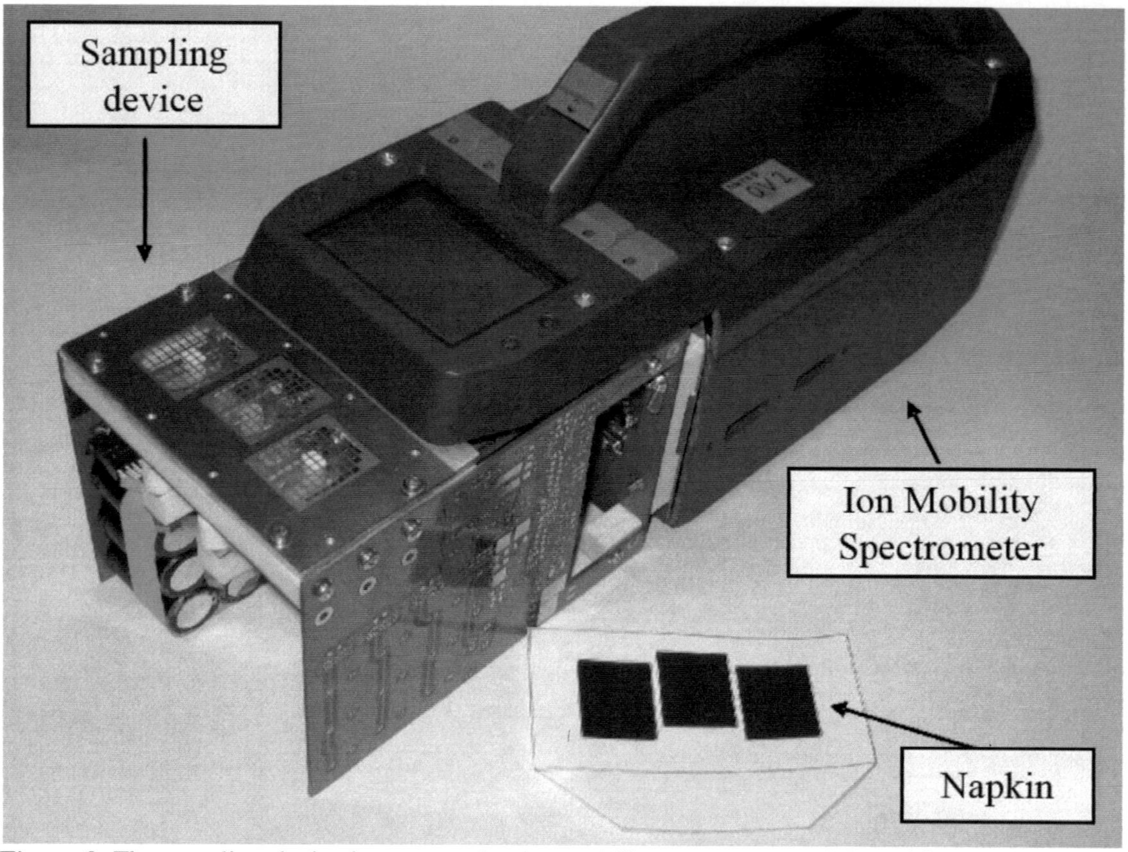

Figure 3. The sampling device is connected to ion mobility spectrometer. Adapted from Ref. [4] with permission.

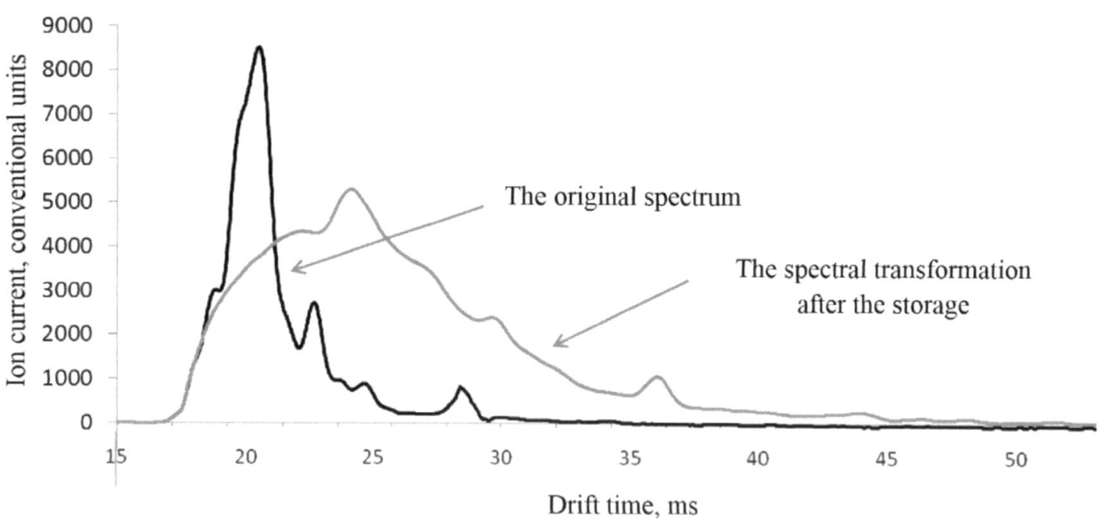

Figure 4. The spectrum before and after a week of storing the device.

The figure shows the spectrum of degradation:
- peaks of substances have bad resolution;
- there are new peaks of additional substances;
- the sensitivity decreases.

According to the figure, it can be resumed that initially clean parts of construction have a clean surface, but they may contain the scope of the "polluter" is diffused within. It evaporates slowly and steadily when heating starts. It tends to in the following processes:

Penetration of "pollutant" substances through porous materials. It includes:

- dissolving the solute in the surface layer of the material;
- diffusion of dissolved solute through the material as a result of the concentration gradient;
- the evaporation of the adsorbate from the space between assembled details;
- the rate and depth of penetration of the adsorbate are increased respectively through high concentration, temperature or pressure. Moreover, besides, these parameters affect crystal structure of the material.

Teflon is popular inert fluorplastic which has "semi-crystalline" structures with both crystalline and amorphous domain structures. Various domain structures are determined by the chemical composition of the polymer and its production technology. Increased crystallinity reduces the intermolecular volume.

2. The deposition (adsorption) on the surface of materials with low porosity

The dense crystalline structure prevents penetration of the adsorbate and diffusion through the volume. Proper materials prevent adsorption on materials surfaces such as borosilicate glass, quartz glass, steel. The surface of the quartz glass has the low adsorptive capacity for various gasses and moistures.

Silanization method reduces surface adsorption by protective coating for metal and glass. The method is based on silanol group (-Si-OH). It is presented on the glass surface with the silanizing agent for chemically sewn hydrophobic siloxane film formation. This reaction eliminates of free chemical bonds.

An applicability comparison of various materials for inlet channel is held. The experiment consists of blowing through the investigated samples of TNT vapor under constant conditions (time, concentration, flow, temperature). The results showed (Fig. 5) that the use of Teflon tubes as the pipeline is highly undesirable due to effect described earlier.

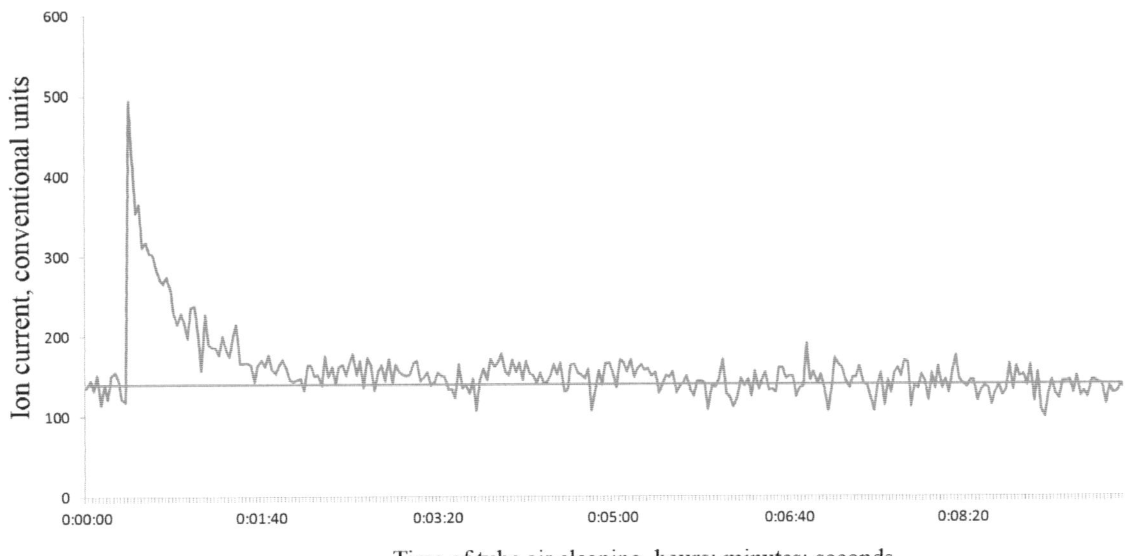

Figure 5. Signal amplitude of trinitrotoluene trace concentrations while cleaning process in the processed quartz tube sample (blow of clean air at a certain temperature) versus time of tube air cleaning.

In order to avoid contamination diffusion effect the following procedure is applied:

- using standard heat treatment. A sample of material is packed in foil. Contamination materials evaporate and condense immediately because of substances being of low volatile. This foil is analyzed on the spectrometer;
- using minimization of materials list up-to several components which are characterized by stable physicochemical characteristics from certain manufacturers.

Silanized glass at low (room) temperatures reduces a level of contamination, but at operating temperatures (about 100 degrees) the difference is irrelevant. A similar situation may be observed with tubes of steel with protective coatings. The quartz tubes are collected the least amount of contamination from components mentioned above. However, the quartz high melting point makes it difficult to manufacture tubes with special shape (bented). Glass tubes is an applicable option, with improved dynamics and sensitivity at low concentrations of the sample.

3. Conclusion

The complex solutions for acceleration and purity increase of IMS probing are discussed. The paper summarizes different approaches used in recent developments: sampling automation, adsorption fabric study, fast cleaning and multiple reuses of the collecting particles surface. Using of chemically inert materials and passivation of transport tube surface allows to solve the problem of contamination sorption. On the one, hand the best characteristics were demonstrated by quartz and silanized glass tubes having minimum TNT sorption. On the other hand, it is concluded that the using of Teflon tubes as the pipeline leads to undesirable growth of analyzed substances in material. Nowadays the methods of illicit substances' transportation systematically improves that is why the additional development of IMS methods is needed

Acknowledgements
This work was supported by the Competitiveness Program of NRNU MEPhI.

References

[1] Eiceman G A, Karpas Z and Hill H 2013 *Ion Mobility Spectrometry (3rd ed.)* (Taylor & Francis Group) **406**(11) p 2493–2494

[2] Borsdorf H, Mayer T, Zarejousheghani M and Eiceman G A 2011 *Appl. Spectrosc. Rev.* **46** 472-521

[3] Sysoev A, Poteshin S, Chernyshev D, Karpov A, Tuzkov Y, Kyzmin V and Sysoev A 2014 *Eur. J. Mass Spectrom.* **20**(2) 185-192

[4] Samotaev N, Vasilyev V, Malkin E, Gromov E, Belyakov V, Golovin A, Pershenkov V, Ivanov I, Shaltaeva Y and Matusko M 2015 *Proc. Eurosensors* **5** p 10-15

[5] Vasilyev V, Pershenkov V, Belyakov V, Samotaev N, Golovin A, Malkin E, Gromov E, Ivanov I, Matusko M, Ivanova A, and Lipatov D 2015 *Proc. 3rd Int. Conf. Nanotechnologies and Biomedical Engineering (Chisinau)*(Springer: IFMBE Proceedings) **55** 515-519

[6] Samotaev N, Pershenkov V, Belyakov V, Vasilyev V, Golovin A., Pershenkov V., Ivanov I, Shaltaeva Y and Matusko M 2015 *Procedia Engineering* **87** 436-439

[7] Matsaev V, Gumerov M, Krasnobaev L, Pershenkov V, Belyakov V, Chistyakov A and Boudovitch V 2002 *International Journal for Ion Mobility Spectrometry* **5**(3) 112-114

[8] Samotaev N, Golovin A, Vasilyev V, Malkin E, Gromov E, Shaltaeva Y, Mironov A and Lipatov D 2014 *Lecture Notes in Electrical Engineering* **268** 447-451

IOP Conf. Series: Materials Science and Engineering **151** (2016) 012004 doi:10.1088/1757-899X/151/1/012004

Combined methods of tolerance increasing for embedded SRAM

L A Shchigorev and I I Shagurin

National Research Nuclear University MEPhI (Moscow Engineering Physics Institute), Kashirskoe shosse 31, 115409 Moscow, Russian Federation

Corresponding author e-mail address: l.shchigorev@module.ru

Abstract. The abilities of combined use of different methods of fault tolerance increasing for SRAM such as error detection and correction codes, parity bits, and redundant elements are considered. Area penalties due to using combinations of these methods are investigated. Estimation is made for different configurations of 4K × 128 RAM memory block for 28 nm manufacturing process. Evaluation of the effectiveness of the proposed combinations is also reported. The results of these investigations can be useful for designing fault-tolerant "system on chips".

1. Introduction

One of the most important challenges in the design of electronic systems is ensuring the validity and reliability of storage and transfer of information (instructions, addresses, data). This problem is topical for using chips containing embedded memory cores, as it is the main source of failures especially with the influence of ionizing radiation. Most modern chips are manufactured on commercial sub-100-nm processes, which are often unique and extremely specialized. Memory cells of these devices have increased sensitivity to the effects of external factors, such as a strike of heavy charged particles, which cause the occurrence of failures [1].

The increase of radiation tolerance of chips produced on commercial CMOS technology is achieved by using special circuit design and algorithmic techniques – Radiation Hardening by Design (RHBD) [2].

Detection and error correction during data storage and transfer are made by special error detection and correction codes (EDAC) [3-4] and redundancy. EDAC can detect and/or correct the errors associated with the emergence of faults under the influence of heavy charged particles or electromagnetic noise, but in the case of failed memory elements, their corrective abilities are greatly reduced. Thus, single error correcting code (SEC) will not be able to correct failures under the influence of heavy charged particles, or electromagnetic noise if one faulty element already exists.

Double or triple redundancy of memory blocks requires considerable hardware costs, which increases the area of VLSI and reduces its technical and economic characteristics. More economical is the use of redundant elements (column and/or rows) in the matrix of memory elements that is connected instead of failed ones. This method is used at the final stage of production of memory chips to increase the yield rate of the chip [5].

This article is devoted to the investigation of the possibility of combined use of EDAC and the redundant elements of SRAM for memory blocks failure tolerance improvement.

2. Used methods

To identify single-bit errors occurring in the transmission of information, the parity bits of the transmitted data words are used. Error correction is implemented using a Hamming code that provides double detection and correction of single errors. EDAC encoders and decoders are introduced into the composition of VLSI.

To improve the yield of chips the redundant memory elements (columns and/or rows) are used. To implement this method, built-in self-test (BIST) and built-in self-repair (BISR) units are included. The BIST can conduct unit testing and determine the position of the damaged elements; then the BISR unit replaces the faulty element by the redundant one. Redundant rows and columns that are not used for production control can be used during the operation of the VLSI by replacing the failing memory elements.

To estimate the number of redundant elements required for manufacturing repairing, we define the probability of obtaining faulty memory elements, depending on the area A, occupied by the memory block on the chip, and the density of defects D. Calculation of the probability of k faulty memory elements is made according to the equation (1) obtained for the distribution of defects in accordance with the Poisson distribution [6].

$$P(k) = \frac{\lambda^k \times e^{-\lambda}}{k!}. \quad (1)$$

The value of λ is calculated as the product of chip area A and the density of defects per unit area D:

$$\lambda = A \times D. \quad (2)$$

Table 1 shows the probability of k (0, 1, 2 and 3) number of error in the memory block for different areas: 1 mm^2, 10 mm^2 and 100 mm^2. The value D is 1.395×10^{-3} faults/mm^2 according to the experts ITRS for 2016 [7].

Table 1. Probability distribution of the number of errors in the SoC memory blocks

A (mm^2)	P(0)	P(1)	P(2)	P(3)
1	0,9986	0,0014	0	0
10	0,9861	0,0138	0,0001	0
100	0,8697	0,1213	0,0085	0,0004

The results show that even for large memory blocks (A = 100 mm^2), the probability of failure due to crystal defects does not exceed 12%. With an area of 10 mm^2 and less than this probability does not exceed 1.4%. Thus, a significant part of the redundant elements remains unused for production control and can be used to perform self-repairing during the further operation of the chip.

To evaluate the possibility of using redundant elements as the factors of fault tolerance increasing SRAM 4K × 128 was designed. The results are shown for different ways of memory block dividing: 1 bank 4K × 128 bit, 2 banks 4K × 64 bit, 4 banks 4K × 32 bit, 8 banks 4K × 16 bit and 16 banks 4K × 8 bit. Each of the banks contains 2 redundant columns.

To evaluate the proposed method of reserving two columns, the simulation of repair rate of redundant elements was performed. It is defined as the ratio of repaired memory elements to the total number of faulty elements. The simulation was performed for 10000 SRAM blocks using programs in the System Verilog language. According to Poisson probability distribution k faulty elements errors occurring repair ratio for manufacturing repairing is calculated for different memory dividing. The evaluation was performed for 100mm^2 area. The result is shown in Figure 1

These obtained results allow making the conclusion that the repair ratio for all memory dividing is more than 99%.

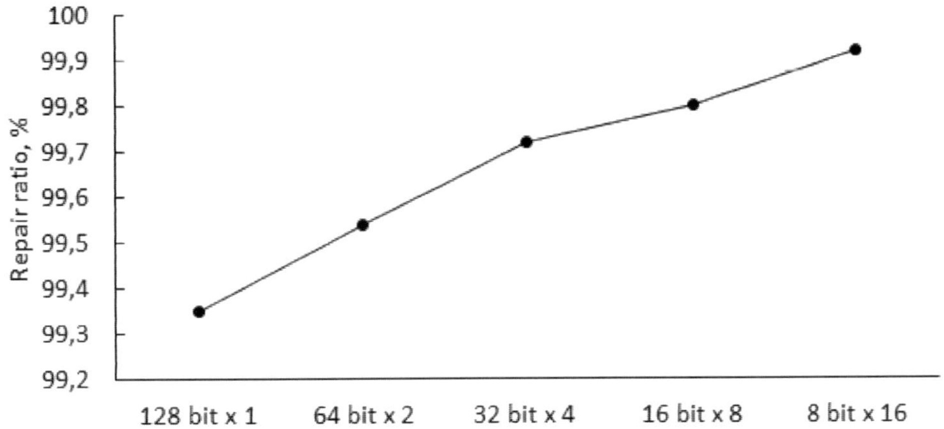

Figure 1. Repair ratio for manufacturing repair.

The next step is to evaluate a number of redundant elements which will be available after yield improvement. For this, the estimation of the percentage of unused redundant elements (URR: unused redundancy ratio) was made. This value is calculated by the equation (3), where MRE (made redundancy) means redundant elements that were designed and URE (used redundancy) means redundant elements which were used in the process of yield improvement at the semiconductor factory.

$$URR = \frac{MRE-URE}{MRE} \times 100\%. \quad (3)$$

Result is shown in Figure 2.

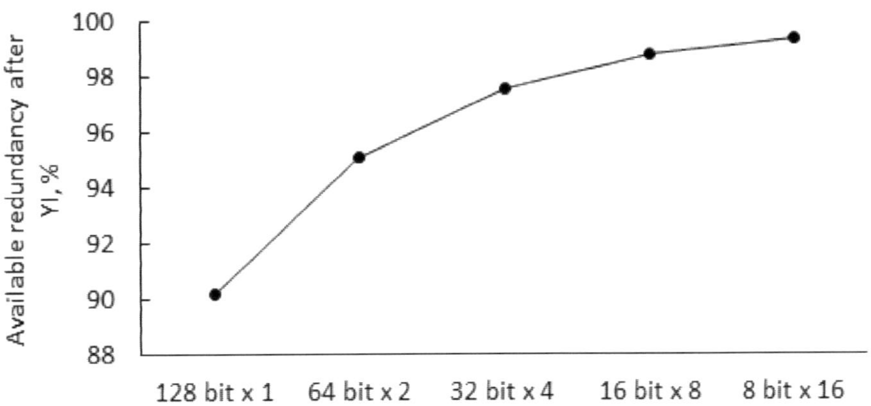

Figure 2. Available redundancy after yield improvement.

To evaluate the effectiveness of the method of reserving two columns the simulation of repair ratio for different numbers of faulty elements per block of memory was made. In Figure 3 the dependence of repair ratio from the number of faulty elements per block of memory is shown.

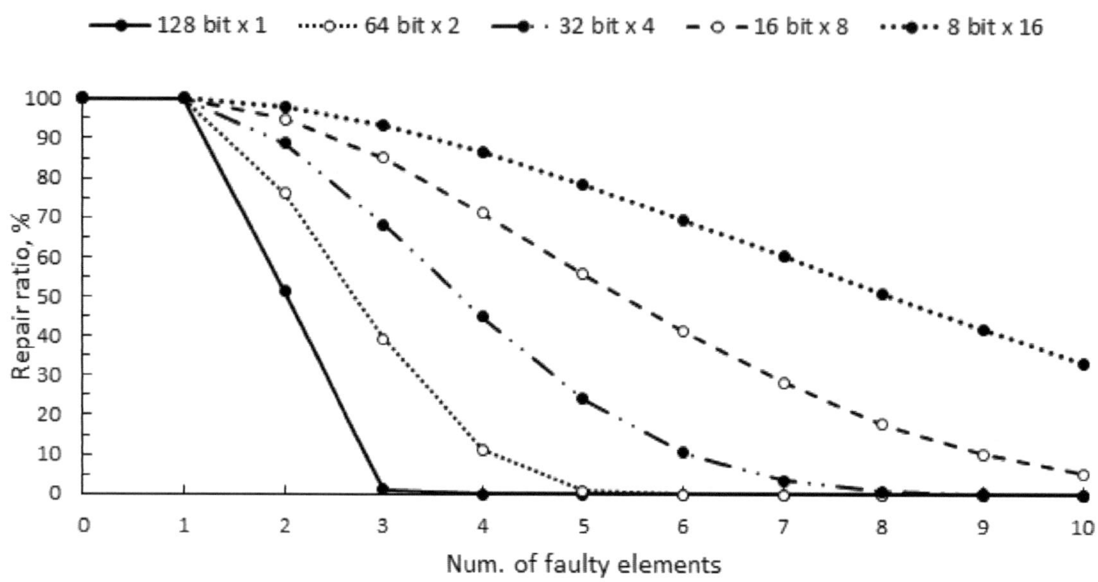

Figure 3. Repair ratio as a function of number of errors.

These obtained results allow that the implementation of memory blocks in multiple banks provide a high probability of correction (90% or more) for both single and double failures occurrence during the operation.

3. Combined methods of fault tolerance increasing

The use of redundant elements is proposed as one of the methods of increasing the fault tolerance for SRAM. This way can be applied alone or in combination with EDAC. The proposed variants are combinations of the three methods: byte parity checking (BPC), single error correction (SEC) and 2 redundant columns per bank (RC). Different combinations can contribute to the total fault tolerance effect differently. For example, the combination of methods SEC and BPC allows to correct the read data words in the case of failures of storage elements and to detect errors occurring on the READ and WRITE buses. The combination of RC and SEC, in its turn, allows to activate the redundant columns with addition of saving the error symptom and address after the READ operation for this address (without interrupting access to the memory and preservation of the available information), followed by periodic sessions of self-repair after detection of SEC the damaged items (with the temporary termination of access to the memory and loss of information).

4. Area penalties estimation

However, all these methods require additional hardware and costs. So EDAC requires additional area for additional bits, encoders, and decoders. Redundant elements require additional area not only for themselves but also for the built-in self-test (BIST) and built-in self-repair (BISR) units. Thus, the challenge of increasing the fault tolerance leads to finding the balance between the additional costs of hardware and the achieved level of data protection.

For estimation of hardware penalties different configurations of 4Kx128 RAM memory block with additional units such as EDAC encoders-decoders, BIST and BISR for 28 nm manufacturing process were produced.

Area estimation was provided in the following situations:

- without fault tolerance increasing (WOFT);
- with per byte parity bits (PB);
- with single error correction (SEC);

- the combination of PB and SEC (PB+SEC);
- the combination of SEC and redundant columns (SEC+RC);
- the combination of SEC, RC and PB (SEC+RC+PB).

Result is shown in Figure 4.

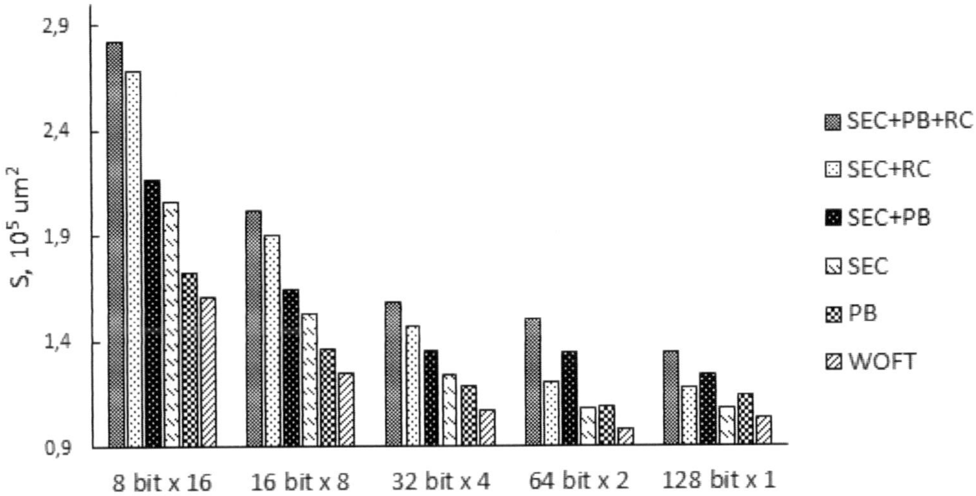

Figure 4. Area for different ways of fault tolerance increasing as a function of different arrays dividing.

These obtained results allow to make the following conclusions:

- The combination of all three considered methods (SEC+PB+RC) leads to increasing required area from 30 to 75% when compared variants have the same dividing;
- The dividing of SRAM block into the banks increases the total area up to 1.6 – 2.1 times depending on the number of banks and method of increasing the failure tolerance;
- The most protected case is obtained by dividing a block into 16 8-bit banks and the use of SEC. In this embodiment, the number of detected errors increases by 16 times. Also the number of redundant elements increases by 16 times (with 2 to 32).

Conclusion

The results of the investigation show that the use of combined techniques for increasing the fault tolerance of the embedded SRAM blocks requires 1.6 – 2.1 times area penalties.

The efficiency of array dividing is shown by evaluating the repair ratio. The number of detected errors increases in proportion to the number of used banks. The probability of correction single or double errors is more than 90% for the case of all combined methods using.

The obtained results can be used in the development of embedded memory blocks included in the design of "systems on a chip" and the evaluation of their characteristics.

References

[1] Uznanski S, Gaisot G, Roche P and Autran J 2010 *IEEE T. Nucl. Sci.* **57** 1876-83
[2] Cannon E, Tostenrude J, Cabanas-Holmen M, Meaker B, Neathery C, Carson M and Brees R 2013 *IEEE T. Nucl. Sci.* **60** 4207-13
[3] Krasnyuk A and Petrov K 2013 *Russian Microelectronics* **42** 53-8
[4] Hamming R 1950 *Bell Syst. Tech. J.* **29** 147-60
[5] Chang D Li J and Huang Y 2008, *J. of electronic testing-theory and applications* **24** 181-92
[6] Horiguchi M, Itoh K 2011 *Nanoscale Memory Repair* (New York: Springer) p 215
[7] International Technology Roadmap for Semiconductors 2007 *Yield Enhancement* Available at: http://www.itrs2.net/itrs-reports.html.

Electrochromic thin-film components for information representation systems

V A Shakhnov, A I Vlasov and S V Tokarev

Bauman Moscow State Technical University, 2-ya Baumanskaya st. 5/1, 105005 Moscow, Russian Federation

Corresponding author e-mail address: shakhnov@iu4.bmstu.ru

Abstract. The solutions in the sphere of energy efficient technologies of information representation systems development are overviewed in this paper. The main emphasis is on "electrochromism" phenomenon, which is a reversible change of material optical characteristics during the application of an external electrical potential. The possible use of electrochromism phenomenon in energy-efficient electromagnetic radiation modulation systems is shown. The generalized model of the universal electrochromic cell construction is presented. The materials with electrochromic characteristics, the field of electrochromes use and the developments in the sphere of electrochromic technologies are analyzed.

1. Introduction

Energy efficient information representation systems development is a real challenge to modern science. A promising possibility in this sphere is a technology based on 'electrochromism' phenomenon. This phenomenon is reflected in thin-films structures in the electromagnetic radiation absorption spectrum change during the application of an electrical potential. The electrical potential is a result of electrons transmission in oxidation-reduction chemical reaction during the application of an external electrical field [1,2]. The absorption spectrum could be located in the visible, infrared or ultraviolet scope. The spectrum location depends on the used electroactive material. Electrochromic device principle of action is based on oxidation-reduction chemical reaction. The electrochromic device contains electrochromic (electroactive) material which consists of electroactive particles. The electroactive material is directly involved in oxidation-reduction chemical reaction. It also changes the adsorbing power during the application of an external electrical potential. The electrochromic material is an electroactive thin film or electroactive solution so that it can be in the solid or liquid state. The term "chromism" means the ability of the substance to change the color reversibly during different external exposures [3,4].

The first reference to electrochromic materials dates from 1706. The first material with electrochromic characteristics (color change from dark blue color to the discolored state during the application of the external voltage) was Prussian blue (Heinrich Diesbach). Wolfram oxide electrochromic characteristics showed up in experiments conducted by Jacob Berzelius (1815). The experiment showed wolfram oxide color change in reduction chemical reaction in gaseous hydrogen. The electrochromism phenomenon in thin films of wolfram oxide was first described by Craus in 1953. The starting point to electrochromism phenomenon studying was the publication of S.K. Deba (1969). In his paper, he presented the first controlled electrochromic device based on the wolfram oxide. The application interest in electrochromism phenomenon is related to the use of this effect in the smart glass, windows and displays [5,6]. The use of photochromic and thermochromic technologies is energy efficient during the room lighting or cooling respectively [7,8].

The advantage of the electrochromic display is low power consumption because energy consumes only at the moment of information updating. Forward-looking electrochromic displays are retro-reflective as well as LCD displays (unlike light-emitting OLED displays). A static image doesn't consume any power because the electrochromic material has a memory effect. The main disadvantage which limits the use of electrochromic displays is long response time. Big electrochromic display panels may have uneven color because the electrodes near the external contact have a higher voltage that the other electrodes.

2. The concept of the universal electrochromic cell construction

Electrochromic devices may have different configurations, but the base configuration is similar to universal electrochemical cell configuration.

Electrochromic devices usually consist of seven layers which are shown in figure 1 [2]. Three inner active layers (counter electrode, ion conducting layer or electrolyte, working electrochromic layer) are located between two conductive transparent electrodes. These five layers are located between protective layers, which are usually made of glass or flexible polymeric material. Oxidation-reduction chemical reaction proceeds between two electroactive materials - counter electrode and working electrode in the electrolyte.

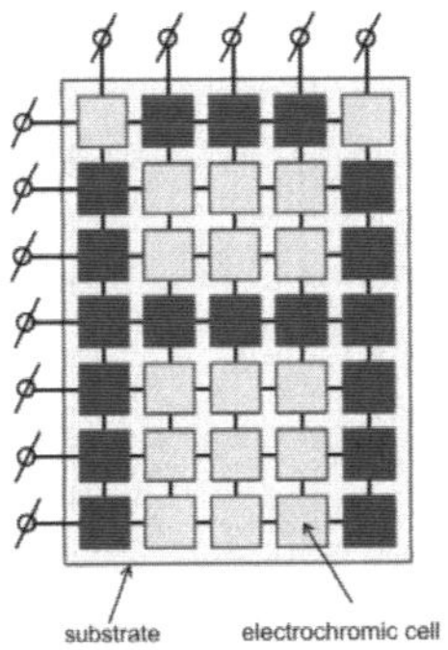

(a)

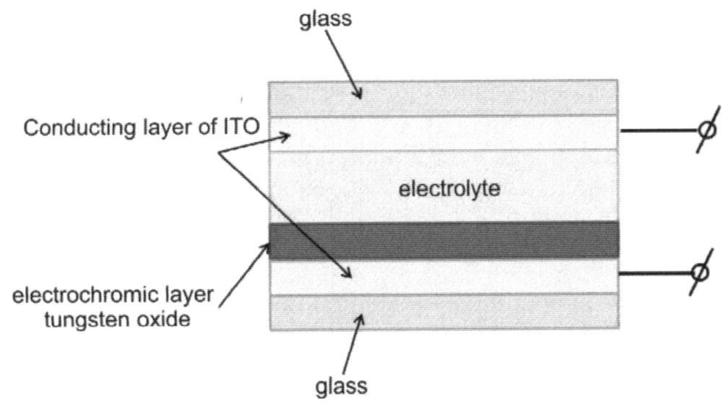

(б)

Figure 1. Universal electrochromic cell construction. (a) – a variant of an electrochromic indicator construction; (б) – a variant of an electrochromic cell construction.

There are four types of electrolytes for electrochromic devices – aqueous electrolytes, organic liquid electrolytes, ionic liquid electrolytes, solid polymer electrolytes [3,4]. Aqueous and organic liquid electrolytes are durable and easy to manufacture. Due to these advantages, these electrolytes are often used in electrochromic smart glass. The disadvantages o these electrolytes are the possibility of their leakage and low chemical stability. Organic liquid electrolytes contain propylene carbonate or ethylene carbonate with the addition of salts such as lithium perchlorate ($LiClO_4$) or natrium perchlorate ($NaClO_4$) to improve their electrochemical characteristics. Ethyl ammonium nitrate ($EtNH_3NO_3$) is related to ionic liquid electrolytes [9].

Solid polymer electrolytes are matrixes of polymer or gel with absorbed liquid electrolyte solution, where only one type of ions can move in the polymer matrix. Another realization of solid polymer electrolytes is an ion-related system consisting of the mixture of ion-containing polymer and salt, where cations and anions can move in the polymer structure.

Counter electrode in electrochromic device structure is used as a layer for ions keeping. The counter electrode layer can be made from any material which has a proper electron and ion conductivity similar to the working electrode. The counter electrode may also have electrochromic characteristics as well as the working electrode with the only difference in the type of colouration (cathode/anode).

Glass or polyethylene terephthalate (PET) substrates can be the base of multilayer electrochromic device structure. Both substrates should be transparent to electromagnetic light emission to be used in smart glass technology. For electrochromic displays and mirrors, only one transparent substrate is needed.

The high light transmitting ability and low resistance allow using transparent conductive materials as an electric current collector in solar panels and as electrodes in the smart glass, electrochromic and LCD displays and sensor panels.

Today the best combination of light transmitting and conductive characteristics has indium oxide doped with tin (ITO) at a ratio of 90% indium oxide In_2O_3 and 10% tin oxide SnO_2. ITO cover is non-organic which combining with low resistance coefficient (0,0001 Ohm/sm), the high light transmitting coefficient (more than 90%) and infrared radiation reflection makes it suitable for use in different devices and systems. The disadvantage of ITO is relatively high cost because indium is rare earth element. As an alternative to ITO cover other high-alloy conductive materials oxides can be used, e.g. fluorine-tin oxide SnO_2:F (FTO), zinc-aluminum oxide ZnO:Al (AZO), zinc and gallium oxide ZnO:Ga (GZO) and titanium-niobium oxide TiO_2:Nb (TNO). It is expected that 90% of capacitive sensor panels and smartphones market will use ITO and its alternatives in 2016 [10]. Other alternatives to transparent conductive oxides are organic polymers and nanomaterials.

PEDOT or poly(3,4-ethylenedioxythiophene) is an organic conductive polymer. This polymer has several advantages over ITO covers such as flexibility, low cost and ability to use PEDOT in printing processes. PEDOT chemical compounds polyaniline PAN1 and polypyrrole PPY have required physical characteristics to use them as a transparent conductive layer.

Carbon nanotubes usage is a forward-looking alternative due to their high conductivity and light transmitting coefficient about 90% at wavelengths from 440 nm to 22 μm. Carbon nanotubes have been already used as a conductive material in electrochromic devices instead of ITO cover. Carbon nanotubes use helped to improve the electrochemical stability of the device, increase the contrast and decrease the response time. Graphene also can be used as a transparent conductive material due to the high mobility of charge carriers and absorption of only 2.3% of visible light.

3. Electrochromic materials classification

Electrochromes are classified by a phase condition of a substance before and after the process of an absorption band change. The phase condition defines the time current characteristic of the electrochromic device. This characteristic is related to substance coloring time characteristic. There are three types of electrochromes: Type I – Liquid-phase, Type II – Transitional, Type III – Solid-phase.

Electrochromic devices of the first type use the electrochromic substance in the liquid phase. The substance stays in this phase throughout the device working period (e.g. liquid methyl viologen (1,1'-dimethyl-4,4'-bipyridyl)).

Electrochromic devices of the second type – initially electrochrome is in the liquid phase and then changes to solid phase precipitating on the electrode in the process of electron charge transfer. This phase change increases the efficiency of write-erase cycles and reduces the response time during discoloration. The examples of second type electrochromes are heptyl or benzyl viologen [11], methoxyflurane in a solution of acetonitrile [12]. The examples of non-organic second type electrochromes are bismuth, lead or silver that recover on the electrode from the liquid solution of ions or cations in combination with joined organic or non-organic molecules (ligands).

Electrochromic devices of the third type are characterized by a solid phase of the electroactive substance throughout the whole working cycle. The most of the non-organic electrochromic materials are materials of the third type. Non-organic electrochromic materials of the third type are phthalocyanine complexes and metal hexacyanoferrates ("Prussian blue"). Organic electroactive conductive polymers are polypyrrole, polythiophene, polyaniline which are monomers and form a solid electrochromic structure in polymerization reactions. The most important electrochromic substances are transition metal oxides (*TMO*), materials on the basis of Prussian blue (*Prussian blue*) that is a mixture of hexacyanoferrates, viologens, lanthanides complexes, metal-polymers and metal-phthalocyanines.

The researches in the sphere of electrochromism phenomenon are continuing. Electrochromic materials are promising for energy-efficient electromagnetic radiation modulation systems constructing [13-17].

4. Experiments analysis, conclusions and recommendations

For universal electrochromic cell prototype forming a conductive layer of ITO was formed on a glass substrate by constant current sputter deposition [2,15]. The source for each prototype was a sputter target from indium and tin oxides alloy at a ratio of 90% and 10% respectively. Sample 1 was not annealed after sputter deposition. Samples 2, 3 and 4 were annealed for 20 minutes at 200, 300 and 400 °C respectively. The resistance of thin-film conductive surface was measured by van der Pauw method with Signatone 1160 scanning probe station. Figure 2 is a plot of a surface layer resistance and the annealing temperature.

The graph on fig. 2 shows that the smallest resistance is achieved after prototypes with ITO cover annealing at the temperature close to 200 °C. The bigger resistance in different annealing conditions is a result of the big number of structural defects.

Prototypes light transmitting coefficient in visible scope was measured by a spectrophotometer. The results are presented on fig. 3. The graph on fig. 3 shows that the increase of the annealing temperature leads to the increase of glass with ITO cover transmitting coefficient. For further experiments prototypes with conducting coating annealed at 200 °C were used. The resistance of the conductive coating layer should be minimized to improve the working parameters of electrochromic indication systems. The results of the experiments are shown on voltammogram of electrochromic cover prototype without annealing (fig.4) and on cyclical voltammogram of the annealed prototype with electrochromic cover (fig.5) [2,15].

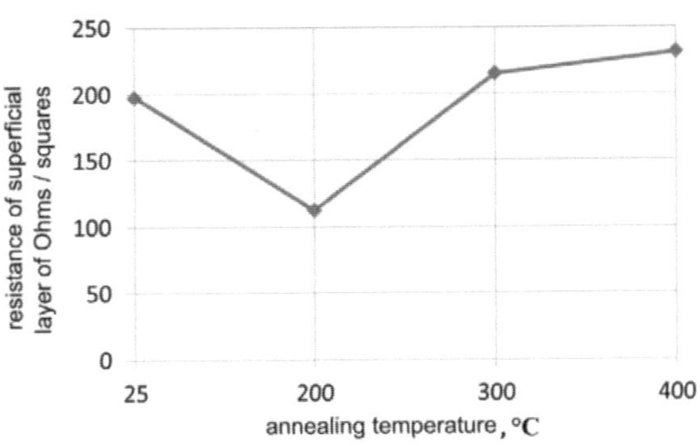

Figure 2. The resistance of ITO thin-film conductive surface on the glass substrate at different annealing temperatures.

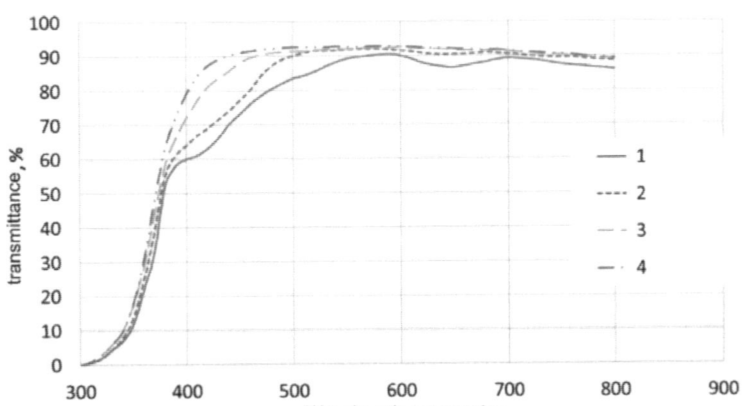

Figure 3. Prototypes with ITO cover transmitting coefficient in visible scope. 1-4 – sample number.

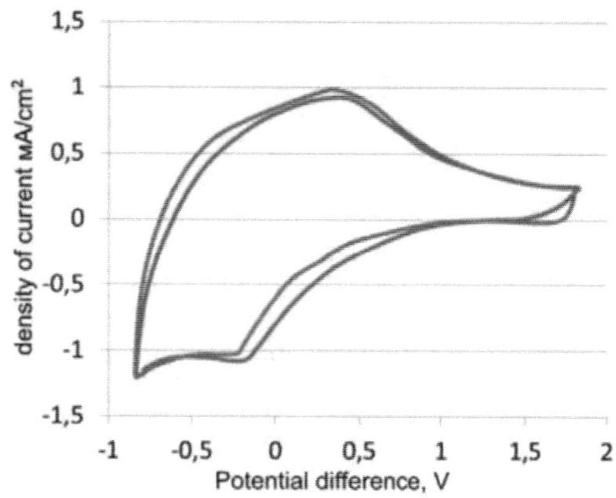

Figure 4. Cyclical voltammogram of electrochromic cover prototype without annealing based on wolfram oxide formed by chemical precipitation from colloidal suspension

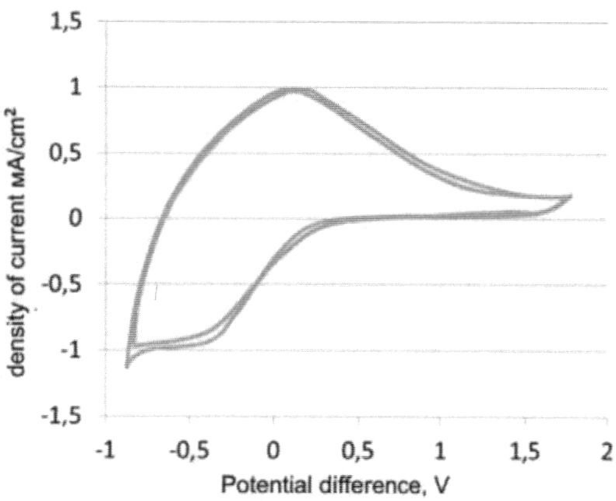

Figure 5. Cyclical voltammogram of the annealed prototype with electrochromic cover based on wolfram oxide formed by chemical precipitation from colloidal suspension

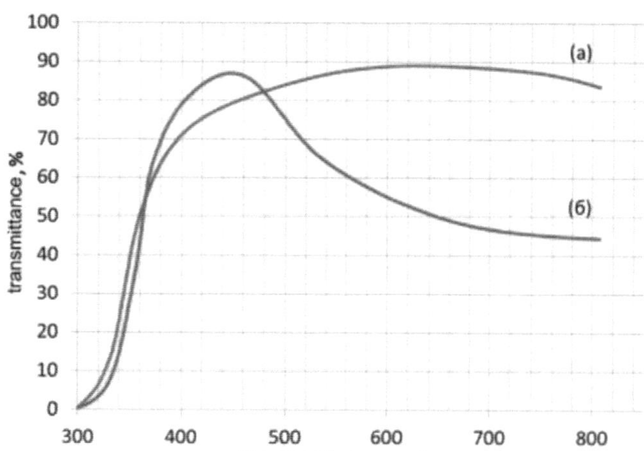

Figure 6. Electrochromic cell with wolfram oxide cover transmission spectrum in discolored (a) and colored state (b)

Graphs on fig.4 and fig.5 have two current peaks – one towards the external voltage increase and one backward. These peaks demonstrate that reduction and oxidation processes proceed towards and backward respectively. The number of peaks suggests that the system consists of only one oxidation-reduction pair without any impurities. Voltammogram of electrochromic cover prototype without annealing also shows the dispersion of the measured values after 10 cycles of measurements. Voltammogram of the annealed prototype with electrochromic cover shows smaller measurements dispersion. In addition, the annealed prototype has more pronounced current peaks than the cover

IOP Conf. Series: Materials Science and Engineering **151** (2016) 012005 doi:10.1088/1757-899X/151/1/012005

prototype without annealing. This difference reveals that the annealed prototype structure is more homogeneous comparing to amorphous structure of the non-annealed prototype.

The research on optical characteristics of the formed electrochromic cell was undertaken using the annealed cover prototype. An electrochromic cell with wolfram oxide cover transmission spectrum is shown in fig.6. Generally, the prototype in the discolored state is transparent in the whole visible scope. The average transmitting coefficient is 82%.

The advantages of electrochromism phenomenon are low power consumption due to color saving effect (saving color without applying any external voltage), low cost of the basic materials, an opportunity to use simple printing technologies to manufacture electrochromic devices and integrate them with other printed electronics and an extensive palette of color materials which simplify color devices construction by eliminating color filters. The disadvantages of electrochromic devices are relatively long time for optical characteristics switching and a possible degradation of some materials.

References

[1] Besedina K N, Vlasov A I, Tokarev S V, Moiseev K M, Panfilov Y V, Shakhnov V A 2011 *Sensors and systems* **7** 69-77 (*in Russian*)

[2] Tokarev S V 2014 Proc. *Youth Int. Scientific and Technical Conf. "Knowledge-intensive technologies and intellectual systems"* (*Moscow*) (BMSTU) p 282-297 (*in Russian*)

[3] Belousov A L and Patrusheva 2014 T N *Journal of Siberian Federal University. Engineering & Technologies* **7** 154-166 (*in Russian*)

[4] Granqvist C G 2002 *Handbook of inorganic Electrochromic Materials* (Elsevier)

[5] Makaryan I A, Efimov O N and Gusev A L 2014 *Alternative energy and ecology* **3**(143) 81-93

[6] Monk P M S, Mortimer R J and Rosseinsk D R 2007 *Electrochromism and electrochromic devices* (Cambridge University Press) p 488

[7] Avendano E, Berggren L, Niklasson G A, Granqvist C G and Azens A 2006 *Thin Solid Films* **496** 30-36

[8] Granqvist C G, Lansaker P C, Mlyuka N R, Niklasson G A and Avendano E 2009 *Sol. Ener. Mat. Sol. C.* **93** 2032-2039

[9] Rosseinsky D R and Mortimer R J 2001 *Advanced Materials* **13** 783-793

[10] Glen A 2009 *Indium Tin Oxide and Alternative Transparent Conductor Markets* (NanoMarket LC)

[11] Monk P M S 1998 *The Viologens: Physicochemical Properties, Synthesis and Applications of the Salts of 4,40-Bipyridine* (Chichester: Wiley)

[12] Grant B, Clecak N J and Oxsen M 1980 *J. Org. Chem.* **45** 702-705

[13] Rowley N M and Mortimer R J 2002 *Science Progress* **85** 243-262

[14] Gabrusenoks J V, Cikmach P D, Lusis A R, Kleperis J J and Ramans G M 1984 *Solid State Ionics* **14** 25-30

[15] Terentiev D S, Vlasov A I and Tokarev S V 2013 *Caspian Sea journal: monitoring and high technology* **1** (21) 016-026

[16] Terentiev D S and Vlasov A 2013 *Proc. 1st Int. Sci. Conf. Science progress in European countries: new concepts and modern solutions (Stuttgart)* (ORT Publishing) p 100-104

[17] Terentiev D S and Vlasov A 2014 *Sensors and systems* **4** 45-55 (*in Russian*)

Multichannel analog front-end and analog-to-digital converter ICs for silicon photomultipliers

Y I Bocharov and V A Butuzov

National Research Nuclear University MEPhI (Moscow Engineering Physics Institute), Kashirskoe shosse 31, 115409 Moscow, Russian Federation

Corresponding author e-mail address: yibocharov@mephi.ru

Abstract. Integrated circuit (IC) of multichannel analog front-end and a mixed-signal chip of multichannel analog-to-digital converter are presented. A chipset of these two ICs is intended for readout, analog preprocessing and analog to digital conversion of silicon photomultiplier array signals. The number of channels of the analog front-end IC as well as the types of their input stages depends on the application. The current test version of the chip contains three current-input channels and three voltage-input channels. Each of the channels includes a programmable pre-amplifier, integrator with baseline-holder, code-controlled amplifier, amplitude discriminator, two programmable timers, pulse-shaping low-pass filter, peak detector, and an output buffer with baseline tuning circuitry. The analog IC has code-configurable architecture. The mixed-signal IC includes nine main channels and one auxiliary channel, containing 10-bit analog-to-digital converter in each channel. It also has a buffer memory and a voltage reference. The chip features low power consumption, which is less than 0.5 mW per channel at a sampling rate of 100 kHz. Both ICs are implemented in 0.35 μm CMOS technology.

1. Introduction

A silicon photomultiplier (SiPM) is an optical sensor [1], which is widely used in medicine, in particular in positron emission tomography (PET) and gamma cameras [2], as well as in experimental physics facilities [3] and the other areas of science and technology. SiPM cell (macrocell) includes a great number of microcells, each of which contains an avalanche photodiode and a quenching resistor limiting the Geiger discharge (Figure 1). SiPM arrays contain multiple cells with independent or multiplexed outputs. New, rapidly growing use of these devices is the use them in portable devices for detection of ionizing radiation and radioactive nuclides. In this application, their main advantage over photomultiplier tubes is a much lower level of power consumption.

A readout of SiPM array signals requires a use of dedicated multichannel analog front-end integrated circuit (IC). A multi-channel analog-to-digital converter (ADC) with a buffer memory is needed as an interface between the analog front-end and a digital processing unit. A complete system for readout, analog processing and analog-to-digital conversion of SiPM array signals is shown in figure 1. A number of ICs have been developed for readout and processing signals of SiPM cells and arrays [4,5]. Most of them are designed for application in medical scanners and multi-channel detectors of physical installations. The use of such ICs in portable applications, where the number of SiPM array elements does not exceed a dozen, is not effective, because of excessive power consumption and functionality.

Known application specific IC for the portable devices [6] do not provide a comprehensive solution to the problem of readout, analog processing and analog-to-digital conversion signals of SiPM arrays.

To solve this problem a chipset of two ICs – analog and mixed-signal have been designed and manufactured. By using this chipset, it is possible to build a fully functional system without additional elements.

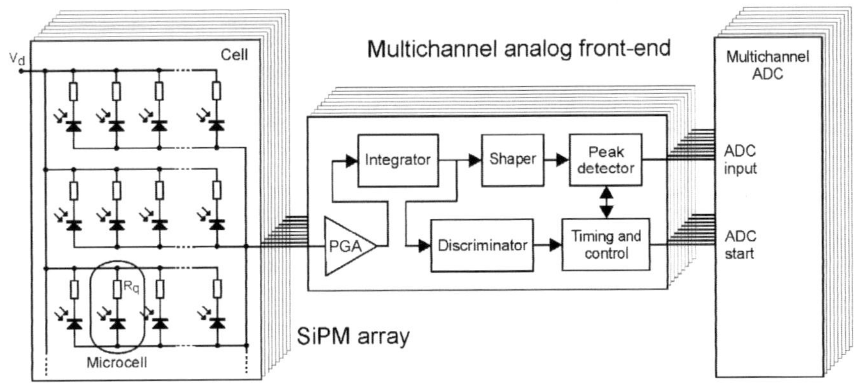

Figure 1. The architecture of a system that performs analog processing and analog-to-digital conversion signals of SiPM array.

2. Analog front-end

2.1. Architecture

Figure 1 shows a block-diagram of a multichannel system, which contains considered in this paper application specific ICs of analog front-end and ADC. The number of channels of the analog front-end IC depends upon the application. The current evaluation version of this chip contains six channels. There are three current-input channels and three voltage-input channels to evaluate IC performance in different applications. The mixed-signal chip comprises ten ADC channels.

2.2. Programmable gain amplifiers

Each channel has a programmable gain amplifier (PGA). All PGAs are identical except for the input stage. Figures 2 and 3 show the simplified schematic diagrams of amplifiers, which have a current-input stage and voltage-input stage, correspondingly.

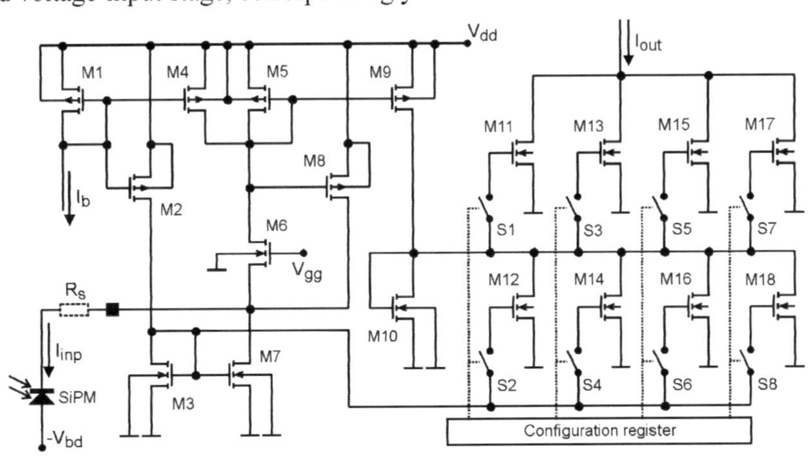

Figure 2. Schematic diagram of the current-input programmable gain amplifier.

A current feedback in the current-input stage reduces the input impedance and stabilizes the current transfer ratio. A programmable stage has a current transfer ratio of 0.5, 1, 2 and 4. It is based on a current mirror circuit. The current gain is a ratio of the channel width of MOS transistors M11 - M17, selected in accordance with the code in the configuration register, to M10 channel width.

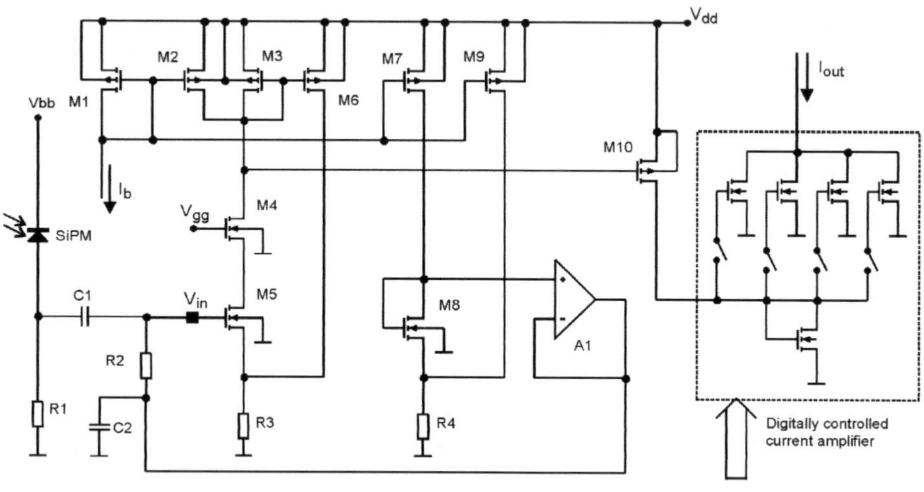

Figure 3. Schematic diagram of voltage-input programmable gain amplifier.

The voltage-input programmable amplifier operates as follows. The current pulse of SiPM is converted to a voltage pulse across the resistor R1. This pulse through a transmission line, loaded on the terminating resistor R2, passes to the input of voltage to the current converting stage. It uses a cascode circuit implementation. A negative feedback stabilizes the circuit's transconductance. The operating points of all input stages are set with the same bias circuit (M7 – M9, A1). The operating point of M5 is the same as the operating point of M8 in a replica-stage. The voltage from its output through the buffer operational amplifier A1 and terminating resistors is supplied to the inputs of all voltage-input stages, setting their operating points.

2.3. Integrator and baseline holder

The schematic diagrams of integrator and baseline holder circuits are shown in figure 4. The current-input integrator provides a continuous reset with two selectable time constants. Their choice is determined by the type of scintillator and by the rate of input events.

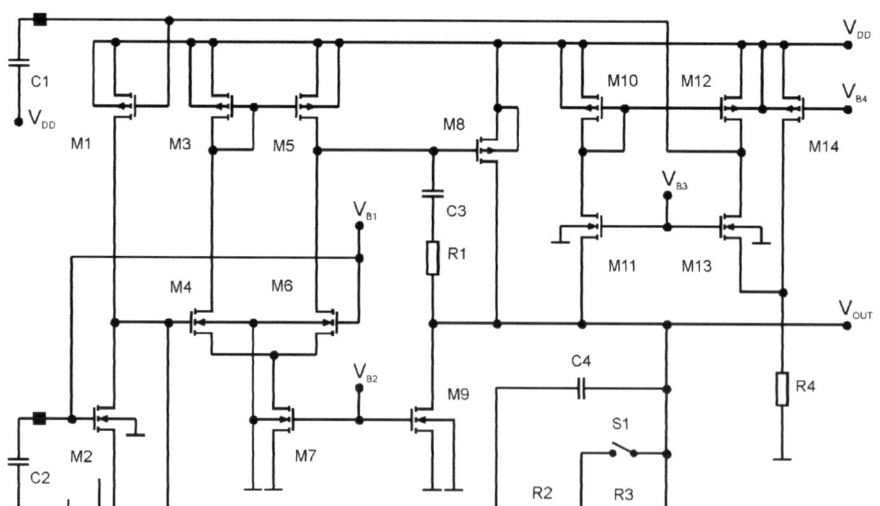

Figure 4. Schematic diagram of integrator and baseline holder.

The integrator is implemented on an operational amplifier (M3 - M9), which is inside of the two feedback loops. The first contains R2, R3 and C4, and the second contains baseline holding circuit components [7]. The baseline holder (DC-level stabilizer) includes a current-output differential amplifier (M10 - M13), an external capacitor C1, and M1 as a controlled current source.

Integrating type circuit in a negative feedback loop generates a zero in a complex transfer function of the signal processing path, forming a high pass filter frequency response that suppresses low-frequency noise components.

2.4. Analog post-processor
Figure 5 shows the main blocks that perform an analog signal processing after integration. The post-processor comprises a time and amplitude channels.

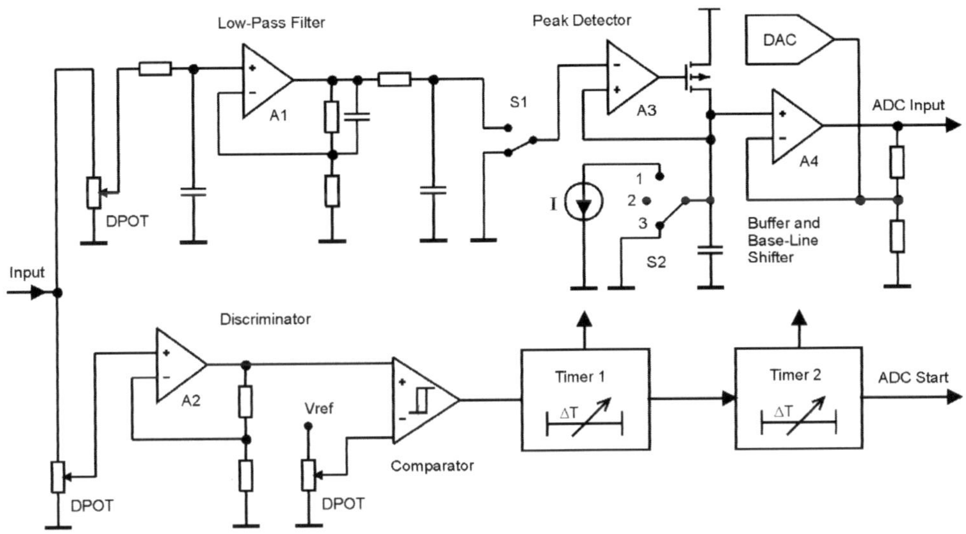

Figure 5. A functional diagram of code-controlled amplifiers, low-pass filter, amplitude discriminator, peak detector, output buffer and timers.

The time channel is used to detect the appearance of the input pulses and for generating signals that control the operation of other blocks. It also generates ADC triggers. The channel includes a digitally controlled amplifier, consisting of the digital potentiometer (DPOT) and fixed-gain amplifier. It also contains amplitude discriminator, consisting of a comparator with a programmable threshold, two programmable timers and a control unit. DPOT has a string architecture, which is based on a tapped resistive divider. The comparator consists of a two-stage preamplifier and Schmitt trigger. Its response time does not exceed 50 ns. Comparator threshold is set by a programmable voltage divider, which consists of an 8-bit DPOT and an external voltage reference. Timers have a resolution of 8 bit.

Amplitude channel comprises a programmable amplifier, similar to the amplifier of the time channel but supplemented by elements forming the frequency response of a third-order low-pass filter (LPF), a peak detector [8] and an output amplifier with a baseline tuning circuit. LPF is used to attenuate high-frequency noise and to ensure the effective operation of the peak detector.

The peak detector has several modes of operation (figure 5). When the switch S1 is in the upper position, and switch S2 is set to the position 1, it operates in a direct transmission (transparent) mode. When the switch S1 is in the upper position, and switch S2 is stored in the position 2, it operates in amplitude (peak) detection mode. When the switch S1 is in the lower position, and switch S2 is stored in the position 2, it operates in a hold mode of fixing the instantaneous value of the signal. When the switch S1 is in the lower position, and switch S2 is set to the position 3, the detector is in a reset mode.

The output buffer is implemented with a rail-to-rail operational amplifier that is able to drive a large capacitive load. It serves as an ADC driver. Baseline adjustment scheme (programmable DC-level shifter) is a 6-bit current-output digital-to-analog converter (DAC). It serves for matching the amplitude range of the processed signal to the input range of the ADC.

3. Mixed-signal IC

3.1. Architecture

Figure 6 shows a block-diagram of ASIC. It contains 9 main channels, each of which includes 10-bit ADC and "data ready" indicator. All ADCs are identical. Each of them has the input for the analog signal and the input for a signal which initiates a conversion process.

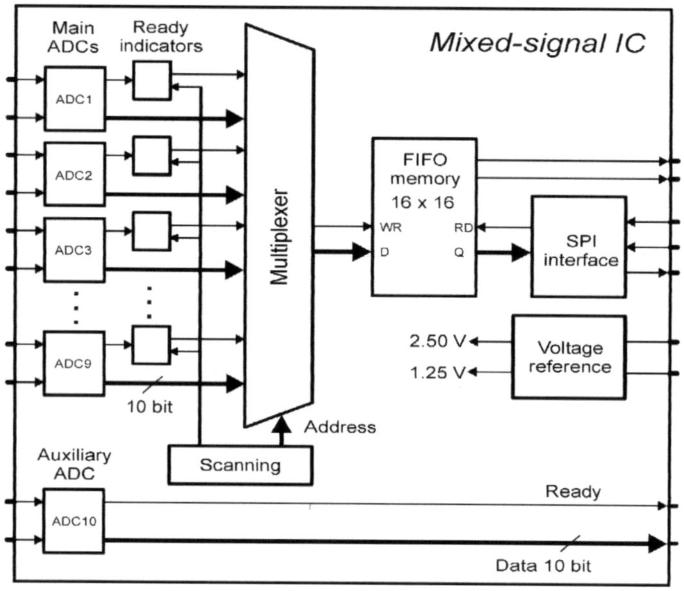

Figure 6. Block-diagram of mixed-signal IC.

The polling system scans the channels, checking the availability of the data. If the indicator of some channel is set, then the system writes output data of the appropriate ADC into the memory along with the channel number. FIFO buffer memory has 16 cells of 16 bits. Data is read from the memory via a serial interface. The chip also includes a 10-th auxiliary ADC that can be used to measure temperature, for monitoring supply voltages and for other purposes.

3.2. ADC details

ADC employs conventional successive approximation register (SAR) architecture (figure 7). One of its main functional blocks is a capacitor array, which serves as a digital-to-analog converter as well as a sample-hold circuit.

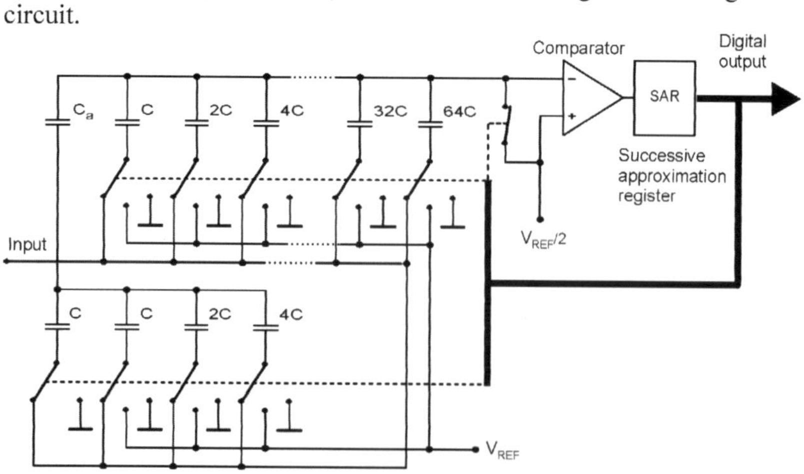

Figure 7. The architecture of SAR ADC.

Previously proposed non-linearity reduction techniques use ADC calibration [9]. They are not effective when designing chips, containing many ADCs, because of the increase in the power consumption and chip area. We proposed a new method of differential non-linearity (DNL) error reduction. It is based on the introduction of additional capacitive elements into capacitor array [10].

4. Implementation and measuring results

4.1. Technology

Both ICs have been manufactured at X-FAB foundry in a mixed-signal 0.35 μm CMOS technology. The capacitors have a structure of "metal–insulator–metal" (MIM). The total chip area is 11.25 mm^2. A layout of designed ICs is shown in figure 8.

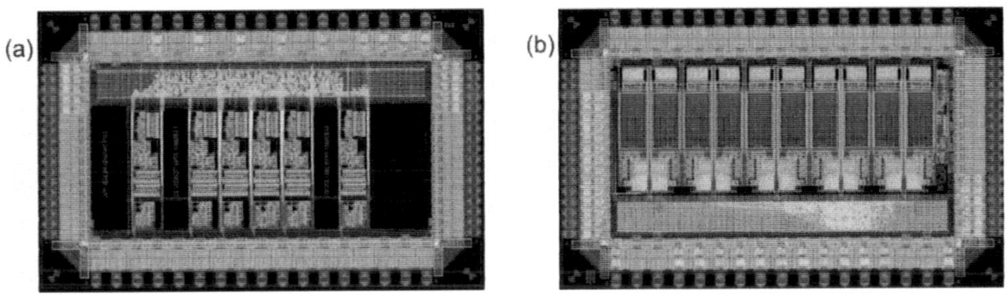

Figure 8. The layout of designed ICs, (a) analog front-end and (b) multichannel ADC.

4.2. Results

In analog IC current-input and voltage-input channels demonstrate a similar performance. Figure 9 shows diagrams, which may be used to estimate their conversion ratio normalized to the current gain of 1.

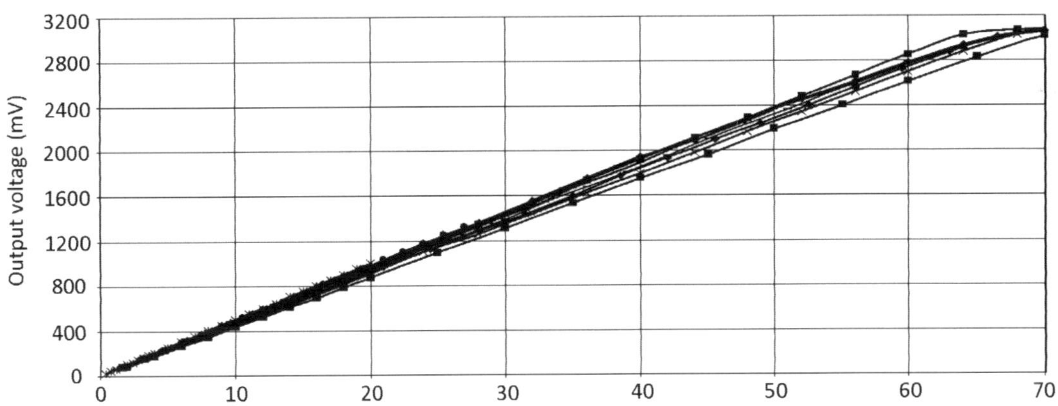

Figure 9. The measured output voltage of analog IC samples versus input charge.

Measurements of the static performance of the mixed-signal IC have shown that the proposed technique of ADC linearity improvement achieves DNL that is less than 0.5 of the least significant bit (LSB). Integral non-linearity (INL) is less than 1.6 LSB. Figure 10 shows diagrams of measured typical differential non-linearity and integral non-linearity of ADC, depending on output digital codes.

Table 1 provides a summary of designed ICs performance parameters.

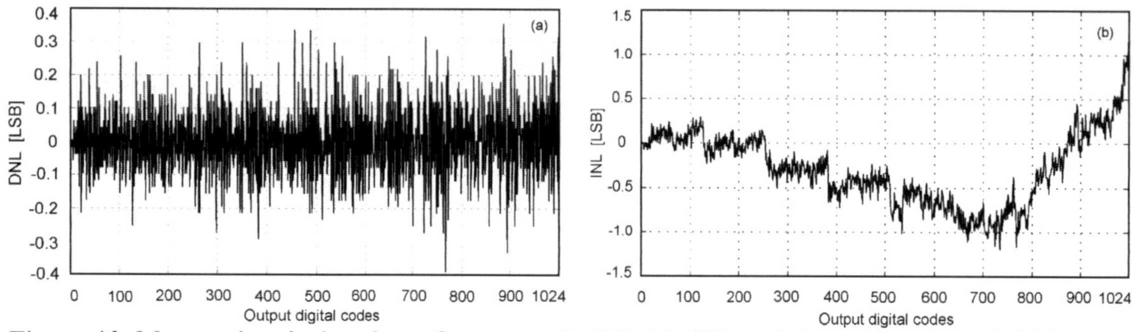

Figure 10. Measured typical static performance of ADC, (a) differential non-linearity and (b) integral non-linearity.

Table 1. Designed ICs performance summary

Analog front-end IC		Mixed-signal IC	
Number of channels	6	Number of channels	10
Current gain control range	0.5 – 4	ADC resolution (bit)	10
Voltage gain control range (bit)	8	Differential nonlinearity (LSB)	0.5
Resolution of timers (bit)	8	Integral nonlinearity (LSB)	1.6
Conversion ratio, max. (mV/pC)	400	Conversion rate, max. (ksps)	125
Dynamic range (pC)	0.1 – 60	Buffer memory capacity (samples)	16
Power consumption per channel (mW)	5.5	Power consumption per channel (mW)	0.45

5. Conclusion

A chipset of two multichannel CMOS ICs has been developed and manufactured. It contains an analog front-end and a low power analog-to-digital converter. The ICs are dedicated for readout, analog processing and analog-to-digital converting of SiPM signals. The chipset is intended for use primarily in portable battery-powered devices. It can also be used in multi-channel data acquisition systems for scientific experiments and in medical equipment such as a gamma camera.

Acknowledgments

The Russian Science Foundation (Grant 16-19-00167) supported this work.

References

[1] Bondarenko G, Dolgoshein B, Golovin V, Ilyin A, Klanner R and Popova E 1998 *Nucl. Phys.* B **61B** 347–352
[2] Roncali E and Cherry S R 2011 *Ann. Biomed. Eng.* **39** 1358–77
[3] Sabet H et al. 2012 *IEEE T. Nucl. Sci.* **59** 1841–49
[4] Rolo M D, Bugalho R, Goncalves F, Mazza G, Rivetti A, Silva J C, Silva R and Varela J 2013 *JINST* **8** C02050
[5] Ahmad S, Fleury J, de la Taille C, Seguin-Moreau N, Dulucq F, Martin-Chassard G, Callier S, Thienpont D and Raux L 2015 *IEEE T. Nucl. Sci.* **62** 664–668
[6] Ghelman M, Paperno E, Ginsburg D, Mazor T, Cohen Y and Osovizky A 2011 *Nucl. Instrum. Meth. A* **652** 866–869
[7] Corsi F, Foresta M, Marzocca C, Matarrese G and Del Guerra A 2009 *JINST* **4** P03004
[8] Kruiskamp M W and Leenaerts D M V 1994 *IEEE T. Nucl. Sci.* **41** 295–298
[9] Yingying C and Dongmei L 2013 *J. Semicond.* **34** 045007
[10] Bocharov Yu I, Butuzov V A and Simakov A B 2015 *Instrum. Exp. Tech.* **58** 623–630

IOP Conf. Series: Materials Science and Engineering **151** (2016) 012007 doi:10.1088/1757-899X/151/1/012007

Influence of chip temperature on the hydrogen sensitivity of MISFET-based sensors

B I Podlepetsky[1], A V Kovalenko[2] and M Y Nikiforova[1]

[1]National Research Nuclear University MEPhI (Moscow Engineering Physics Institute), Kashirskoe shosse 31, 115409 Moscow, Russian Federation
[2]Induko Ltd., Seslavinskaia str. 32/2, Moscow, Russian Federation

Corresponding author e-mail address: bipod45@gmail.com

Abstract. The chip temperature influence on the hydrogen sensitivity of the metal-insulator-semiconductor field-effect transistor (MISFET) with structure Pd-Ta_2O_5-SiO_2-Si was investigated. MISFET sensing element was integrated on silicon chip together with (*p–n*)-junction temperature sensor and heater-resistor. There were measured the hydrogen responses of the MISFET threshold voltage for room and higher chip temperatures. The threshold voltage V_T as a function of hydrogen concentration C was determined for different temperatures T. The models of hydrogen and temperature sensitivities based on the experimental dependencies of $V_T(C,T)$ are presented in this work.

1. Introduction

The hydrogen-containing gases sensors are being used in the fire-explosion safety and in the environmental monitoring systems [1]. The development of integrated sensors with elements fabricated by microtechnology is a promising area for the creation of small-sized gas-analysis devices and systems. The sensitive elements based on metal-insulator-semiconductor transistor structures (MISFETs) possess the best compatibility with the standard elements of integrated circuits. So these sensitive elements seem promising to develop the integrated gas sensors.

The sensors based on MISFETs have been studied by many researchers [2-7]. The studies have shown that the sensor characteristics depend on technological parameters, chip temperature, performance time and external factors (other gases, irradiation). The motivation of this paper is to investigate the chip temperature influence on the hydrogen sensitivity of MISFET with structure Pd-Ta_2O_5-SiO_2-Si and to develop the hydrogen and temperature sensitivities' models of these sensors.

2. Experimental and results

The sensors chip fabricated by conventional MIS-technology consists of Pd-resistor, Pd-gate MISFET, thermosensor and heater (figure 1). Technological processes details are presented in [4-5]. The main parameter of MISFET is the threshold voltage V_T. To measure the threshold voltage, there was used the special circuit (figure 2). The circuit provides the constant drain current I_D = 0.1 mA and source-drain voltage V_D =1.0 V. The constant chip temperature ($T \pm 2$°C) was supported by the temperature-stabilization circuitry with feedback loop using on-chip thermosensor and heater [4-7].

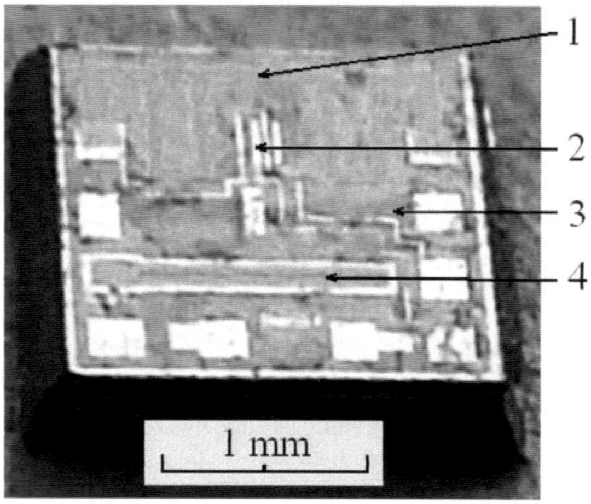

Figure 1. The sensor chip photograph: (1) – Pd-resistor; (2) – (*p*–*n*)-junction temperature sensor; (3) – heater-resistor; (4) – *n*-channel Pd-gate MISFET.

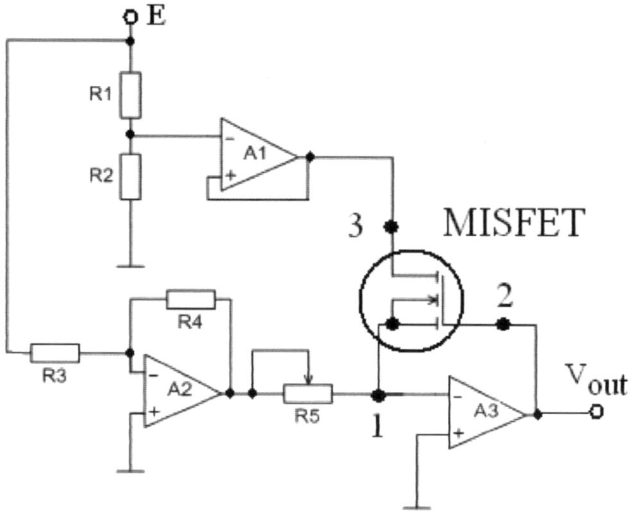

Figure 2. The measuring circuit: (1), (2), (3) are respectively outputs of source, gate, drain of MISFET).

For this circuit's mode, the voltage V_{out} is equal the gate voltage V_G. Therefore, the threshold voltage V_T can be calculated in accordance with the MISFET electrical model [8] as:

$$V_T = V_{OUT} - \left\{2\left[I_D / b - (kT/q)^2\right]\right\}^{1/2}. \quad (1)$$

where b is the MISFET specific transconductance, k and q are the Boltzmann constant and the electron charge respectively. The threshold voltage change ΔV_T is equal ΔV_{out} because of the output voltage V_{out} is equal V_G, and parameters I_D, b, T are constant.

There were tested the five sensors. The sensors was consecutively exposed to hydrogen pulses at duration (25–30) s and concentrations C being equal 0.05, 0.1, 0.15 and 0.2 % vol. (percentage per volume) for different chip temperatures. There were determined the initial threshold voltage V_{T0} and amplitude of hydrogen response ΔV_{TC} for each sensor. The average response amplitudes ΔV_{TC} as a function of concentration C for different chip temperatures are demonstrated in figure 3.

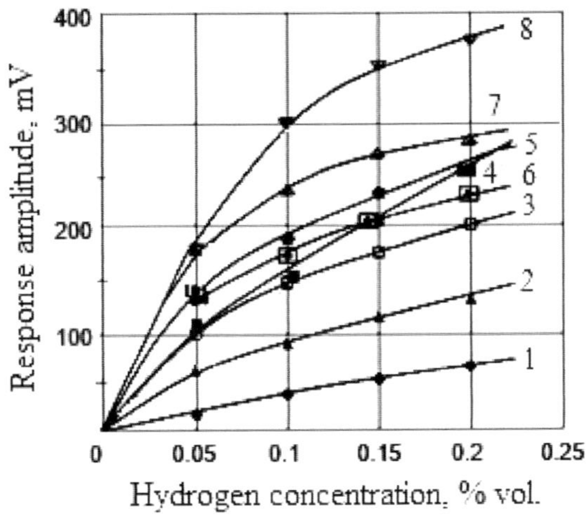

Figure 3. Responses amplitudes ΔV_{TC} as functions of hydrogen concentration C for different chip temperatures: 1 – 25 °C; 2 – 60 °C; 3 – 100 °C; 4 – 130 °C; 5 – 140 °C; 6 – 150 °C; 7 – 160 °C; 8 – 180 °C (symbols are experimental dots, lines are approximating curves).

3. Models of hydrogen and temperature sensitivities

According to the MISFET models [8] the threshold voltage of V_T versus (C; T) is presented as

$$V_T(C,T)=V_{T0}(T)-\Delta V_{TC}(C,T)=V_{T0}(T)-V_M(T)\{1-\exp[-k_C^{'}(T)\cdot C]\} \quad (2)$$

where the values of parameter V_M was determined by extrapolation of characteristic ΔV_{TC} (C) shown in figure 3 for $C \approx 1.0$ % vol. The average value k_C was calculated as the arithmetic mean of

$$k_C=(1/C)\cdot \ln\{V_M/[V_M-\Delta V_{TC}(C)]\} \quad (3)$$

The hydrogen responses' and model parameters are demonstrated in table 1.

Table 1. Average experimental values of the hydrogen response parameters (values of ΔV_{TC} are given for hydrogen concentration 0.1 %) and model parameters of MISFET hydrogen sensitivity.

Parameters	Temperature (°C)							
	25	60	100	130	140	150	160	180
V_{T0} (V)	1.75	1.68	1.46	1.35	1.31	1.27	1.23	1.14
ΔV_{TC} (mV)	35	86	150	158	182	176	235	300
V_M (mV)	82	174	276	415	320	284	330	465
k_C ((%)$^{-1}$)	7.3	6.7	6.6	6.4	7.8	8.2	9.3	9.9

According to (2) the differential hydrogen sensitivity of MISFET threshold voltage

$$S_C(C,T)=|dV_T/dC|=k_C\cdot V_M(T)\cdot \exp(-k_C\cdot C) \quad (4)$$

In accordance with experimental data the maximum values of hydrogen sensitivity ($S_{CM}=k_C\cdot V_M$) are increasing from 0.6 to 4.6 (V/%), if the chip temperature is rising from 25°C to 180 °C. There was supposed the linear approximations for parameters of the chip temperature MISFIT sensitivity model

$$V_{T0}(T)=V_{T00}(1+\alpha_0\Delta T) \text{ and } V_M(T)=V_{M0}(1+\alpha\Delta T) \quad (5)$$

where V_{T00} and α_0 are equal 1.75 V and ($-$ 2.23·10^{-3} (°C)$^{-1}$) for temperatures from 25°C to 180 °C. The parameters of V_{M0} and α are presented in table 2 for the fourth temperature ranges. The values ΔT are ($T - T_0$), and T_0 are initial temperatures for each temperature range.

Table 2. Average calculated model parameters.

Parameters	Temperature ranges (°C)			
	25-60	100-130	140-150	160-180
k_C	7	6.5	8	9.6
V_{M0} (mV)	80	280	410	280
α ((°C)$^{-1}$)	$3.33\cdot10^{-2}$	$1.55\cdot10^{-2}$	$-1.59\cdot10^{-2}$	$2.2\cdot10^{-2}$

According to (2) and (5) the temperature sensitivity of the threshold voltage for different temperature ranges is presented as

$$S_T(T,\tilde{N}) = |dV_T / dT| = \alpha \cdot V_{M0}(T) \cdot \{1 - \exp[-k_C(T) \cdot C]\} - \alpha_0 V_{T00} \quad (6)$$

Equation (6) can be used to estimate the influence of chip temperature fluctuations on the hydrogen response parameters. For example, if T=130 °C, C=0.1 % and chip temperature fluctuations are ±1°C, that absolute errors for ΔV_{TC} and V_{T0} are equal 2.2 mV/°C and 3.8 mV/°C respectively. Such temperature errors should be taken into account in the practical applications of MISFET based hydrogen sensors.

4. Conclusions

The experimental studies of the chip temperature influence on the hydrogen sensitivity of MISFET with structure Pd-Ta_2O_5-SiO_2-Si have shown that the hydrogen sensitivity tends to increase, if chip temperature is rising from 25°C to 180 °C. The optimal operating temperature is about 130 °C. Presented models can be applied to estimate the errors due to chip temperature fluctuations in MISFET-based sensors used to measure hydrogen concentration.

Acknowledgements
This work was supported by the Competitiveness Program of NRNU MEPhI.

References

[1] Hübert T, Boon-Brett L, Black G and Banach U 2011, *Sensors and actuators* **B 157** 329-52
[2] Lundström I, Shivaraman S, Svensson C and Lundkvist L 1975, *Applied Physycs Letters* **26** 55-7
[3] Lundström I, Armgarth M, Spetz A and Winquist F 1986, *Sensors and actuators* **3-4** 399-421
[4] Fomenko S, Gumenjuk S, Podlepetsky B, Chuvashov V and Safronkin G 1992, *Sensors and actuators* **B 10** 7-10
[5] Podlepetsky B, Gumenjuk S and Fomenko S 1996, *Proceedings of Eurosensors X (Leuven)* Belgium (Elsevier) **3** 637-640
[6] Lundström I, Sundgren H, Winquist F, Eriksson M, Krants-Rülcker C and Lloyd-Spets A 2007, *Sensors and actuators* **B 121** 247-62
[7] Podlepetsky B 2015, *Procedia Engineering* **120** 979-82
[8] Podlepetsky B 2015, *Automation and Remote Control* **76** 535-47

IOP Conf. Series: Materials Science and Engineering **151** (2016) 012008 doi:10.1088/1757-899X/151/1/012008

ELDRS in a wide range of total doses

V S Pershenkov, D V Savchenkov and V A Telets

National Research Nuclear University MEPhI, (Moscow Engineering Physics Institute), Kashirskoe shosse 31, 115409 Moscow, Russian Federation

Corresponding author e-mail address: dvsav@spels.ru

Abstract. The ELDRS enhancement factor versus total absorbed dose and dose rate is considered in a wide range of doses.

1. Introduction

The low dose rate effect in bipolar technologies or Enhanced Low Dose Rate Sensitivity (ELDRS) consists of more severe degradation of a bipolar structure's current gain for a given total dose at low dose rate. The ratio of the sensitive electrical parameter's degradation at low dose rate to that at some fixed high dose rate for a given total dose is termed as low dose rate enhancement factor (EF) [1]. EF can reach more than 10. Detailed analysis of ELDRS has been performed in early works [1] typically related to space electronics for relatively small total doses. However using modern integrated circuits for the front end electronics of High Energy Physics experiments (such as the ATLAS Detector at the Large Hadron Collider at CERN) changes the situation [2,3]. Total dose in the ATLAS Upgrade Inner Detector's Middle Region gives dose value near 50 Mrad (SiO_2) for the front-end electronics operation during 10 years [4]. This corresponds to the dose rate near 0.25 rad (SiO_2)/s taking into account shutdown and inactive periods. The authors of [4] investigating the behavior of the silicon-germanium heterojunction bipolar transistors for ATLAS Detector applications observed the effect of the anomalous recovery of transistor current gain after the total absorbed dose of 30 Mrad (SiO_2): reducing the dose rate leads to the decrease of current gain degradation. This means that the enhancement factor EF becomes less than 1. The same result was obtained in [5]. Therefore devices can suffer from ELDRS for small total doses and show recovery dependence for high doses.

The goal of this paper is to describe the enhancement factor EF versus dose-dependence in a wide range of total absorbed doses.

If the excess base current of bipolar transistor is a linear function of excess interface-trap density [6-8], EF can be written as:

$$EF = {\Delta I_b(low\ dose\ rate)}\Big/{\Delta I_b(high\ dose\ rate)}, \tag{1}$$

where $\Delta I_b(low\ dose\ rate)$ and $\Delta I_b(high\ dose\ rate)$ are the excess base current values at low and high dose rates for a given total dose.

2. Physical model

The most developed model of radiation-induced interface-trap buildup is a two-stage "hydrogen" model [9, 10]. Another model (the so-called "conversion" model [11]) is based on the assumption that generation of interface traps is related to the neutralization of positive charge by the substrate electrons. Experiments [12, 13] have shown that the interaction between hydrogen species and the electrons from substrate is a necessary and important component of the interface-trap build-up. Those results extend the hydrogen model and allow for speaking about a hydrogen-electron model of radiation-induced interface-trap buildup [13].

The concentration of hydrogen in a given microelectronic structure is unknown and depends on a particular manufacturing process. The conversion model allows us to consider the effect of substrate electrons on the interface-trap buildup at any specified hydrogen concentration in the structure. This can be very important for low dose rate applications during a long term device operation, for example, in space. In this work, we assume that the generation of interface traps relates to the oxide trapped charge annealing.

Surface states generation and annihilation can be expressed as follows:

$$\frac{dN_{it}}{dt} = (K_{oi})_{sub}\frac{Q_{ot}}{(\tau_{ann})_{sub}} - \frac{N_{it}}{(\tau_{ann})_{it}}, \tag{2}$$

where N_{it} is the density of interface states; Q_{ot} is the oxide trapped charge; $(\tau_{ann})_{sub}$ is the time constant of the annealing of oxide space charge by electrons from the substrate; $(K_{oi})_{sub}$ is the fraction of space charge converted to interface traps due to the interaction with substrate electrons; $(\tau_{ann})_{it}$ is the time constant of N_{it} annihilation.

The first term on the right-hand side of (2) represents the interface traps buildup through the annealing of trapped charge by the substrate electrons. The second term is responsible for the interface traps annihilation (annealing) [14,15].

Interface traps annihilation is considered from the view of AD-center model [16]. According to [17] time constant of N_{it} annihilation is

$$(\tau_{ann})_{it} = \frac{K_{AD}}{P}, \tag{3}$$

where K_{AD} is a function of thermal velocity, capture cross-section of AD center, generation rate and the radiation induced electrons yield; P is the dose rate.

We suppose that the accumulation and annealing of the positive oxide trapped charge Q_{ot} are described by the following equation:

$$\frac{dQ_{ot}}{dt} = (K_{acc})_{ot}P - \frac{Q_{ot}}{(\tau_{ann})_{sub}}, \tag{4}$$

where $(K_{acc})_{ot}$ is a coefficient characterizing the accumulation of trapped charge; P is the dose rate.

Time constant of the annealing process is a function of temperature described by the Arrhenius law:

$$(\tau_{ann})_{sub} = \tau_0 e^{E_A/kT}, \tag{5}$$

where k is the Boltzmann's constant; T is the temperature; E_A is the activation energy of the oxide trap thermal excitation; τ_0 is the pre-exponential coefficient.

3. Shallow and deep traps

According to the ELDRS conversion model [18,19] the inverse S-shaped form of the excess base current versus dose rate [1,6] assumes that there are two kinds of oxide trapped charge at the Si-SiO_2 interface: shallow one Q_S laying near the edge of Si conduction band, and deep one Q_D, laying near the middle of Si bandgap. The shallow trapped charge has small conversion time constant $(\tau_{ann})_{sub} = \tau_S$ so it is converted to N_{it} in the first place. Deep trapped charge has relatively large conversion time constant $(\tau_{ann})_{sub} = \tau_D$, so it converts to interface traps much later than the shallow one does. When a device is irradiated with high dose rate just shallow charge has time to convert to interface traps. When the device is irradiated with low dose rate both shallow and deeply trapped charges, convert into interface traps causing more degradation than that at high dose rate.

We assume that the processes of shallow and deep trapped charge conversion are independent of each other, that leads us to two separate sets of differential equations (2) and (4) each one having its values of $(K_{acc})_{ot}^{S}$, $(K_{acc})_{ot}^{D}$, $(\tau_{ann})_{sub}^{S} = \tau_S$ and $(\tau_{ann})_{sub}^{D} = \tau_D$.

4. Fitting parameters extraction

Practical use of the model requires extraction of its fitting parameters [18-21]. The model described above has six of them: $(K_{acc})_{ot}^{S}$, $(K_{acc})_{ot}^{D}$, $(K_{oi})_{sub}$, τ_S, τ_D and K_{AD}. The extraction procedure includes the following approaches:

- Irradiation at different doses and dose rates at room temperature. The set of experiments can include, for example, three doses and two dose rates.
- Post-irradiation annealing performed in order to extract the activation energy and the pre-exponential coefficient in (5).
- Irradiation at elevated temperature.

Measurements number (i.e. number of equations for extraction procedure) must exceed the number of fitting parameters. This allows for obtaining averaged and thus more reliable values of the fitting parameters.

In Table 1 fitting parameters are presented that we have extracted for LM111 from [22,23].

Table 1. Fitting parameters for LM111 from [22,23].

$(K_{oi})_{sub}$	1
K_{AD}	10^4 rad
τ_S	$4.5 \cdot 10^2$ s
τ_D	$3.2 \cdot 10^6$ s
$(K_{acc})_{ot}^{D}/(K_{acc})_{ot}^{S}$	5.7

5. Enhancement factor in a wide range of total doses

Fig. 1 shows the calculated EF versus total dose for LM111. The calculations were made using equation (1) for four dose rate values: 100 rad(Si)/s (reference dose rate), 0.1 rad(Si)/s, 10 mrad(Si)/s and 1 mrad(Si)/s. EF stays approximately constant at doses in the range $10^5 - 10^7$ rad(Si) for dose rates 10 mrad(Si)/s and 1 mrad(Si)/s. At large doses, EF monotonically decreases finally approaching 1. For 0.1 rad(Si)/s and 1 rad(Si)/s dose rates EF has the maximum at 10^6 rad(Si) and 10^7 rad(Si) respectively.

6. Discussion

In general, EF versus total dose is a non-monotonic curve. For LM111 at small dose rates (10 mrad(Si)/s and 1 mrad(Si)/s) the enhancement factor stays constant up to 10^7 rad(Si) and then falls to 1 at higher doses. For dose rates 0.1 rad(Si)/s EF has a plateau for doses $10^6 - 10^7$ rad(Si), becoming smaller outside this range. EF for dose rate 1 rad(Si)/s has the maximum at 10^7 rad(Si). Decrease of EF at doses smaller than 10^6 rad(Si) is due to the fact that at these doses exposure time becomes smaller than the deep traps conversion time ($3.2 \cdot 10^6$ s, see Table 1) and during the exposure time deep traps convert to interface states only partially, that leads to less base current degradation. For 1 rad/s dose rate, the behavior of EF is the same.

EF decrease at high doses is caused by the interface states annealing. At very large absorbed doses the accumulation of interface states saturates. Note that the saturated N_{it} value doesn't depend on dose rate [14,15], and therefore at doses greater than 10^8 rad(Si) all EF curves converge. The obtained data tell us that at very large doses the influence of ELDRS is not that critical. The interface states build-up saturation can occur for relatively small doses [24,25]. Therefore the role of ELDRS can reduce even for space electronics devices. This issue requires special analysis.

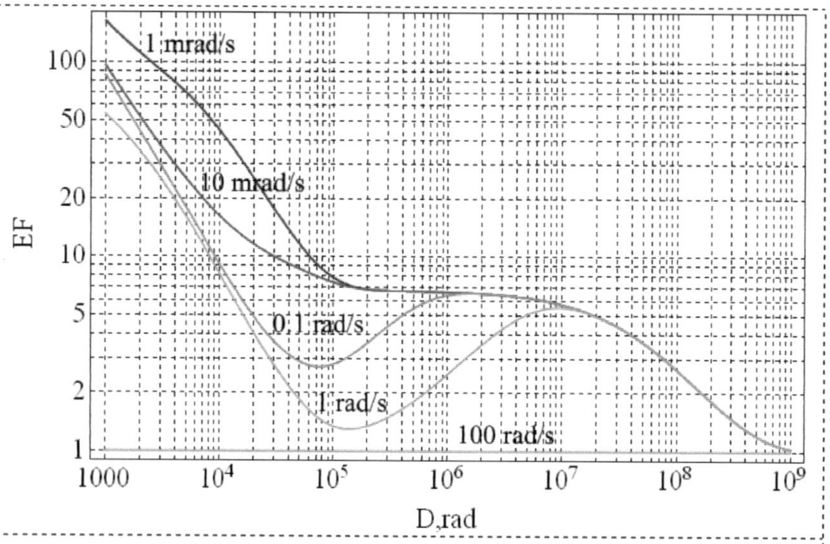

Figure 1. Enhancement factor versus total dose at different dose rates for LM111. Fitting parameters were extracted from [22] (see Table 1).

7. Conclusion

The conversion model of surface traps accumulation and annihilation has been described and based on this model the analysis of enhancement factor versus total dose has been made in a wide range of doses (10^3 – 10^9 rad(Si)).The model has several fitting parameters which can be extracted from experimental data. By the example of LM111, we found that EF versus total dose is a non-monotonic curve. At large doses inherent in the front end electronics of High Energy Physics experiments, the enhancement factor approaches 1 which leads to the lack of ELDRS. The obtained results improve our understanding of the ELDRS physical mechanism and can be useful for bipolar devices' radiation tolerance estimation for the front end electronics in High Energy Physics experiments.

Acknowledgements
This work was supported by the Competitiveness Program of NRNU MEPhI.

References

[1] Pease R L, Schrimpf R D and Fleetwood D M 2009 *IEEE T. Nucl. Sci.* **56** 1894
[2] Ullán M, Rice J, Brooijmans G, Cressler J D, Damiani D, Díez S, Gadfort T, Grillo A A, Hackenburg R, Hare G, Jones A, Kierstead J, Kononenko W, Mandic I, Martinez-McKinney F, Metcalfe J, Newcomer F M, Parsons J A, Phillips S, Rescia S, Sadrozinski H F-W, Seiden A, Spencer E, Spieler H, Sutton A K, Tazawa Y, Wilder M and Wulf E 2009 *Nucl. Instrum. Meth.* A **604** 668
[3] Díez S, Lozano M, Pellegrini G, Mandic I, Knoll D, Heinemann B and Ullán M 2009 *IEEE T. Nucl. Sci.* **56** 2449
[4] Ullan M, Wilder M, Spieler H, Spencer E, Rescia S, Newcomer F M, Martinez-McKinney F, Kononenko W, Glillo A A and Diez S 2013 *Nucl. Instrum. Meth.* A **724** 41
[5] Zebrev G I, Petrov A S, Useinov R G, Ikhsanov R S, Ulimov V N, Anashin V S and Elushov I V 2014 *IEEE T. Nucl. Sci.* **61** 1785
[6] Boch J, Saigne F, Schrimpf R D, Vaille J-R, Dusseau L and Lorfevre E 2006 *IEEE T. Nucl. Sci.* **53** 3655
[7] Pershenkov V S, Savchenkov D V, Bakerenkov A S and Egorov A S 2009 *Russian Microelectronics* **38** 17
[8] Adell P C, Rax B, Esqueda I S and Barnaby H J 2015 *IEEE T. Nucl. Sci.* **62** 2476

[9] McLean F B 1980 *IEEE T. Nucl. Sci.* **27** 1651
[10] Oldham T R and McLean F B 2003 *IEEE T. Nucl. Sci.* **50** 483
[11] Lai S K 1983 *J. Appl. Phys.* **54** 2540
[12] Sogoyan A V, Cherepko S V, Pershcnkov V S, Rogov V I, Ulinov V N and Emelianov V V 1997 *Proc. Eur. Conf. on Radiation and Its Effects on Components and Systems (Cannes)*(IEEE) p 69
[13] Sogoyan A V, Cherepko S V and Pershcnkov V S 2014 *Russian Microelectronics* **43** 162
[14] Benedetto J M, Boesch H E and Mclean F B, 1988 *IEEE T. Nucl. Sci.* **35** 1260
[15] Baze M P, Plaag R E and Johnston A H 1989 *IEEE T. Nucl. Sci.* **36** 1858
[16] Pershenkov V S, Cherepko S V, Sogoyan A V, Belyakov V V, Ulimov V N, Abramov V V, Shalnov A V and Rusanovsky V I 1996 *IEEE T. Nucl. Sci.* **43** 2579
[17] Pershenkov V S, Bakerenkov A S, Solomatin A V, Belyakov V V and Shurenkov V V 2014 *Appl. Mech. Mater.* **565** 142
[18] Pershenkov V S, Savchenkov D V, Bakerenkov A S, Ulimov V N, Nikiforov A Y and Chumakov A I 2009 *Proc. Eur. Conf. on Radiation and Its Effects on Components and Systems (Bruges)*(IEEE) p 286
[19] Pershenkov V S, Savchenkov D V, Bakerenkov A S and Ulimov V N 2010 *Russian Microelectronics* **39** 91
[20] Skorobogatov P K and Nikiforov A Y 1998 *Russian Microelectronics* **27** 1
[21] Akhmetov A O, Boychenko D V, Bobrovskiy D V, Chumakov A I, Kalashnikov O A, Nikiforov A Y and Nekrasov P V 2014 *Proc. Int. Conf. on Microelectronics (Belgrade)* (IEEE) p 329
[22] Johnston A N, Rax B G and Lee C I 1995 *IEEE T. Nucl. Sci.* **42** 1650
[23] Bakerenkov A S 2016 *Proc. Int. Conf. on Nanotechnologies and Biomedical Engineering (Chisinau)*(Springer) p 520
[24] Boch J, Saigne F, Ducret S, Schrimpf R D, Fleetwood D M, Iacconi P and Dusseau L 2004 *IEEE T. Nucl. Sci.* **51** 3225
[25] Boch J, Velo Y G, Sainge F, Roche N, Schrimpf R D, Vaille J, Dusseau L, Chatry C, Lorfevre E, Ecoffet R and Touboul A D *IEEE T. Nucl. Sci.* **53** 3347

Competing nucleation of islands and nanopits in zinc-blend III-nitride quaternary material system

K M Gambaryan, V M Aroutiounian, A K Simonyan and L S Yeranyan

Department of Physics of Semiconductors and Microelectronics, Yerevan State University, 1 Alex Manoukyan, Yerevan 025, Armenia

Corresponding author e-mail address: kgambaryan@ysu.am

Abstract. The growth mechanism of quantum dots (QDs), nanopits and collaborative QDs-nanopits structures in GaN-InN-AlN material system is theoretically investigated using the continuum elasticity model. The islands energy versus their volume, as well as the critical energy and volume versus the island and wetting layer lattice constants relative mismatch ratio (strain ε), are calculated. It is shown that when the zinc-blend GaN is used as a substrate and when the strain between the wetting layer and a substrate overcomes critical $\varepsilon^* = 0.039$ value, instead of QDs nucleation, the formation of nanopits becomes energetically preferable. Revealed feature is critical and has to be taking into account at QDs engineering in GaInAlN material system.

1. Introduction

During the last two decades, the use of quantum dots (QDs), corresponding semiconducting materials and their band gap engineering, opens up entirely new functionalities of traditional devices as well as new challenges for the fabrication of devices with unique properties. In particularly, single photon sources for quantum cryptography, quantum dot lasers, single photon detectors, single electron transistors, resonant tunneling diodes, etc [1-4]. Indeed, the physical properties of QDs depend on QDs size and shape, as well as on the mechanism of their formation. The most useful approach for the fabrication of QDs is Stranski–Krastanow growth mode, where the sum of the surface free energy and the interface free energy is about the same as the substrate free energy. In this case, the wetting layer is compressively strained in a few percent. Interestingly note, that in the original publication by Stranski and Krastanow, no strain effects were considered. The strain relaxation leads to the formation of coherent (dislocation free) islands on top of a thin wetting layer. Depending on the strain value and its sign, the growth of QDs, the formation of nanopits or even QDs–nanopits cooperative structure can be achieved.

Binary III-V compound semiconductors, especially nitrides and their ternary and quaternary alloy are very attractive for several applications [5]. For instance, GaInN alloys are used for fabrication of blue and green light emitted diodes, as well as for violet and blue lasers [6]. Since the band gap of GaInN can be varied from 2.0 to 3.5 eV by increasing of GaN concentration, the potential operating wavelengths cover nearly the entire visible spectra range [7,8]. High-speed field effect transistors, high-temperature electronic devices, UV and blue light emitters, detectors and gas sensors were made of GaN [9]. Among III-nitride semiconductors, InN has lowest effective mass and small band gap. Therefore, InN-related solid solutions can extend the emission or absorption from the UV to near infrared regions. The photovoltaic (PV) and thermo-PV cells were also fabricated using InN [9].

Additionally, the sufficient lattice mismatch between the III-N binary compounds allows growing of nanostructures in Stranski–Krastanow growth mode.

In this paper, the growth mechanism of QDs, nanopits and collaborative QDs-nanopits structures in GaN-InN-AlN material system is theoretically investigated using the continuum elasticity model proposed by J. Tersoff (IBM) [1,10].

2. Total energy of island–pit structure in GaN-InN-AlN material system

Here, according to [10] we assume that the zinc-blend GaN substrate has only discrete orientations and therefore only one angle can be used [10]. We also assume that islands and pits have a shape as schematically presented in figure 1(*a*). High-resolution SEM images of the InAsSbP composition pyramidal island and a nanopit [4] grown on InAs(100) substrate are presented in Figures 1(*b*,*c*).

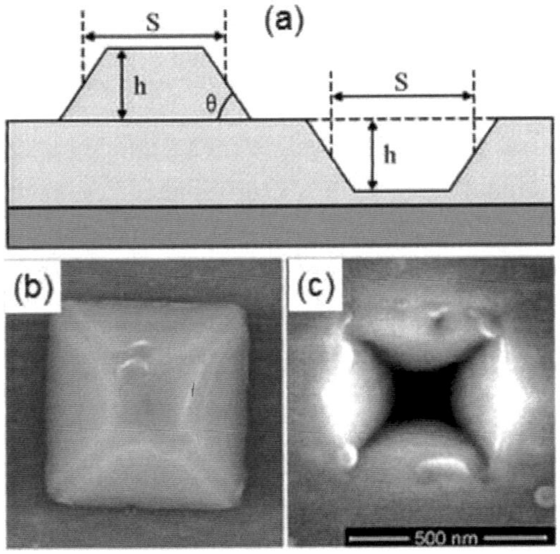

Figure 1. Schematic view of the island-nanopit structure's cross section – (*a*), high-resolution SEM images of the InAsSbP composition pyramidal island and a nanopit – (*b, c*).

As it is known from [10], the total Gibbs free energy for the formation of an island (or a pit) can be written as $E = E_S + E_R$, where E_S and E_R are the change in surface energy the reduction of the strain energy by elastic relaxation, respectively. Considering island's volume as a constant, in the case of $s = t = h \cot\theta$, where s, t, h and θ are the length, width, height (depth) and contact angle, as in figure 1(*a*), the energy is equal to:

$$E = 4\Gamma V^{2/3} \mathrm{Tan}^{1/3}\theta - 6cV\mathrm{Tan}\theta \qquad (1)$$

where $\Gamma = \gamma_e \mathrm{Csc}\theta - \gamma_s \mathrm{Cot}\theta$. For the crystals with a cubic symmetry, $\gamma_s = \frac{1}{2}\cdot\varepsilon^2(C_{11} + C_{44})d_{wet}$, $c = \sigma_b^2 \frac{(1-\nu)}{2\pi\mu}$, $\sigma_b = \varepsilon(C_{11} + C_{44})$. Here γ_s and γ_e are the surface Gibbs free energy of mixing per unit area for the normal orientation and the beveled edge, respectively, $\varepsilon = \frac{\Delta a}{a}$ is the lattice mismatch ratio (strain) and d_{wet} is the wetting layer thickness. The value for γ_e can be found from Young equation $\gamma_{sl} = \gamma_s - \gamma_e \mathrm{Cos}\theta$ [11], where for Stranski–Krastanow growth mode $\gamma_{sl} = 0$ is the surface Gibbs free energy of mixing corresponding to the solid-liquid interface, $\nu = \frac{\lambda}{2(\lambda+\mu)}$ is the Poisson ratio, μ, λ and C_{ij} are the Lame coefficients and the elastic modulus of the substrate. Finally, the expression for the total energy can be written as:

$$E = 4\left(\gamma_e Csc\theta - \frac{1}{2}\varepsilon^2 (C_{11} + C_{44}) d_{wet}(\varepsilon) Cot\theta\right) V^{2/3} Tan^{1/3}\theta - $$
$$- 3\varepsilon^2 (C_{11} + C_{44})^2 \cdot \frac{(1-\nu)}{\pi\mu} \cdot V\, Tan\theta \qquad (2)$$

Next, we performed a mathematical approximation of experimental data [12] in order to evaluate for the GaInAlN material system an analytical expression for the dependence of wetting layer thickness versus strain. In our calculations we used the following expressions for d_{wet} in monolayers (ML): (*i*) if the deformation strain is positive, then $d_{wet} = 0.05\varepsilon^{-3/2}$ at $\varepsilon > 0.03$ [12] and $d_{wet} = 24.181 e^{-31.034\varepsilon}$ at $0 < \varepsilon < 0.03$ (accuracy of approximation $R^2 = 0.9635$), (*ii*) if the deformation strain is negative, then $d_{wet} = 0.15|\varepsilon|^{-3/2}$ at $|\varepsilon| > 0.035$ [12] and $d_{wet} = 45.162 e^{-23.03|\varepsilon|}$ at $0 < |\varepsilon| < 0.035$ (accuracy of approximation $R^2 = 0.9934$).

Dependence of the GaInAlN strain-induced islands and pits total energy versus volume, calculated at $\gamma_e = 10.15 \cdot 10^{-5}$ J/cm^2, $\mu = 30.34 \cdot 10^4$ J/cm^3, $C_{11} = 272.3 \cdot 10^3$ J/cm^3, $C_{44} = 130.3 \cdot 10^3$ J/cm^3, $\nu = 0.361$ and $\theta = 0.785$ (45°), is presented in figure 2(*a*) at different strains.

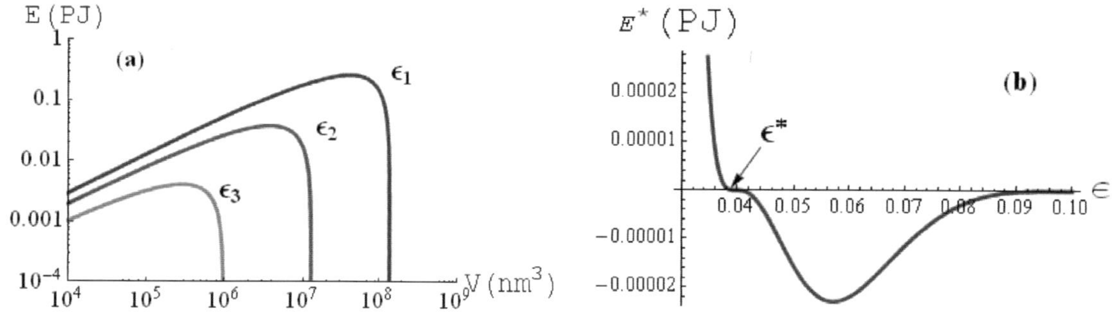

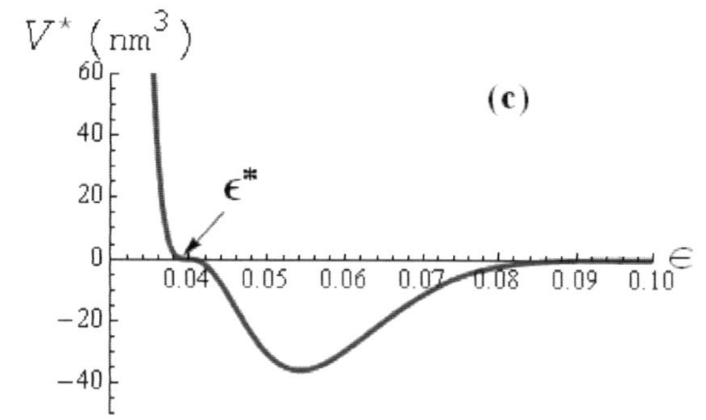

Figure 2 (color online). (*a*) – dependence of the GaInAlN composition islands energy on volume at different strain (ε_1=0.02, ε_2=0.025, ε_3=0.03). (*b,c*) – critical energy and critical volume versus strain, respectively.

From Figure 2(*a*) it is quite visible that an island has to overcome the energy barrier E^* which reached at volume V^*. Figures 2(*b*) and 2(*c*) show the dependences of the critical energy and critical volume versus strain, respectively. Our calculations show the evident result, i.e. both E^* and V^* depend on the strain and that at some critical strain of $\varepsilon^* = 0.039$ the sign of energy and volume is changed. According to the islands configuration (figure 1(*a*)), as well as question (2), we conclude that ε^* is the edge of the strain for island nucleation, after that the formation of a pit becomes energetically preferable.

3. Conclusion

Thus, the growth mechanism of QDs, nanopits and collaborative QDs-nanopits structures in GaN-InN-AlN material system was investigated using the theoretical approach proposed by J. Tersoff (IBM). The islands energy versus their volume, as well as the critical energy and volume versus the strain, are calculated. When the zinc-blend GaN is used as a substrate, it was shown that when the strain between the wetting layer and a substrate overcomes critical $\varepsilon^*=0.039$ value, the formation of nanopits becomes energetically preferable instead of QDs nucleation. Revealed feature has to be taking into account not only at QDs engineering but also at the growth of bulk crystals and epitaxial thin films in GaInAlN material system.

Acknowledgments

This work was performed in the frame of the project № 15T-2J137 funded by the State Committee of Science of Armenia.

References

[1] Tersoff J and Le Goues F K 1994 *Phys. Rev. Lett.* **72** 3570
[2] Gambaryan K M, Aroutiounian V M and Harutyunyan V G 2011 *Infrared Phys. and Technol.* **54** 114
[3] Gambaryan K M 2010 *Nanoscale Res. Lett.* **5** 587
[4] Aroutiounian V M, Gambaryan K M and Soukiassian P G 2010 *Surf. Sci.* **604** 1127
[5] Ihsiu H and Stringfellow G B 1996 *Appl. Phys. Lett.* **69** 18
[6] Nakamura S, Senoh M, Iwasa N and Nagahama S 1995 *Jpn. J. Appl. Phys.* **34** L79
[7] Nakamura S, Senoh M, Nagahama S, Iwasa H, Yamada N T, Matsushita T, Kiyoko H and Sugimoto Y 1996 *Jpn. J. Appl. Phys.* **35** L74
[8] Strite S and Morkoc H 1992 *J. Vac. Sci. Technol. B* **10** 1237
[9] Kim H W, Kim H S, Na H G, Yang J C, Kim S S and Lee C 2010 *Chemical Engineering Journal* **165** 720–727
[10] Tersoff J and Tromp R M 1993 *Phys. Rev. Letters* **70**(18) 2782-2785
[11] Żenkiewicz M 2007*J. Achievements in Materials and Manufacturing Engineering* **24**(1) 137-147
[12] Biehl M, Much F and Vey C 2005 *Int. Series of Numerical Mathematics* **149** 41

IOP Conf. Series: Materials Science and Engineering **151** (2016) 012010 doi:10.1088/1757-899X/151/1/012010

THORON-SCOUT – first diffusion based active Radon and Thoron monitor

W Wagner, T Streil, V Oeser, G Horak and M Duzynski

SARAD GmbH, Wiesbadener Str. 10, 01199 Dresden, Germany

Corresponding author e-mail address: w.wagner@sarad.de

Abstract. THORON-SCOUT is a stand-alone diffusion based active Radon and Thoron monitor for long term indoor measurements to evaluate the human health risk due to activity concentration in the breathing air. Alpha-particle spectroscopy of Po isotopes, being the progeny of the decay of the radioactive noble gas Radon, is applied to separately monitor activity contributions of ^{222}Rn and ^{220}Rn (Thoron) as well. In this work we show that the portion of Thorpn (Tn) may locally be remarkable and even dominating and cannot be neglected as often has been assumed up to now. Along with tobacco consumption, Rn radioactivity turned out to be a dangerous cause of lung cancer, especially in older badly vented buildings situated in regions of radioactive geological formations. THORON-SCOUT allows a precise examination of the indoor atmosphere with respect to Rn and Inactivity concentration and, therefore, a realistic evaluation of corresponding health risk.

1. Basic considerations

Two radiologically important isotopes of the noble gas Radon occur in nature: ^{222}Rn (isotopic ratio ≈90%) and ^{220}Rn (≈9%; called Thoron (Tn)). Together with their short-lived radioactive progeny, they are members of the natural decay series of ^{238}U and ^{232}Th, respectively. Migrating from environmental bedrock and soil into buildings, radioactive Rn gas and its aerosol-attached and free progeny, in this way, become health-relevant constituents of the breathing air.

The evaluation of human health risk due to inhalation of radioactive Rn and subsequent damage of lung tissue by energy-rich alpha particles from the decay of absorbed progeny is mainly based on activity concentration measurement of ^{222}Rn. The risk due to Thas usually been ignored up to now because of putative less abundance. As shown below, this must not be the case in general but strongly depends on the environmental conditions.

The ratio of the so called potential alpha energy (PAE) per decay for Tn and Rn progeny with short half-life amounts to PAE/Bq (^{216}Po⇐Tn) / PAE/Bq (^{218}Po ⇐Rn) ≈8.75%.

Summing up over the entireTn and Rn decay series, this ratio increases considerably and, consequently, the incorporated equivalent dose due to Tn alpha decay. The precise evaluation of the local health risk, therefore, requires long-term monitoring of both the Rn and Tn activity concentrations in indoor air.

2. THORON-SCOUT

The THORON-SCOUT developed by SARAD GmbH Dresden is the first diffusion based device on the market appropriate for long term active Rn and Tn monitoring[1] (figure 1).

Figure 1. Layout of the THORON-SCOUT.

It applies fast gas diffusion through a special membrane into a measurement volume which is arranged outside of the corpus of the device to guarantee a higher sensitivity. The measurement status and actual results are visualized on a small display.

The ionized decay products from Rn and Tn alpha radioactivity are collected by an electrical field on the surface of a Si semiconductor detector. Alpha-particle spectroscopy within four regions of interest (ROI) is applied to register the decay of $^{218,\ 216,\ 214,\ 212}$Po isotopes as well as to distinguish between Rn and Tn progeny (figure 2).This enables the quantitative evaluation of the relevant activity concentrations in the local environment.

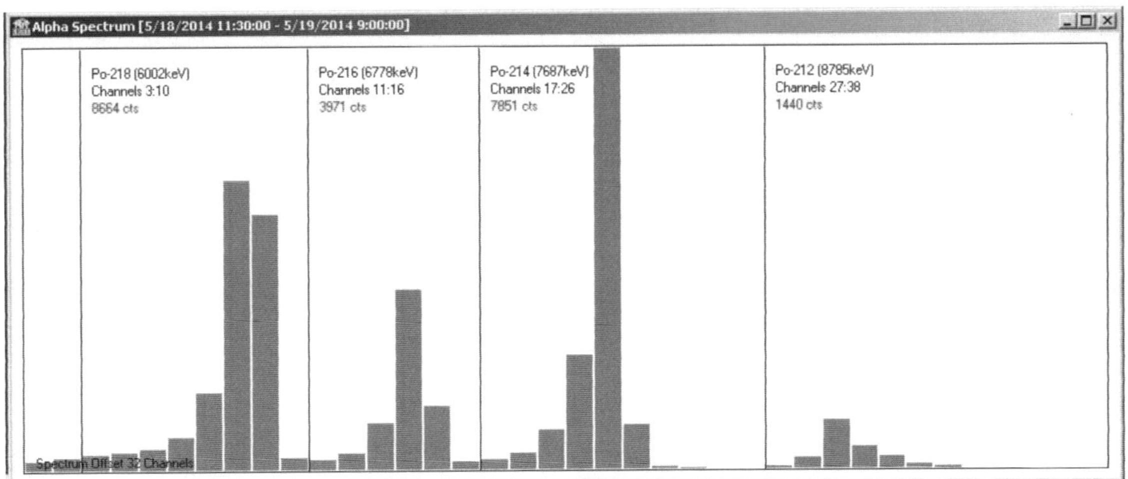

Figure 2. Registration of alpha particle spectra within four ROIscalibrated to ^{218}Po (Rn "fast" mode), ^{214}Po (Rn "slow" mode) and ^{216,212}Po (Tn) decay, respectively.

One of the principal advantages of THORON-SCOUT is its self-sustaining measurement regime without any pump, i.e., without possible vibrations or disturbing noise. Powered by conventional mono-cells, the device works autonomously over about one month. An internal clock controls the user adjusted measurement cycle which may be chosen between 1 min and 4 h. Two measurement modes are possible for Rn detection where the "fast" mode relates to only ^{218}Po decay detection, and the "slow" mode relates to both ^{218}Po and ^{214}Po decay detection. In the "slow" mode, the registration sensitivity is, therefore, doubled. Simultaneously, the Tn activity concentration measurement is realized by registering ^{216}Po and ^{212}Podecay with a sensitivity of 0.67 cpm per kBq/m^3[1].

The alpha spectra shown in figure 2 reveal another problem which had to be solved during the design of THORON-SCOUT. This concerns the optimal assessment of the four ROIs for alpha particle counting from the decay of the corresponding isotopes and the necessary determination of correction factors for the low-energy tails of the alpha spectra which partly fall into neighboring ROIs. Furthermore, the alpha decay branch of the isotope ^{212}Bi (36.2 %), which belongs to the Th decay series and, therefore, is also a Tn progeny, energetically overlaps with the ROI for ^{218}Po decay, which is used for the determination of the Rn activity concentration, especially in the "fast" mode. When the latter problem has been solved by additionally registering the alpha particles from ^{212}Po decay following the β-decay branch of ^{212}Bi (63.8 %), the former one requires a special calibration procedure of every device inRn and Tn gas atmosphere, respectively. The absolute sensitivity with respect to Rn and Tn is obtained by comparison with a reference device [1].

The corresponding correction factors as well as all other measurement parameters concerning the duration of a measurement cycle, the measurement mode, etc. can be written into the device using special set-up software. The ROI counting data for every measurement cycle are automatically converted into concentration activity units of Bq/m^3 for Rn and Tn, respectively, and stored in the internal memory of volume of 2047 records (2 Gbyte). Additionally, the actual temperature, barometric pressure, and relative air humidity are registered and stored for every cycle. An internal movement sensor registers any position change of the device during the measurement period.

The device set-up and the data read-out are realized via USB or RS232 interface. Wireless data transfer via 2.4 GHz ZigBee is also possible what allows a remote measurement control.

3. Indoor Rn and Tn monitoring using THORON-SCOUT

Exemplarily, some results of Rn / Tn monitoring with THORON-SCOUT carried out in the cellar of an older apartment building during 63 hours are shown in figure 3 [2].

The dotted lines represent the actual measurement data obtained within a cycle of 1 hour. The full lines are obtained by smoothing over 3 h to reduce the statistical data scattering. The right ordinate of figure 3 indicates the difference between indoor and outdoor temperatures. The daily change of this temperature difference is also given in the graph (black line) together with the activity concentration data. In this way, it is rather clearly seen in figure 3 how the indoor Rn gas concentration changes with this temperature difference during the routine of the day. Since the atmospheric pressure as well as the indoor temperature (about ±1 degree) and air humidity (not given in the graph)changed only marginally during the measurement period, one may conclude that the determining factor for the actual indoor Rn / Tn activity concentrations is the outdoor temperature which of course strongly depends on the actual weather. Obviously, at increasing or decreasing temperature difference, the velocity of diffusion of Rn gas through the grounding of the building remarkably changes, and the indoor Ra / Tn activity concentrations change adequately, however, with a delay of several hours.

This behavior seems to be plausible from a consideration of the so-called chimney effect where warm air rises upward. The density of warm indoor air is slightly less than that of colder outdoor air. Convection within the building leads to the built up of some suck through the grounding into the building. In other words, Rn gas diffusion into the building is influenced by a gradient of specific pressure which for his part is caused by the mentioned temperature difference.

Figure 3 also reveals that the indoor activity concentrations of Rn and Tn show a similar trend during the routine of the day. Since the source concentrations of Rn and Tn in the environmental bedrock and soil of the building are governed by the natural radioactive decay of ^{238}U and ^{232}Th being constituents of the locally present geological formation, the ratio of the corresponding indoor activity concentrations should, therefore, be nearly constant and characteristic for the location of the building. The more it is important at new building activities that the environmental ground is investigated with respect to geogenic Rn gas sources prior to final decisions. SARAD GmbH offers the necessary technical equipment and corresponding monitors for such measurements [3]. At considerably higher

Rn / Tn concentrations in the environmental, geological formation, special measures have to be planned and carried out to eventually meet national standards of indoor activity concentration.

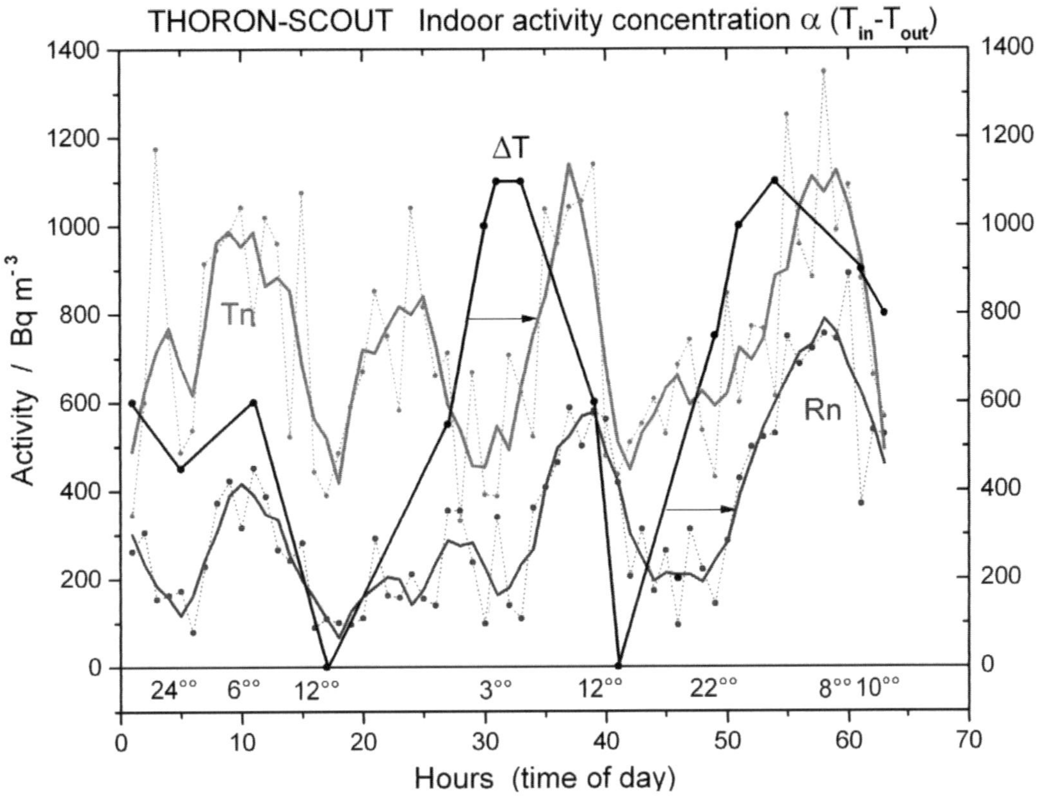

Figure 3. Results of indoor Rn / Tn activity concentration monitoring carried out by means of THORON-SCOUT in the cellar of an older apartment building situated in the hill country near the town Dresden in Germany [2].

The specific result of Rn / Tn monitoring given in figure 3 consists in the fact that in this case, the indoor Tn activity concentration exceeds that of Rn by roughly a factor of two. This impressively proves the statement given above. The realistic evaluation of the health risk due to indoor radioactive Rn gas necessarily requires Rn and Tn monitoring as well. One cannot and must not neglect the locally in no way the negligible contribution of Tn.

References

[1] Horak G, Streil T, Oeser V, Sabol J and Duzynski M 2014 *8th Int. Conf. on High Levels of Natural Radiation and Radon Areas(Prague, Czech Republic)* (priv. comm.)

[2] Streil T, Oeser V, Horak G, Duzynski M and Wagner W 2015 *Third Int. Conf. on Radiation and Applications in Various Fields of Research (Budva, Montenegro)*ed G Ristić(Nis, Serbia: RAD Association, Book of Abstracts) p 47

[3] http://www.sarad.de

Evaluation performance of digital integrated circuits while exposed to radiation

V M Barbashov and N S Trushkin

National Research Nuclear University MEPhI (Moscow Engineering Physics Institute), Kashirskoe shosse 31, 115409 Moscow, Russian Federation

Corresponding author e-mail address: VMBarbashov@mephi.ru

Abstract. The methods of functional-logical simulation of digital integrated circuits (ICs) exposed to radiation are observed. It is shown that in a number of cases functional and electrical deterioration of ICs performances have both deterministic and non-deterministic nature. Methods for simulating IC failure exposed to radiation based on the model of fuzzy digital machine and Brauer probabilistic reliability machine are proposed.

1. Introduction

Digital integrated circuits (ICs) are widely used in electronic systems [1], in some cases under high radiation level conditions. Analysis of the radiation test results shows there is a statistical dispersion of threshold levels of failure for functional and parametric failures for similar ICs [2].

In our radiation IC reliability studies, we use the digital machine model involving apparatus of probability theory. In Boolean reliability networks internal IC models are used in state-space terms in models where the axioms of a Boolean lattice [2,3] are applicable. In the situation in where the physical mechanisms of IC failure must be taken into account, the construction of a functional-logical model assumes transition from axioms of Boolean lattice to the axioms of a vector lattice with the corresponding change of algebraic operations to operations "minimum", "maximum" and "complement" $\chi \in X$ [3].

Thus, the real nature of the IC radiation behavior is determined by the specific ratio of the radiation-sensitive parameters of its internal elements, taking into account the impact of the spread. The ratio between the probability-density function of the spread and criterion function [4,5] should be determined by the usefulness of IC functional-logical radiation behavior models for a particular occasion. It should be noticed the parameters of the distribution functions characterizing statistical uncontrollable factors are themselves dependent on radiation. Moreover, the nature of their changes during irradiation depends on many factors, including the type of radiation, its intensity, and range, type a criteria parameter characterizing the radiation resistance of ICs and work mode. Therefore, in different ranges of levels or intensities of the impact of the IC model can be either of fuzzy or probabilistic nature. Carrying out such a comparison is an essential step in the overall analysis of the IC radiation resistance.

2. Model environment

With the radiation dose increasing the value of logic zero "0" and "1" in the IC changes. This leads to amplitude distribution spread of logic voltage levels "0" and "1", as shown in Fig. 1. In this case, to assess the nature of distributions it is necessary to expand the set of modalities (level of logic signals) to the continuum $\chi \in [0,1]$. In this case, it is preferable to describe a continuum set by fuzzy logic (fuzzy sets). Here we add it by apparatus of probability theory [5,6]. With increasing doses of ionizing radiation, these distributions come together (indicated by arrows), making the difference logic signal levels decreases. Operating capacity of the logic element (LE) can be characterized by the magnitude of the probability distribution which can be calculated as follows.

Let the distribution of logical "0" zero indicates a function $F(x)$ being like:

$$F(x)=\begin{cases} f(x), & x\in\Delta_0 \\ 0, & x\notin\Delta_0 \end{cases}, \tag{1}$$

Moreover, the distribution of logic "1" unit denotes a function being like:

$$\Phi(x)=\begin{cases} \varphi(x), & x\in\Delta_1 \\ 0, & x\notin\Delta_1 \end{cases}, \tag{2}$$

where Δ_0 is the ultimate distribution area «0»; Δ_1 is the ultimate distribution area "1" while $0\le\Delta_0<1/2$, $1/2\le\Delta_1<1$ (see Fig. 1).

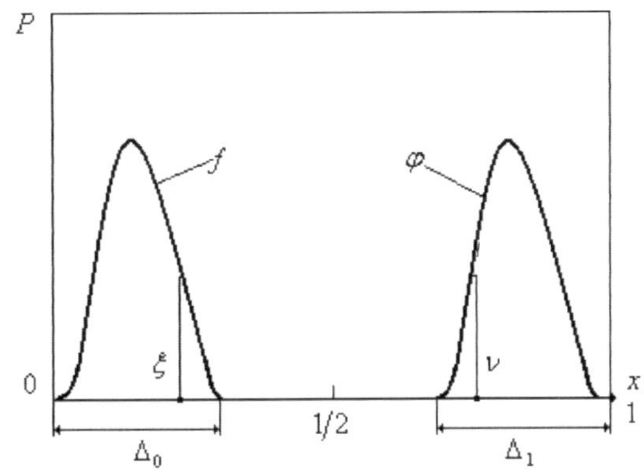

Figure 1. The probability distribution of the voltage amplitudes of logic levels "0" and "1" depending on the radiation dose.

In this case, it is obvious that:

$$\int_0^{\Delta_0} f(\xi)d\xi=1 \tag{3}$$

$$\int_{1-\Delta_0}^{1} \varphi(1-\nu)d\nu=1 \tag{4}$$

If ξ and ν are random variables, the difference $\nu-\xi$ is also a random variable from the logical pulse switching elements, and the probability that the value of the product $f(\xi)\cdot\varphi(\nu)$ [7,8].

Then the fuzzy probability value of the switching pulse, respectively, equal to:

$$\tilde{P}=\iint_\Omega(\nu-\xi)\cdot f(\xi)\varphi(\nu)d\xi d\nu \tag{5}$$

where $\Omega=[0,\ \Delta_0]\times[1-\Delta_1,\ 1]$.

As a result of the transformation, the integral (5) can be written as follows:

$$\tilde{P}=\int_{1-\Delta_1}^{1}\nu\varphi(1-\nu)d\nu-\int_0^{\Delta_0}\xi f(\xi)d\xi \tag{6}$$

The second integral in equation (6) describes the fuzzy probability of "0" (logic zero), consequently:

$$1-\int_0^{\Delta_1} \lambda\varphi(\lambda)d\lambda = 1-\tilde{P}_1 \tag{7}$$

where $\tilde{P}_1$ is the fuzzy probability of logic one "1".

Therefore, from the expressions (6) and (7) finally find the fuzzy probability value of the pulse switching logic element:

$$\tilde{P} = 1-(\tilde{P}_1+\tilde{P}_0) \tag{8}$$

In practice, the actual distribution asymmetrical, but even that case, the resulting formula (8) is valid for them.

Failures of systems under the influence of radiation factors depend not only on the values of the fuzzy probability but also on their probability of switching elements themselves in LSI [9,10]. The amount of each switching LSI logic element per unit time is the multiplicity of nodes connected to this element. Critical nodes in the LSI are fuzzy sets of LE with a large multiplicity of nodes. When LSI Logic degradation probability of switching can either decrease or increases in a certain range. Therefore, despite the fact that the fuzzy probability for logic elements is decreasing, the logic element, however, can recover after a failure in a certain range of radiation exposure. In this case, the physical processes are contributing largely in explaining the emergence of the systematic slot effect.

3. Conclusions

The actual nature of ICs performance under ionizing radiation is determined by the specific ratio between determined and non-determined radiation-sensitive parameters of the logic elements considering effects of their statistical spread. In this case, the relationship between the density-probability distribution function of the spread and criterion function determines in the end, the feasibility of using functional and logic ICs radiation behavior patterns in relation to each particular case. Carrying out such a comparison is an essential step in the overall analysis of ICs radiation resistance. It should be noted that in different ranges of levels and intensities of the impact the ICs model can be either of fuzzy or of probabilistic nature. That structure of the model is a probabilistic one with operators such as probability, and a criterion function is a superposition of statistical and deterministic criterion function. At the same time, the relationship between fuzzy logic and probability is defined, which makes it highly accurate method to assess the quality of the IC operation when exposed to radiation.

Acknowledgements
This work was supported by the Competitiveness Program of NRNU MEPhI.

References

[1] Pershenkov V S, Podlepetsky B I, Bocharov Yu I and Shagurin I I 2015 *Sensors & Systems.* **1** 3 (*in Russian*)
[2] Frank M J 1975 *Aeguationes Math* **12** 121
[3] Gratzer G 1982 *General theory of lattices* (Academic Press, Inc., New York)
[4] Yagera R R 1986 *Fuzzy sets and the theory of opportunities* (Moscow: Radio and Communications) p 408 (*in Russian*)
[5] Pospelova D A 1986 *Fuzzy sets in management models and artificial intelligence* (Moscow: Nauka) p 312 (*in Russian*)
[6] Denisenko V V 2010 *Compact models of MOSFETs for SPICE in micro- and nanoelectronics* (Moscow: FIZMATLIT) p 408 (*in Russian*)
[7] Barbashov V M and Trushkin N S 2009 *Russian Microelectronics* **38** 34
[8] Barbashov V M 2011 *Sensors & Systems* **6** 29 (*in Russian*)
[9] Sogoyan A V, Chumakov A I and Nikiforov A Yu 1999 *Russian Microelectronics* **28** 224
[10] Barbashov V M, Trushkin N S and Epifantsev K A 2014 *Russian Microelectronics* **43** 148

Latch up effect under electromagnetic pulse

V V Shurenkov and D V Gromov

National Research Nuclear University MEPhI (Moscow Engineering Physics Institute), Kashirskoe shosse 31, 115409 Moscow, Russian Federation

Corresponding author e-mail address: VVShurenkov@mephi.ru

Abstract. The physics of CMOS Latch-Up (latchup) under high power microwave radiation is discussed.

1. Introduction

There has been a significant interest in the understanding the susceptibility of an electronic system to the electromagnetic interference effects due to the high power microwave (HPM) radiation. In particular, the interest is focused on the potential errors of the electronic system function introduced as a result of the electromagnetic pulse (EMP) radiation attack. A number of prior test reports have indicated that intensive EMP might cause the electronic system failures, as extensive as complete shutdown or abort of operations due to the complex interference effects of the inner circuits, and even a permanent physical hardware damage [1-4].

The electronic device level effects studies are required in order to better understand the electronic circuits effects, including the latchup, under the HPM radiation. For this review, the CMOS inverters are chosen in digital circuits device level effects studies, due to their widespread usage as an electronic component. Any typical CMOS device has an inverter in the input and output stage. The latch-up effects are the traditional inherent destruction phenomena.

2. The latch up physics

The latchup is a CMOS integrated circuit failure mechanism that is characterized by an excessive current flow between the power supply and the ground rails [6-11]. It may be a temporary transitional state that is eliminated upon removal of the exciting stimulus, a catastrophic state, requiring a shut-down of the system in order to clear the state and recover the function, or a fatal state requiring the replacement of the damaged device.

More specifically, a latch-up is the creation of a low-impedance path be-tween the power supply rails of a MOSFET circuit, triggering a follow-up parasitic structure. The parasitic structure is usually equivalent to a thyristor (or Silicon Controlled Rectifier SCR), a PNPN structure which acts as a PNP and an NPN transistors stacked next to each other. In the CMOS technology, there are a number of the intrinsic bipolar junction transistors (Fig.1). These transistors can create problems when the combination of n-well/p- substrate results in the formation of parasitic PNPN structures. Triggering these thyristor-like devices leads to a shorting of the V_{dd} and GND lines, usually resulting in the destruction of the chip or a system failure. This thyristor can be triggered into latchup by the electrical conditions or by the radiation environment. Although these stimuli initiate latchup in a very different ways, the conditions for latch-up are the same.

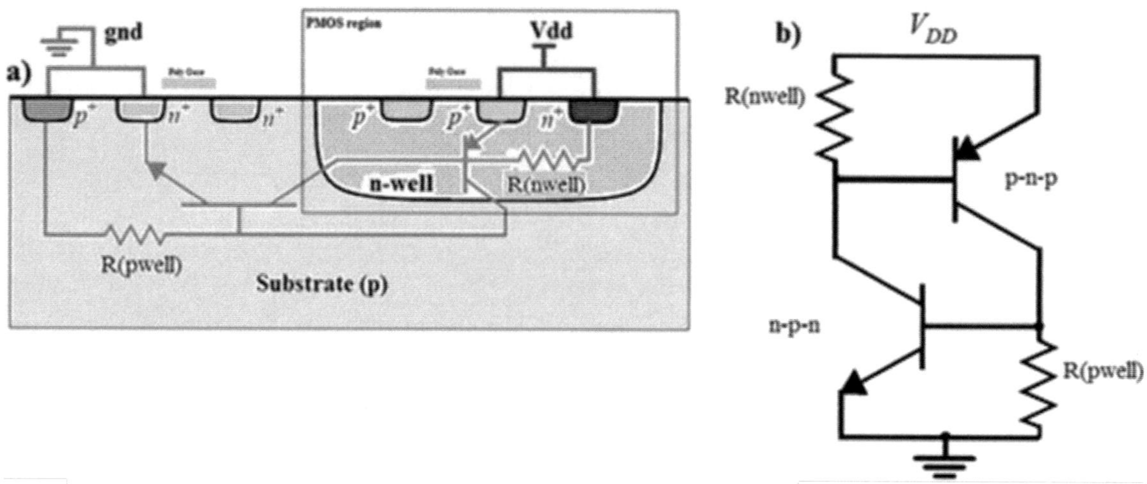

Figure 1. a) Intrinsic BJTs in the CMOS technology b) Equivalent circuit of CMOS latch-up.

Consider the structure in Figure 1 [6]. The n-p-n-p structure is formed by the source n+ of the NMOS, the p-substrate, the n-well and the p+ source of the PMOS. A circuit equivalent is also shown in Fig.1. This simplified circuit can be used to understand how the latchup susceptibility is affected by the operating conditions and structural changes to the PNPN region. The feedback loop shown in Figure 1 is composed of cross-coupled parasitic PNP and NPN bipolar junction transistors (BJTs) and decoupling resistors R(pwell) and R(nwell). Under the normal bias conditions, the parasitic BJTs conduct very little current and regenerative feedback does not occur in this loop. The both junctions of each bipolar junction transistors are reversed biased, $V_{BE} < 0$ and $V_{BC} < 0$, so that very little current goes through the device (a thyristor). This corresponds to the "off" state of the device. However, any BJTs can be triggered first into the forward-active mode, and then in the saturation mode when both junctions of the bipolar transistors are forward biased, $V_{BE} > 0$ and $V_{CB} > 0$. During this triggering a significant current and the regenerative feedback is produced, which introduces a low impedance path between the supply voltage (V_{DD}) and ground (GND). As shown in Figure 1, the base/collector junction of both of the BJTs (i.e., the N-well/P-substrate junction) initially is reverse biased under the normal conditions. Therefore, to trigger the two BJTs into the forward active mode, it is only necessary to forward bias their emitter/base junctions, which are the source/body junctions of the MOSFETs. The n+ and p+ sources are tied directly to the power rails and are heavily doped, so their potentials are fixed. However, the bodies of the NMOS and PMOS transistors shown in Figure 1 are biased through the substrate R(pwell) and R(nwell) resistances, respectively. These resistances shunt the emitter/base junctions of the PNP and NPN transistors. Therefore, the potential drops across these resistances cause the emitter/base junctions to be forward biased. If a triggering event causes only one of these two junctions to be forward biased (0.6-0.7 V), then the corresponding BJT enters the forward active mode and forces current through the base resistance of the other BJT, and can, therefore, cause it to also enter the forward active mode, thereby establishing the latchup. Turn-on of any one of the BJT on and the SCR action will occur when both junctions are forward biased, $V_{BE} > 0$ and $V_{CB} > 0$. This corresponds to the low resistance "on" state.

It is important to remember that the two conditions are necessary for catch-up to occur:

1) both parasitic bipolar transistors must be biased into the saturation state; furthermore, since both transistors are in the saturation state, a significant amount of minority carriers is accumulated in the base region of each transistor;

2) the product of the parasitic bipolar transistor current gains must be greater than or equal to one.

3. Experimental reports

Here is presented a review of some experimental results of the HPM radiation effects and examine the destruction and malfunction of on the CMOS structures caused by the high power microwaves. During the HPM interference, electromagnetic pulse or radiofrequency (RF) radiation energy usually couples into the interior electronics of a system through the following two generic paths: the front-door coupling and the back-door coupling.[5,11]. The coupling is a kind of mechanism whereby microwave energy is delivered to equipment under the test (EUT) through a circuit line. The front-door coupling examples are the intensive EM signal couples into the front-end receivers in the communication and radar systems through the antenna. The back-door coupling means EM radiation couples through the apertures in the system shielding structure, such as ventilation et al. The interconnected wires and power cables are also considered as the strong back-door coupling ways.

The growing attention has been focused on the threat posed by the HPMs, EMPs and RF (EM) radiation, which can be coupled into electronic system intentionally from EM sources or unintentionally due to the proximity to the environmental EM signals. The EM radiation results in temporary or permanent upset, or physical hardware damage. It is necessary to note that the CMOS digital circuits could be triggered not only by EM radiation but in other various ways such as the ESD pulse, cosmic ray, heavy ion particles, etc.

The CMOS digital circuits, as another example of modern information devices, also can experience the serious failures and malfunctions because of the thermal secondary breakdown caused by high power microwaves. As these devices are mostly comprised of the integrated circuits and microelectronics, which are sensitive to the microwaves. This is caused by the over-current when the reverse voltage is biased to the PN junction region [3]. The computer networks, PC mainboards, and basic logic components experiments in the TEM waveguide indicated that the system susceptibility becomes more sensitive as the system complexity increases [2].The test results had shown that the field threshold depends on the type of radiation and the type of circuit component as well, and it changes from 0.2 KV/m for UWB radiation for some computer network up to the 120 KV/m for the EMP radiation of the CMOS logic components.

In the publication [4] the authors have investigated the malfunction of the CMOS IC inverter under a certain critical field strength. The critical value of the field strength is dependent on many factors. First, there are constant influencing factors such as chip fabrication technology and chip layout of the CMOS IC. Also, there are variable factors such as switching states of the transistor. Because of these uncertain outcomes, the critical value of field strength varies randomly. The destruction threshold (DT) values of the studied inverters are changing from 8 kV/m up to 12.5 kV/m. The power supply current of the CMOS IC inverter also changes under high power microwave radiation.

An internal chip analysis of destructed semiconductor was carried out using an optical microscope. The inverter devices were separated into single inverter gates each with an input (A) and output (Y) and two connectors for the supply voltage (V_{CC}) and the ground (GND). The distribution of the destruction on the chip indicated that the primary destruction was to the inverter gates in the V_{CC} and GND sides. A microscopic analysis of the inverter devices statistically shows three different levels of damage, as indicated in Figure 2.

The damage became more severe with the increasing field strength. At a field strength of level 1 the component destruction on the chip had occurred. When field strength of level 2 the melting processes mainly damaged the IC-lines. A field strength of level 3 exceeding 13.1 kV/m led mostly to bond wire destruction.

4. Uncovering the different latchup mechanisms

In reality, the latchup mechanisms discussed earlier are far more complicated. The authors [5,9] have stated that both the MOSFET sources and the drains participate in the latchup. Figures 3, 4 show the cross-section of an N-well CMOS inverter that includes the parasitic BJTs on both the sources and the drains [6,9].

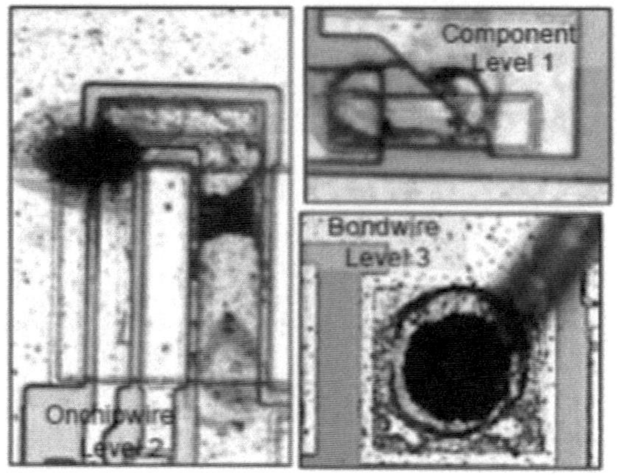

Figure 2. Damage at the field strength level.

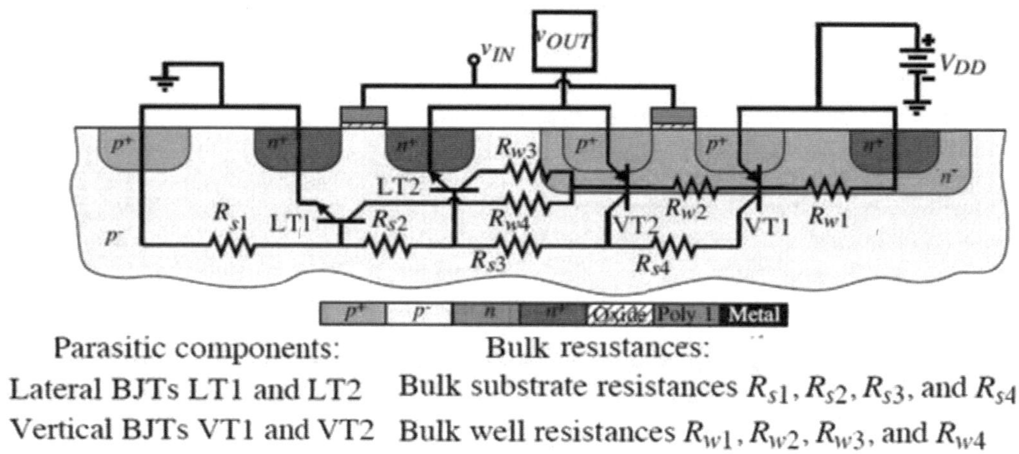

Figure 3. Parasitic bipolar transistors for the n-well CMOS Inverter.

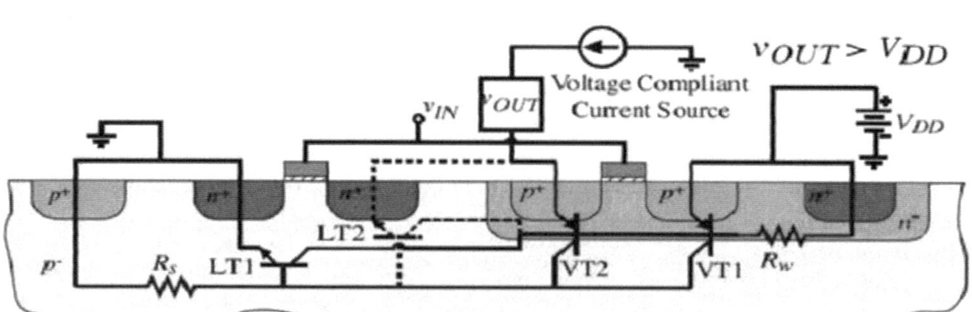

Figure 4. Current source injection.

That MOSFET drains contribute to the formation of BJTs. According to publication [7], the drains can be less susceptible to the latchup than other similarly-positioned sources, for some reasons, but as noted previously, the latchup may be triggered in a drain and then spread to a source, or that the source and drain of the same MOSFET can participate in latchup simultaneously. Figure 5 shows the triggering parasitic thyristors under the output signal. However, an integrated CMOS circuit as mentioned can be triggered in the various conditions:

a) the CMOS logic devices will latch-up when exposed to a strong enough voltage transient on either an input pin or a supply pin;

b) an electrostatic discharge can trigger the parasitic thyristor;
c) at the abnormal applications and misuse of the CMOS ICs at or beyond its rated maximum the voltage range and the presence of inductive transients;
d) the latch-up can be initiated by the dV/dt effect;
e) the high-power microwave interference can also trigger latch-ups.

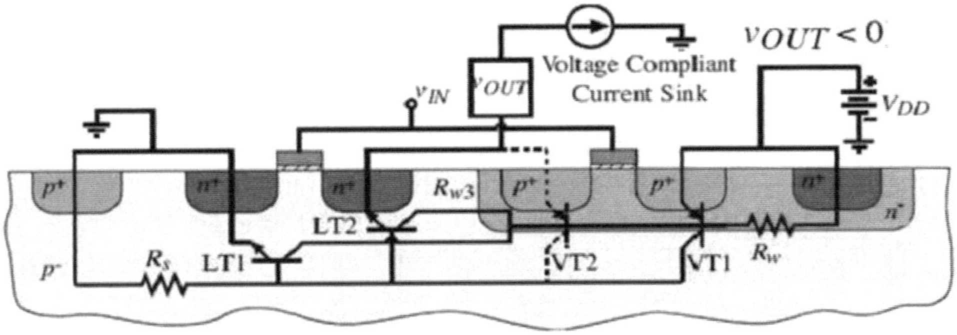

Figure 5. Cross-section of an N-well CMOS inverter showing parasitic bipolar transistors on both the MOSFET sources and drains

It's very important to notice that several triggering conditions from a) to d) inclusive, may be as well produced by the last condition - by the high-power microwave interference. The signals induced at any pin of the IC by the EM radiation may be any polarity, any meaning, and any velocity. So, the EM radiation pulse may induce any triggering condition.

5. Conclusion

The high power microwave radiation on CMOS inverter (the main device of the digital circuits) causes the latchup effects. The high power microwave radiation may be the main factor of the malfunction or irreversible physical destruction of CMOS inverter. The physical mechanisms of latchup are discussed.

Acknowledgements
This work was supported by the Competitiveness Program of NRNU MEPhI.

References

[1] Johnston A H 1996 *IEEE T. Nucl. Sci.* **43** 505-521
[2] Wang H, Li J, Li H, Xiao K and Chen H 2008 *Prog. Electromagn. Res.* **87** 313-330
[3] Hwang S M, Hong J I and Huh C S 2008 *Prog. Electromagn. Res.* **81** 61-72
[4] www.fairchildsemi.com *Understanding Latch-Up in Advanced CMOS Logic*
[5] Dodds N A 2012 *Single event latchup: hardening strategies, triggering mechanisms and testing considerations* (Dissertation presented to obtain the Ph.D degree in Electrical Engineering / Vanderbilt University, Nashville, Tennessee)
[6] Farbiz F 2010 *Modeling And Suppression Of Latchup* (Dissertation presented to obtain the Ph.D degree in Electrical Engineering / University of Illinois at Urbana-Champaign, Urbana, Illinois)
[7] Johnson M, Cline R, Ward S and Schichl J 2015 *Latch-Up* Texas Instruments Incorporated Application Report SCAA124 www.ti.com
[8] Haseloff E 2000 *Latch-Up, ESD, and Other Phenomena* Texas Instruments Incorporated Application Report SLYA014A
[9] Allen P E 2010 *CMOS Analog Circuit Design, 2nd Edition Reference Lecture 080 - Latchup and ESD*
[10] Yu X, Chai C, Qiao L, Yang Y, Liu Y and Xi X 2002 *Semiconductors* **36** 1-6

The comparison of radiation hardness of heterojunction SiGe and conventional silicon bipolar transistors

A S Bakerenkov, V A Felitsyn and A S Rodin

National Research Nuclear University MEPhI (Moscow Engineering Physics Institute), Kashirskoe shosse 31, 115409 Moscow, Russian Federation

Corresponding author e-mail address: ASBakerenkov@mephi.ru

Abstract. The results of the X-ray radiation impact on heterojunction SiGe and conventional silicon bipolar transistors are presented. Oxide thickness over the emitter-base junction depletion region determines the radiation hardness of the bipolar transistors. In this article, the estimation of the rate of radiation degradation of electrical parameters for conventional silicon devices and SiGe-transistors is performed.

1. Introduction

Nowadays in modern data acquisition and processing systems are used high-frequency bipolar transistors, made with SiGe technology. These devices have series of advantages in comparison with conventional silicon transistors, that is explained in [1]. Firstly, the implantation of germanium atoms into silicon crystal structure, which concentration increases in the direction from emitter junction to the collector, creates accelerating electric field, which increases the frequency performance of the transistor significantly. Secondly, the dopant of Germanium leads to a local decrease of the bandgap, which results in the increase of emitter efficiently and current gain of the transistors. Third, SiGe-transistors appear to have good compatibility with common silicon technology, low noise level, and high radiation hardness.

To estimate the effect of construction and technology on radiation hardness, a number of radiation experiments were performed in this work with SiGe and conventional bipolar transistors.

2. Bipolar transistors radiation tests

A static current gain that is considered in the article is a radiation-sensitive parameter which is called ß and equal to a ratio of collector current to base current. Under the impact of ionizing radiation, positively-charged traps accumulate in the oxide (SiO_2). They interact with electrons from substrate and converts to interface traps that can further act as recombination centers. It results in an increase of surface recombination current that in its turn raises base current I_b [2]. Since SiGe-transistors usually have a vertical structure, the charge carriers do not reach the subsurface area on their way from the emitter to collector, and so the collector current does not change during irradiation. Generally, the drift of the parameters and characteristics of integrated circuits (in the article this will be base current I_b and static current gain of the transistor ß) is called degradation. The radiation degradation of a static current gain of the bipolar transistor is determined by a radiation-induced increasing of the base current, which usually depends linearly on the total dose [3-6].

Two types of transistors were selected to compare radiation hardness: SiGe-transistors SGN8343Z and silicon 2N2222A. The irradiation was performed with the 8 keV X-ray source at the dose rate of 10 rad(SiO2)/s. The dependencies of collector and base current on the voltage that was applied to the emitter junction at the zero offset of the collector junction have been measured during the irradiation process. These dependencies are shown in Fig. 1 and Fig. 2 for different total dose levels. The dependencies of static current gain on the emitter-base voltage are shown in Fig. 3 and Fig.4. Irradiation and measurement processes were held at (25 ± 0,1) °C. The temperature stabilization system was used to control the temperature during the experiment. Its theory of operation is described in [6].

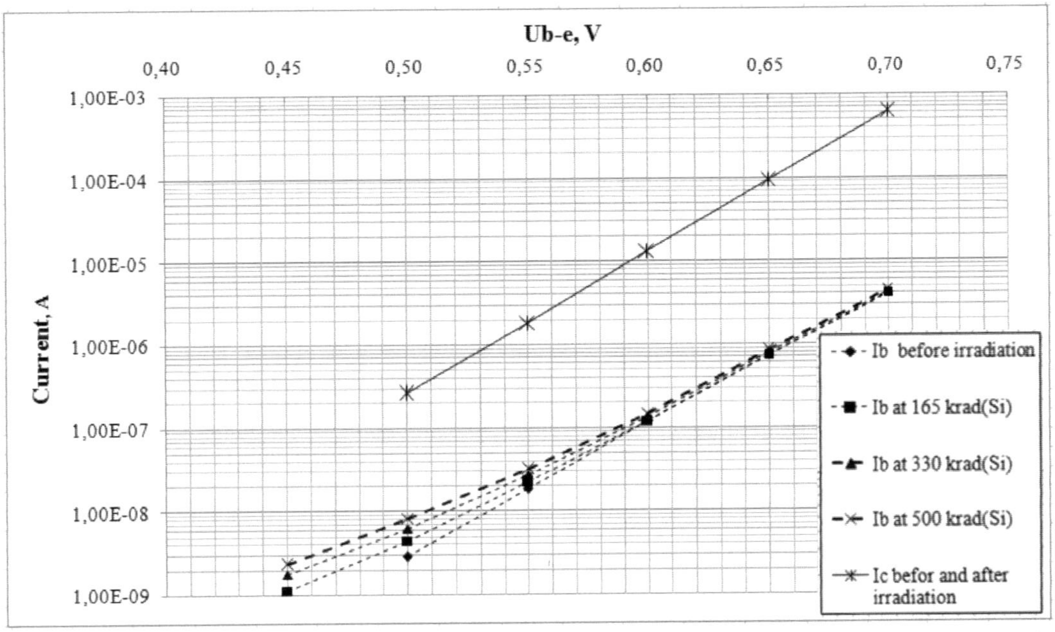

Figure 1. The dependence of the base current I_b and the collector current I_k on the voltage emitter-base at zero offsets of collector junction for the SiGe transistors SGN8343Z before irradiation and for different value of total dose.

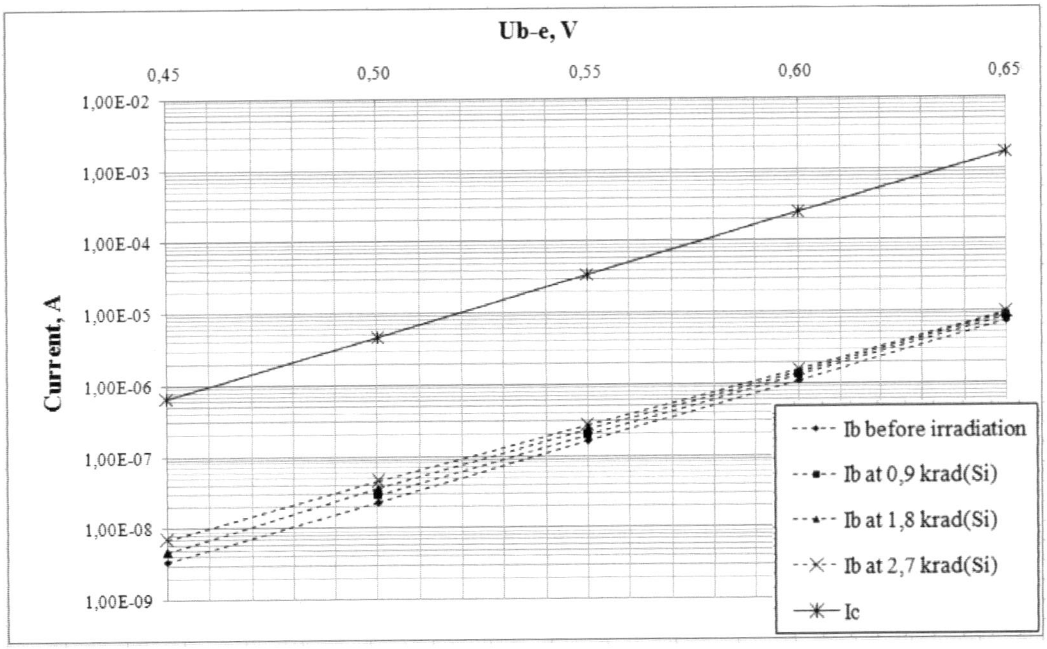

Figure 2. The dependence of the base current I_b and the collector current I_k on the voltage emitter-base at zero offsets of collector junction for the silicon transistors 2N2222A before irradiation and for different value of total dose.

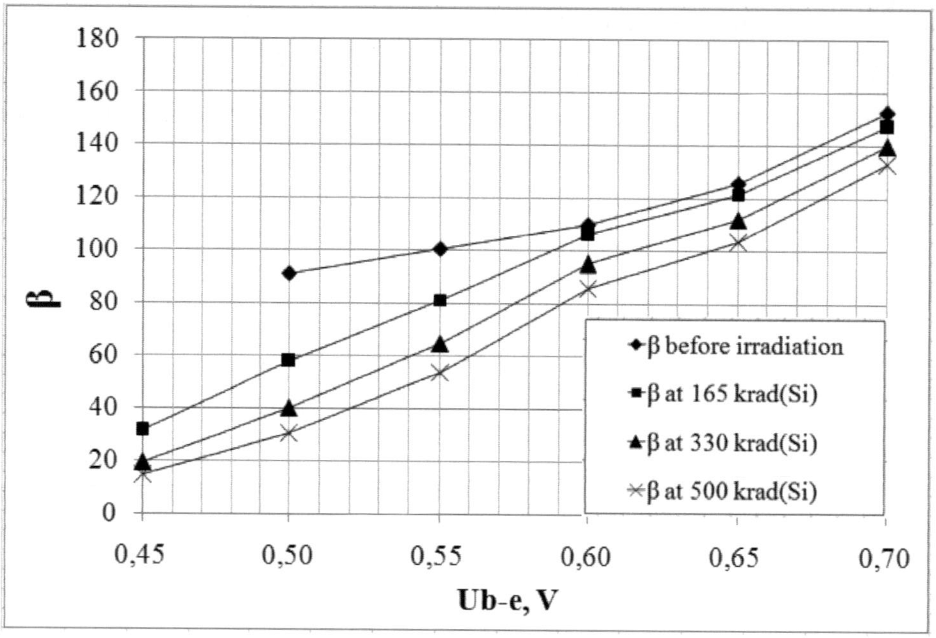

Figure 3. The dependence of the static current gain of the transistor on the voltage emitter-base at zero offsets of collector junction for the SiGe transistors SGN8343Z before irradiation and for different value of total dose.

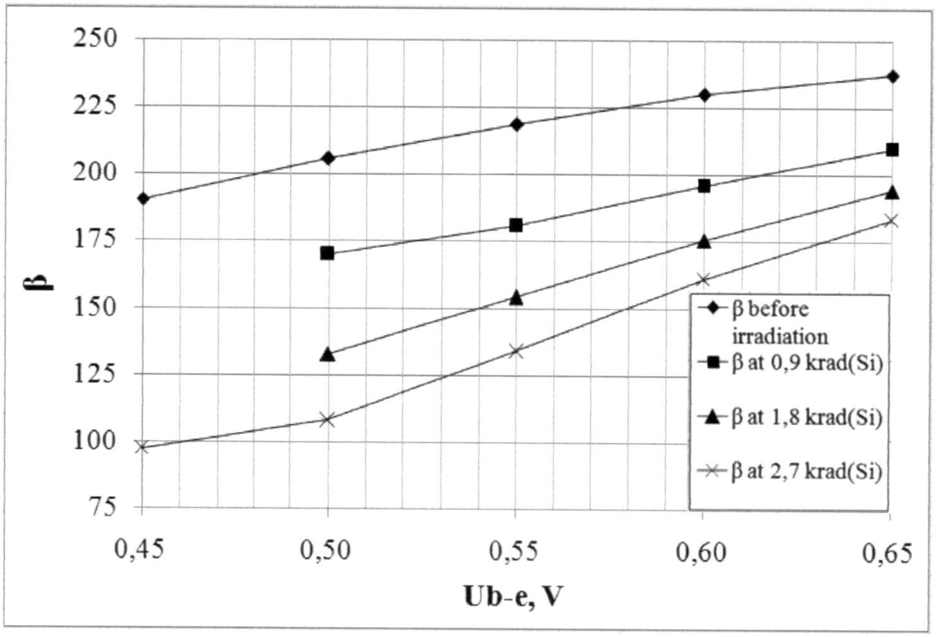

Figure 4. The dependence of the static current gain of the transistor on the voltage emitter-base at zero offsets of collector junction for the silicon transistors 2N2222A before irradiation and for different value of total dose.

The comparison of radiation degradation of base current ΔI_b for different transistors was performed at the emitter-base voltage 0.55 V. The relative increase of base current $\Delta I_b/I_{b_0}$ for SiGe-transistors and silicon devices against its value before the irradiation is shown in Fig. 5. It is easy to see that the radiation degradation of the base current ΔI_b of SiGe-transistors is 2 orders of magnitude less than in silicon transistors.

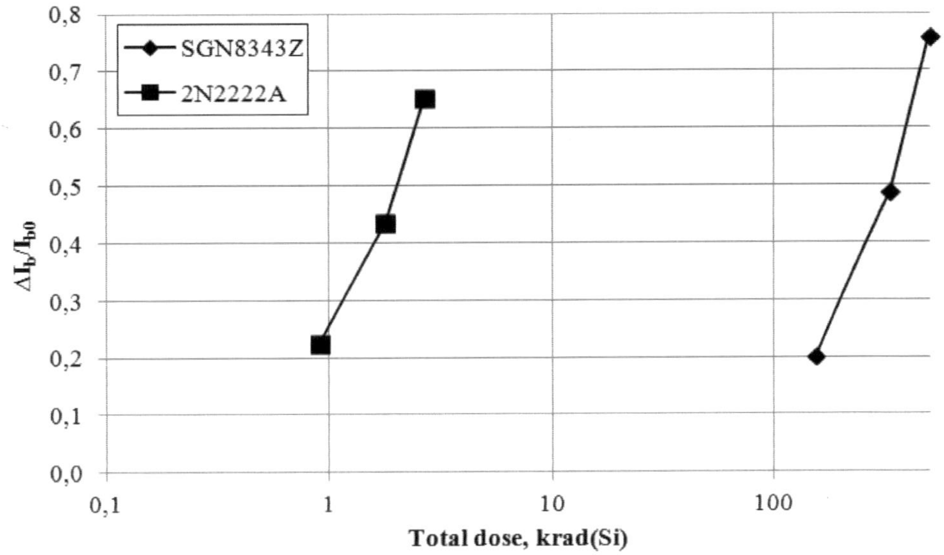

Figure 5. Total dose dependence of relative difference of input bias current of the SiGe transistor SGN8343Z and conventional silicon transistors 2N2222A (I_{b_0}- the value of the base current before irradiation).

3. Physical reasons of increased radiation hardness of the SiGe-transistors

According to the series of work, the main reason of bipolar SiGe-transistors radiation degradation is the formation of positively-charged traps in the oxide with the following conversion to the interface traps. Their build up results in the increase of base current and decrease of a static current gain of the transistor [4-6]. Electron-hole pairs can be most effectively separated during irradiation above the emitter junction border by the edge electric field. There are a number of works that describe the correlation of the dimensions of depleted area and bipolar devices radiation degradation rate [7].

A standard cross-section that shows the structure of bipolar SiGe-transistor is shown in Fig. 6. Above the emitter junction side boundary at the passive area, a thin layer of SiO_2 oxide is located. The layer is many times thinner than the corresponding protective layer in conventional silicon transistors. As shown in Fig. 7, the protective layer in silicon transistors is usually located over the emitter junction on the surface of the silicon substrate.

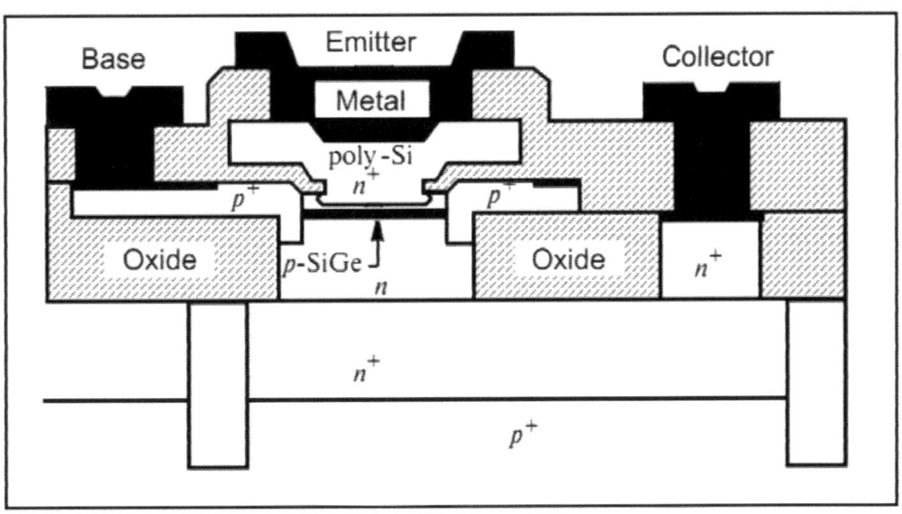

Figure 6. The cross-section that shows the structure of bipolar SiGe-transistor [1]

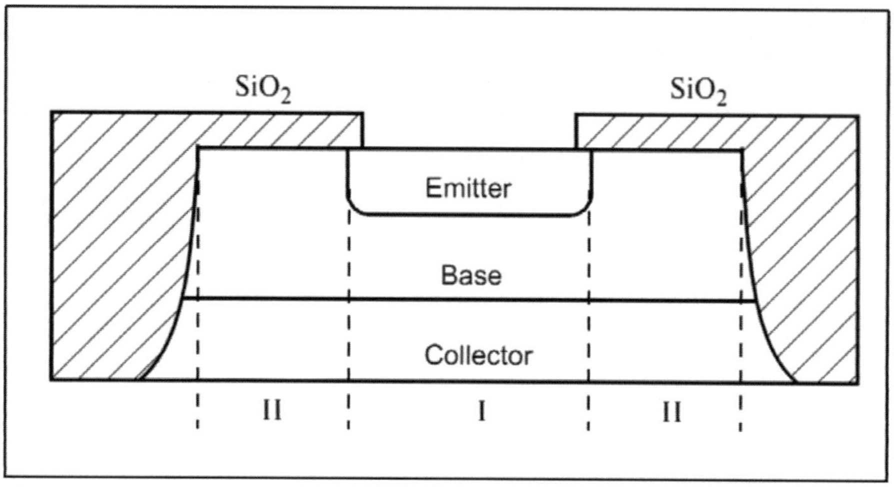

Figure 7. The cross-section that shows the structure of conventional silicon transistor [1]

Positively charged traps are formed on the whole oxide layer during the irradiation. As a result, a lot more holes are accumulated in thick oxides rather than in thin ones. It explains a dependence of the rate of radiation degradation of the base current on the oxide thickness above the depleted emitter junction area.

From the analysis of obtained experimental results (fig. 5), it is clear that the degradation rate of SiGe transistor base current lower than in conventional silicon devices by two orders of magnitude. It approximately corresponds to the ratio of oxide thickness over emitter depletion regions in SiGe and conventional silicon transistors.

4. Conclusion

In this work an experimental comparing of the radiation hardness of SiGe transistors and silicon devices were performed. It was obtained that base currents radiation degradation rate and current gain decreasing of SiGe-transistors are approximately two orders of magnitude less. High radiation hardness of SiGe devices is connected with a small thickness of SiO_2 oxide layer above the depleted emitter junction area. It should be noted that it is also possible to use this technological method in silicon transistors in order to increase their radiation hardness.

Acknowledgements
This work was supported by the Competitiveness Program of NRNU MEPhI.

References

[1] Cressler J D 2013 *IEEE T. Nucl. Sci.* **60** 3
[2] Pershenkov V S, Savchenkov D V, Bakerenkov A S, Ulimov V N, Nikiforov A Y and Chumakov A I 2009 *Proc. Eur. Conf. on Radiation and Its Effects on Components and Systems (Bruges)*(IEEE) p 286
[3] Enlow E W and R L Pease 1991 *IEEE T. Nucl. Sci.* **38** 1342
[4] Pease R L, Schrimpf R D and Fleetwood D M 2009 *IEEE T. Nucl. Sci.* **56** 1894
[5] Pershenkov V S, Savchenkov D V, Bakerenkov A S and Egorov A S 2009 *Russian Microelectronics* **38** 17
[6] Bakerenkov A S, Belyakov V V, Kozyukov A E , Pershenkov V S, Solomatin A V and Shurenkov V V 2015 *Nucl. Instrum. Meth. A* **773** 33
[7] Zebrev G I 2006 *Russian Microelectronics* **35** 177

AMNST2015 IOP Publishing
IOP Conf. Series: Materials Science and Engineering **151** (2016) 012014 doi:10.1088/1757-899X/151/1/012014

Linearity analysis of single-ended SAR ADC with split capacitive DAC

Dmitry Osipov[1], **Evgeny Malankin**[2], **Vitaly Shumikhin**[2]

[1] Institute of Electrodynamics and Microelectronics (ITEM), University of Bremen, Otto-Hahn-Allee NW 1, Bremen, 28359,Germany
[2] ASIC Lab, National Research Nuclear University MEPhI, Kashirskoe highway, 31, Moscow, 115409, Russia

E-mail: osipov@item.uni-bremen.de

Abstract. This paper proposes the design of a 6-bit single-ended SAR ADC with a variable sampling rate at a maximum achievable speed of 50 MS/s. The SAR ADC utilizes the split capacitor array DAC with a non-conventional split-capacitor value. The influence of switches in the capacitive DAC on the ADC's non-linearity is analysed. According to the fulfilled analysis the recommendations for switches and capacitor array dimensioning are given to provide a minimum differential non-linearity (DNL).

At a sampling rate of 50 MS/s, the SAR ADC achieves an ENOB of 5.4 bit at an input signal frequency of 1 MHz and consumes 1.2 mW at a 1.8 V power supply, resulting in an energy efficiency of 568 fJ/conv.-step. The SAR ADC was simulated with parasitics in a standard 180nm CMOS process.

1. Introduction

The SAR ADC architecture provides a compromise for the rising demand on power and area efficient ADC's IP-blocks in up-to-date mixed-signal ASICs. The use of a single-ended architecture instead of a differential one can give the possibility to reduce the block area significantly. But the errors usually not important in differential designs have to be considered to provide the linearity of the converter.

The decrease of circuit elements also leads to the increase of the significance of the errors arising from parasitic circuit elements and elements mismatch. In the DACs with a split capacitor matrix the gain error of LSB bits makes the most contribution in the DAC's non-linearity [1]. In a conventional capacitive DAC the split capacitor equal to $C_S = C_0 2^{N/2}/(2^{N/2}-1)$, which can introduce a LSB gain error leading to differential non-linearity errors when the DAC switches from the MSB half of the array to the LSB one. Several works utilize the split capacitor with the capacitance equal to the unit capacitor one for ease of layout and good matching [2], [3]. In this paper the doubled split capacitor is used as in [4], [5] to improve more the matching properties, utilizing the capacitor connected to the fixed potential in the LSB half of an array to correct gain.

The main purpose of this work was to analyse the errors arising in single-ended SAR ADC with a split capacitor array DAC and to find out the recommendations for optimal switches and unit capacitors of the DAC.

The errors arising in a capacitive DAC were analysed in several works. For example, [6] proposes the SAR ADC's transfer function model taking into account the voltage dependence of the unit capacitors capacitance and mismatch. The non-linearity arising from additional switches capacitance is not taken into account. The cancellation technique of parasitic capacitance at the LSB top plate node is provided in [7]. This work shows that the proper unit capacitor sizing can also eliminate this effect.

The optimal array capacitors and bottom-plate switches sizing, provided in this work, gives an opportunity to design the area efficient SAR ADC's IP-blocks with a desired non-linearity value. The theoretical results are proven by the design and simulation results of a 6-bit SAR ADC in 180nm CMOS with a variable sampling frequency (up to 50 MS/s at a 1.2mW average power consumption at this speed). The dynamic performance simulations (50 MS/s, 1 MHz input) show the 34.5 dBc SNDR value (5.4 ENOB).

2. ADC architecture

The ADC architecture is shown in Fig. 1. The asynchronous clocking was used to eliminate the use of a high frequency external clock (~500 MHz). The internal clock is generated by a comparator after receiving the Start of Conversion (SOC) signal. After the end of conversion cycle the End of conversion (EOC) signal is generated, which also resets the comparator. So the ADC consumes power only during conversion. The digital logic utilises dynamic cells to lower the power consumption.

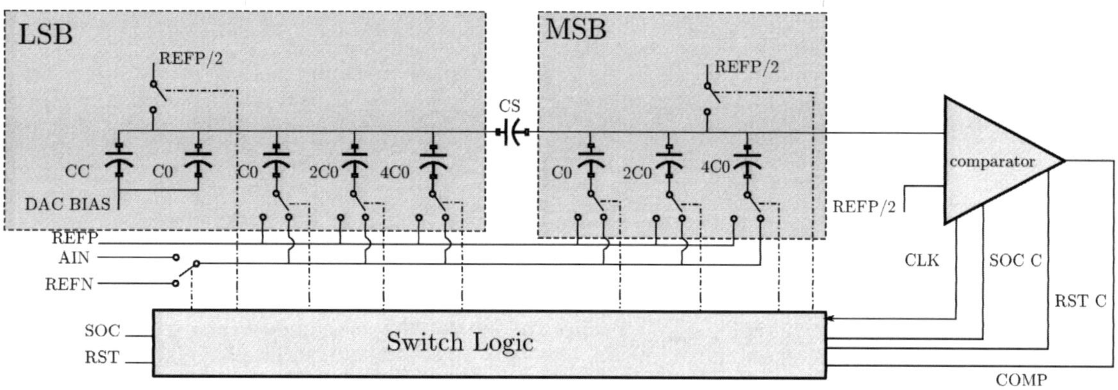

Figure 1. SAR ADC architecture

3. Capacitive DAC

Instead of a conventional binary weighted capacitor matrix with a $\frac{2^{N/2}}{2^{N/2}-1}C_0$ split capacitor C_S, the $2C_0$ split capacitor was used to lower the errors caused by the non-ideal value of split capacitor. The non-conventional split capacitor value can be compensated by a capacitor connected to the LSB-half of array and to some fixed potential [4]. The value of the compensation capacitor C_C depending on the C_S value can be found as follow. Using the notations $\lfloor a/b \rfloor$ for the largest integer not greater than the quotient of a/b division and $(a/b)_{mod}$ for the remainder of a/b division, the output voltage at the MSB half of array can be expressed as follows:

$$V_{out} = V_{ref}\left[m\left(\frac{i}{2^{N/2}}\right)_{mod} + (m+2^{N/2}+\chi)\left\lfloor\frac{i}{2^{N/2}}\right\rfloor\right] \times \\ \times\left[m(2^{N/2}-1)+(m+2^{N/2}+\chi)(2^{N/2}-1)+\chi m+m\right]^{-1}, \quad (1)$$

where i is DAC the input code (from SAR logic) and $\chi = C_C/C_0$, $m = C_S/C_0$. Thus, for a given m, χ can be found from:

$$\chi = m(2^{N/2}-1) - 2^{N/2}. \quad (2)$$

For the 6 bit case with a 3 bit LSB half-matrix for $m = 2$ the equation (2) gives $\chi = 6$. This method introduces the overall gain error of a capacitive DAC, rising from the $\chi m + m$ term in the denominator of equation (1). The overall gain error can be found from equation (1), rewriting (1) for $i = 2^N - 1$:

$$G_{error} = \frac{V_{FULLSCALE}}{V_{FULLSCALE}^{Ideal}} = 1 + \frac{\chi m + m}{[m(2^{N/2}-1)+(m+2^{N/2}+\chi)(2^{N/2}-1)]} \quad (3)$$

For the considered case of a 6 bit ADC the equation (3) gives $G_{error} = 1.1(1)$.

The connection of array capacitors C_C and C_S is also important. In [1] was shown, that only the parasitic capacitors connected to the LSB common node of array have the influence on the code-dependent error of DAC as:

$$DNL(j) = \frac{C_{pL}}{C_S}\delta\left(\left\lfloor\frac{j+1}{2^{N/2-1}}\right\rfloor\right), \quad (4)$$

where C_{pL} is the parasitic capacitance common to LSB capacitors, j – DAC code, δ – Dirac measure. Thus, the array capacitors and C_C should be connected with their top plates to a common node, the split capacitor C_S should be connected with it's top plate to the LSB-half of array to lower the influence of parasitic capacitance on matrix performance. For example, according the to SPICE simulation file for CMIM of the considered 180 nm CMOS technology, parasitic capacitors are connected only to the bottom plate of capacitor.

3.1. Top plate switches

The main effect of the top plate switches arises from their capacitances ($C_{GD}+C_{BD}$). According to (4) only the top plate switch in the LSB part of array has an influence on the ADC's linearity. The capacitance of a single transistor in the switch can be expressed C_{pLS} as:

$$C_{pLS} = C_{GD} + C_{BD}, \quad (5)$$

where C_{GD} and C_{BD} can be found from model equations [8]:

$$C_{GD} = WC_{GD0}, \\ C_{BD} = \frac{C_J(AD)}{\left(1-\frac{V_{BD}}{PB}\right)^{MJ}} + \frac{C_{JSW}(PD)}{\left(1-\frac{V_{BD}}{PB}\right)^{MJ}}, \; V_{BD} \le 0, \quad (6)$$

where AD is the drain area, PD is the drain perimeter, C_J, C_{JSW}, PB, FC, MJ are the BSIM3v3 model related parameters, which can be found in the designkit simulation models. Some features of the ADC's DNL dependence can be obtained from (5-6):

- Firstly, the top plate switch of the LSB half of the capacitor matrix should have the minimum possible size, allowing the recharge of matrix during the sampling phase.
- Secondly, the value C_{pLS} and,respectively, DNL depends on the potential at the top plate of LSB array.

Thus, the parasitic capacitance calculated by the formula (5) can be substituted in (4) to obtain the theoretical expression for the ADC's DNL arising from the influence of the top-plate switch. In Fig. 2 the theoretical dependence and the schematic-level simulation results are compared. The simulation results were obtained by the simulation of ADC with the schematic-level top-plate switch of a LSB-half and with the ideal one (Verilog-A model).

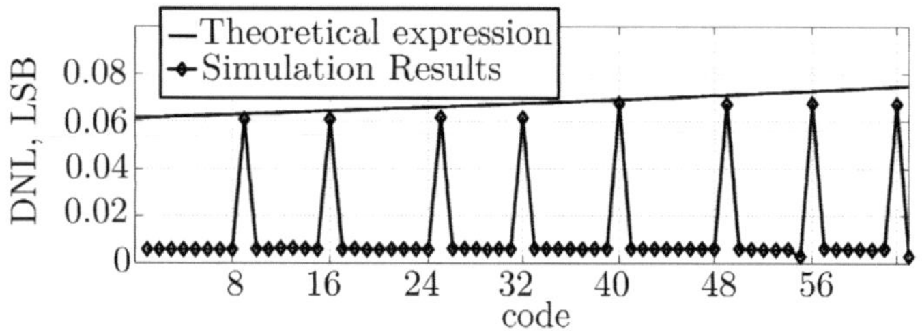

Figure 2. The DNL arising from the influence of the top-plate switch. Complementary switch dimensions: $L_p = 240n$, $W_p = 800n$, $L_n = 240n$, $W_n = 800n$, $C_0 = 37.8fF$

According to (4) the DNL decreases hyperbolically with the increase of unit capacitance. The theoretical curve and the simulation results are plotted in Fig. 3 for the top-plate switch transistors width of 800nm. Thus, starting from some value the further increase of the unit capacitance doesn't provide any significant decrease of DNL.

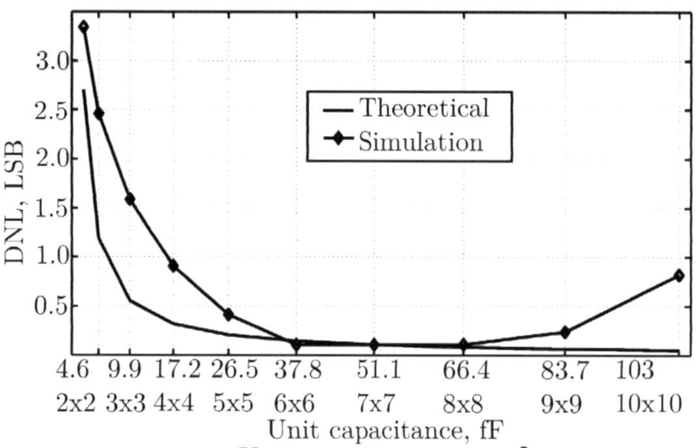

Figure 3. The dependence of DNL on unit capacitor value for a fixed top-plate LSB switch dimension ($W_p = W_n = 800n$)

The simulation results for different top-plate switch dimensions (Fig.4) confirm the linear dependence of parasitic switch capacitance on transistor width.

3.2. Bottom plate switches

There are two most significant error sources when using MOS transistors as switches. Firstly, charge injection, which can be effectively cancelled by using complementary MOS devices as switching elements. Secondly, clock feedthrough, which can cause code-dependent errors in capacitive DACs. The strict formulation of voltage errors arising in sample-and-hold circuits class from clock feedthrough can be found in [9]. For the clarity the simpler equation from [10]

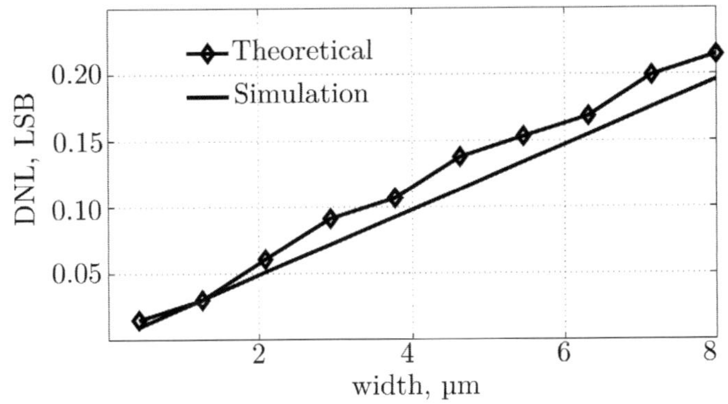

Figure 4. The dependence of DNL on top-plate LSB switch dimension for a fixed unit capacitor value, (C_0 – $6 \times 6\mu m$)

will be used to define the voltage error:

$$V_{error} = V_{dd}\frac{WC_{ov}}{WC_{ov} + C_L}, \tag{7}$$

where W is the effective switching transistor width, V_{dd} – switch control voltage level, C_{ov} – overlap capacitance per unit width, C_L – load capacitance.

For a non-split capacitor array to not to introduce any DNL error the V^i_{error} of every i-th switch should be binary weighed $\frac{1}{2^i}V^0_{error} = V^i_{error}$. The appropriate width of i-th MOS switch can be found from:

$$\frac{1}{2^i}V_{dd}\frac{W^0C_{ov}}{W^0C_{ov} + C_L} = V_{dd}\frac{\zeta W^0C_{ov}}{\zeta W^0C_{ov} + C_L/2^i},$$
$$\zeta = W^i/W^0. \tag{8}$$

Thus,

$$\zeta = \frac{C_L}{2^i\left((2^i - 1)W^0C_{ov} + 2^iC_L\right)}. \tag{9}$$

If $C_L \gg W^0C_{ov}$:

$$\zeta = \frac{1}{2^{2i}}. \tag{10}$$

Thus, if $C_L \simeq W^0C_{ov}$, the optimum W^i value can be found from (9).

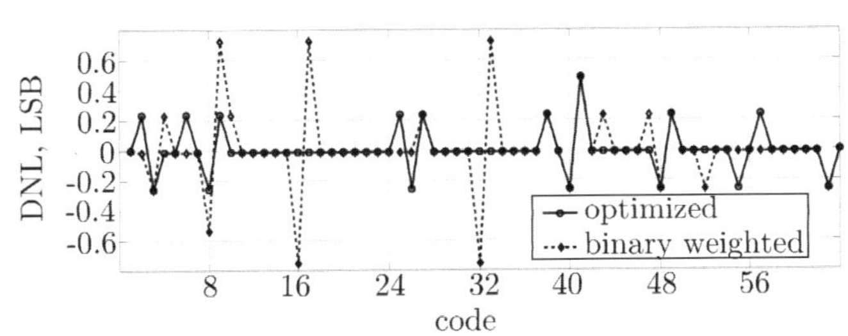

Figure 5. DNLs for binary weighed switches and optimized ones

As an example of such calculation the DNL plots for binary-weighed switches and for optimum width ones is given in Fig. 5. For the binary weighed switches the error value is equal for each bit switch. Thus, the maximum DNL values are reached when the maximum number of MSB array capacitors are recharged. For the case of 6-bit ADC with the capacitor array split into the two equal parts the maximum DNL could be expected at codes 8,16,32 (MSB switches in positions 110, 101, 011). The unit capacitor value used is $6\mu m \times 6\mu m$, $37.8fF$. These plots don't take into account topology related parasitics extraction & errors introduced by the comparator.

3.3. Capacitive DAC design algorithm
Considering the above dependences the following algorithm can be summarized:

- Define the optimum unit capacitor value from minimum switch dimensions allowable in the technology used (4)& (5).
- Calculate the bottom-plate switches dimensions using (9) or (10).

In this design the unit capacitor size of $6\mu m \times 6\mu m$ was used. The common-centroid capacitor array with a dummy capacitors shield occupies $68\mu m \times 60\mu m$.

4. Simulation results

The SAR ADC structure shown in Fig.1 was designed and simulated in the standard 180nm CMOS technology. The non-linearity analysis was proven by transistor level simulations. The overall ADC performance was simulated taking into account the topology parasitics extraction. The ADC's topology shown in Fig. 6 occupies $0.0255\, mm^2$.

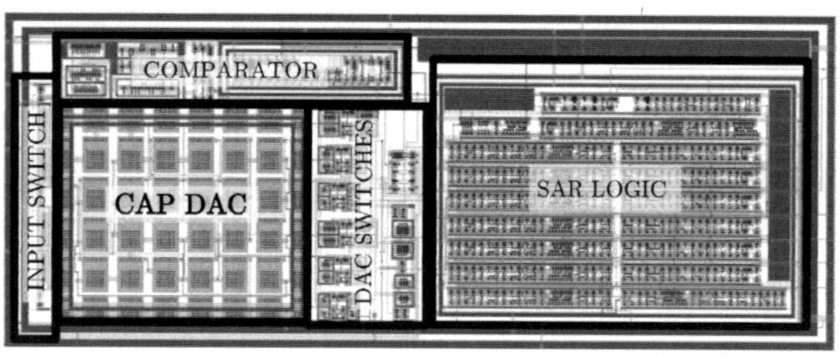

Figure 6. SAR ADC Topology

Fig.7 illustrates the simulated FFT spectrum with a conversion rate of 49.152 MS/s and an input of 1.032 MHz. The SNDR is 34.3 dBc, resulting in 5.4 ENOB.The DNL and INL, as shown in Fig.8, are $+0.6/-0.6$ and $+0.65/-0.64$ LSBs, respectively. The average power consumption measured while dynamic simulation is equal to 1.2 mW.

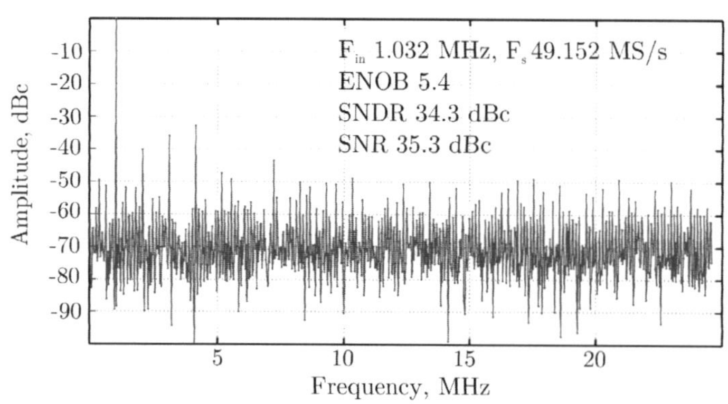

Figure 7. Simulated single-tone spectrum (Post-layout)

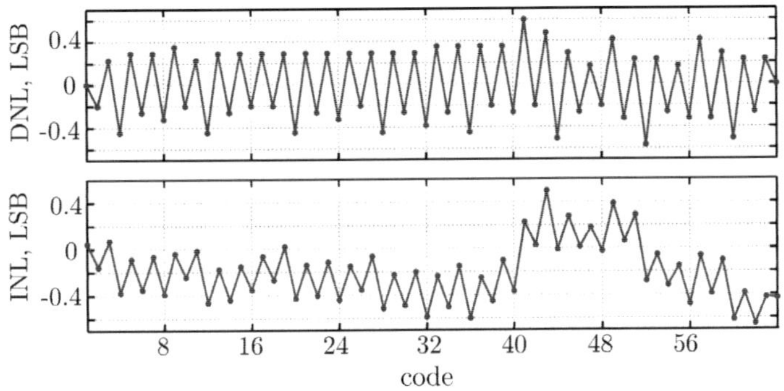

Figure 8. Simulated INL and DNL (Post-layout)

Tab. 1 summarizes the ADC performance and shows the comparison with recently published SAR ADCs with 20-200 MS/s . The achieved performance exceeds the previously published results in 180nm technology and is comparable with the one for more advanced technologies.

Table 1. Comparison with previous art

	This Work	[11]	[12]	[13]	[14]	[15]
Resolution(bits)	6	7	6	8	8	7
Technology (μm)	0.18	0.18	0.18	0.09	0.065	0.065
Sampling frequency (MS/s)	50	26	220	40	100	50
Area (mm^2)	0.0255	0.53	0.031	0.055	-	0.017
SNDR (dBc)	34.3	43.11	32.62	48.4	49.5	43.2
ENOB (bits)	5.4	6.87	5.13	7.75	7.9	6.8
max DNL (LSB)	0.6	0.168	1.15	0.88	-	-
FOM (fJ/conv.-step)	568	480	880	20.6	23.3	28
Year	2016	2012	2009	2011	2013	2013

5. Conclusion

A split capacitive array DAC for SAR ADC errors analysis arising from bottom and top plate switches was presented. It makes possible to design area efficient SAR ADC blocks with a desired non-linearity level. The analysis was proved by the design and simulation of 6 bit SAR ADC with variable sampling rate in the 180 nm CMOS technology.

Acknowledgement

This research was supported by the Ministry of Science and Education of the Russian Federation (RF Governmental resolution No 220).

References

[1] Osipov D and Bocharov Y 2012 *Ph.D. Research in Microelectronics and Electronics (PRIME), 2012 8th Conference on* pp 1–4
[2] Zhang D and Alvandpour A 2012 *ESSCIRC (ESSCIRC), 2012 Proceedings of the* pp 369–372 ISSN 1930-8833
[3] Agnes A, Bonizzoni E, Malcovati P and Maloberti F 2008 *Solid-State Circuits Conference, 2008. ISSCC 2008. Digest of Technical Papers. IEEE International* pp 246–610
[4] Zamprogno M, Minuti A, Girardi F, Confalonieri P and Nicollini G 2012 *Circuits and Systems II: Express Briefs, IEEE Transactions on* **59** 138–142 ISSN 1549-7747

[5] Chen H W, Liu Y H, Lin Y H and Chen H S 2009 *Solid-State Circuits Conference, 2009. A-SSCC 2009. IEEE Asian* pp 153–156
[6] Haenzsche S, Henker S and Schuffny R 2010 *Mixed Design of Integrated Circuits and Systems (MIXDES), 2010 Proceedings of the 17th International Conference* pp 300–305
[7] Zhu Y, Chan C H, Chio U F, Sin S W, U S P and Martins R 2010 *Circuits and Systems (MWSCAS), 2010 53rd IEEE International Midwest Symposium on* pp 889–892 ISSN 1548-3746
[8] Liu W, Jin X, Chen J, Jeng M C, Liu Z, Cheng Y, Chen K, Chan M, Hui K, Huang J, Tu R, Ko P and Hu C 1998 Bsim 3v3.2 mosfet model users' manual Tech. Rep. UCB/ERL M98/51 EECS Department, University of California, Berkeley
[9] Xu W and Friedman E 2002 *ASIC/SOC Conference, 2002. 15th Annual IEEE International* pp 181–185
[10] Razavi B 2002 *Design of Analog CMOS Integrated Circuits* McGraw-Hill higher education (Tata McGraw-Hill) ISBN 9780070529038
[11] Yang T and Li Z 2012 *High Speed Intelligent Communication Forum (HSIC), 2012 4th International* pp 1–4
[12] Liu C C, Huang Y T, Huang G Y, Chang S J, Huang C M and Huang C H 2009 *VLSI Design, Automation and Test, 2009. VLSI-DAT '09. International Symposium on* pp 215–218
[13] Tsai J H, Chen Y J, Shen M H and Huang P C 2011 *VLSI Circuits (VLSIC), 2011 Symposium on* pp 264–265
[14] Wu W L, Zhu Y, Ding L, Chan C H, Chio U F, Sin S W, Seng-Pan U and Martins R 2013 *Circuits and Systems (ISCAS), 2013 IEEE International Symposium on* pp 2239–2242 ISSN 0271-4302
[15] Xu Y and Ytterdal T 2013 *NORCHIP, 2013* pp 1–4

IOP Conf. Series: Materials Science and Engineering **151** (2016) 012015 doi:10.1088/1757-899X/151/1/012015

Technology and characteristics of the transistor with a channel based on graphene

S A Shostachenko[1], G I Zebrev[1], R V Zakharchenko[1], S V Leshchev[1] and I V Komissarov[2]

[1] National Research Nuclear University MEPhI (Moscow Engineering Physics Institute), Kashirskoe shosse 31, 115409 Moscow, Russian Federation
[2] Belarusian State University of Informatics and Radioelectronics, P. Brovki street 6, 220013 Minsk, Belarus

Corresponding author e-mail address: sashostachenko@gmail.com

Abstract. In this article we have studied the current-voltage characteristics of the graphene-based transistor. The test structure of graphene transistor was fabricated with the back gate. Graphene has been produced by chemical vapor deposition, and then transferred to the silicon dioxide on a silicon wafer. The channel of the transistor has been formed by etching in oxygen plasma through a photolithographic mask. Metals electrodes of the drain, source and gate were deposited by resistive evaporation in a vacuum. It was used titanium / gold with a thickness of 20/200 nm. In the case of the back gate, silicon dioxide was used, obtained by thermal oxidation of the silicon substrate. The field effect was demonstrated. With IV curves graphene parameters were defined.

1. Introduction

The unique properties of opened in 2004 2D material - graphene such as high mobility of charge carriers (theoretically up to 100 000 cm^2 / V·s), ideal electron-hole symmetry and line intersecting the zone at the Fermi level, defining the massless behavior of electrons, make it useful for various applications in solid state electronics and photonics [1-3]. The use of graphene as a channel of the transistor allows reaching the terahertz frequency range [4-6], as the threshold of the device is proportional to the frequency of the carrier mobility. To date, individual research groups obtained field-effect transistors with a channel based on graphene frequency range of more than 100 GHz [7-10]. However, in practice, the actual mobility of the samples of graphene films on the substrate is in the range from 100 to 10 000 cm^2 / V·s [9-11]. However, the unique properties of graphene, on the other hand, are a limiting factor in the production of transistors based on it. One of the main problems in the formation of such devices is the problem of preserving the possibility of effective management of the flow of charge carriers - field effect.

In this paper, a study of the current-voltage characteristics of the transistor structures with conduction channel based graphene film in order to obtain its basic parameters.

2. Experimental details

Graphene film was obtained by the method of chemical vapor deposition (CVD) at atmospheric pressure to a substrate of copper foil 60 microns thick, using methane as a hydrocarbon precursor. The synthesis was carried out at 1050 °C as a carrier gas was a mixture of argon and hydrogen. After synthesis process, copper foil is dissolved in a solution of $FeCl_3$. After washing in distilled water graphene film was transferred to the SiO2/Si substrate with a thermally formed oxide thickness of 300 nm [12].

Back gate contact to the silicon oxide film represented by a continuous metal layer to the reverse side of the plate. SiO2 film at the same time played the role of the lower gate dielectric. Graphene film etching to form a transistor structure was conducted in an oxygen plasma (200 W, 6 minutes) by the Diener Nano. Metal contacts drain, source and gate Ti/Au (20/200 nm) deposited by electron beam evaporation in Kurt J. Lescker PVD 250. The operating pressure in the chamber was $2{\cdot}10^{-7}$ torr. The resulting resistance of ohmic contacts defined by the test proposed in [13], was $5.6{\cdot}10^{-5}$ Ohm·cm^2. The resulting structure is shown in Figure 1.

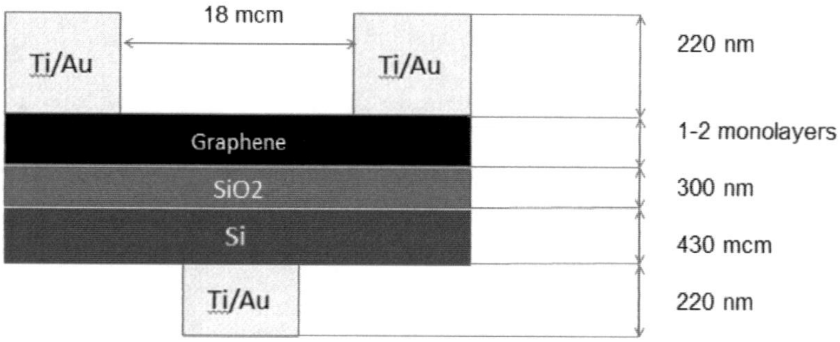

Figure 1. Schematic profile of the test channel transistor based on graphene film

3. Results and discussions

The study was carried out on the current-voltage characteristics of the installation of Agilent B1500, equipped with a manual probe station EP6 company Cascade Microtech.

To determine the necessary parameters for the current expression has been used:

$$I_d = \frac{W}{L} e\mu_0 n_s V_{ds}, \tag{1}$$

where W, L – the width and length of the channel, μ_0– mobility, e – the electron charge, V_{ds} – drain-source voltage, n_s – the concentration of charge carriers in the channel. The concentration of charge carriers in the channel was calculated using the 2 models: approximate the classic formula, and accurate, taking into account the state of the surface. The approximate formula is:

$$n_s \cong \sqrt{n_0^2 + \frac{C_{ox}^2}{e^2}\left(V_g - V_{NP}\right)}, \tag{2}$$

where n_0 – the concentration of charge carriers at the point of electrical neutrality, V_{NP} – of electrical voltage, V_g – gate-source voltage, C_{ox} – the linear capacitance gate dielectric. Substitute equation (2) (1) and comparing with the experimental points can be determined and the mobility of the charge carrier concentration at the origin (Fig. 2).

Same parameters were defined by Hall effects measurements (Table 1). We observe the convergence of the order of magnitude indicating that the applicability of the used model.

Table 1. Graphene parameters from Hall effect and IV curve.

	Mobility, cm2V-1s-1	the concentration of charge carriers, 10^{13}cm^{-2}
Hall	18	1.56
IV curve	33	1.45

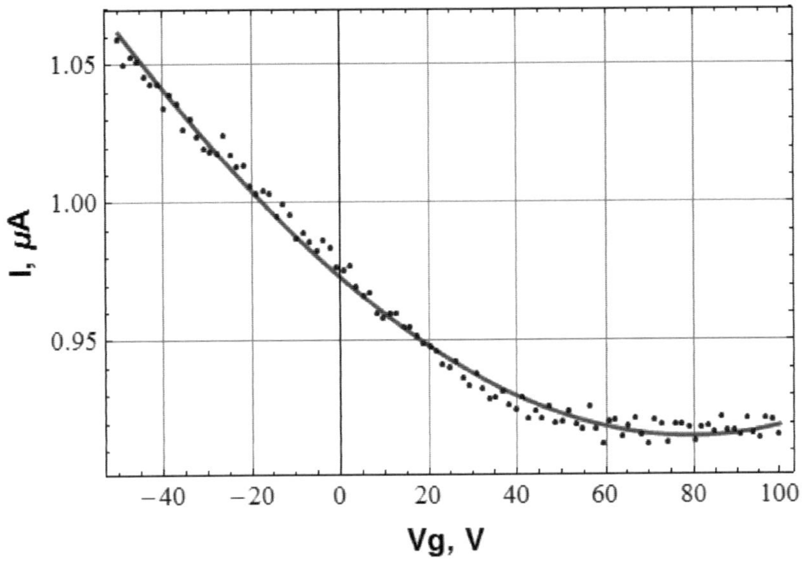

Figure 2. Comparison of the model with experimental data. Points – the experimental data, line – modeling. Model parameters: V_{ds} = 0.01V, μ_0 = 26 $cm^2V^{-1}s^{-1}$, n_0= 1.6 $10^{13}cm^{-2}$

4. Conclusion

Graphene field effect transistor was made. Main parameters of graphene, such as Dirac point, carrier concentration and mobility were determined from IV curves. The main results of the work was developing the technology of forming a channel and an ohmic contact to the graphene and demonstration of field effect. It is shown that the technology does not impair the properties of the graphene film. That is all the further work to improve the parameters of the transistor are reduced to improve film technology of graphene.

Acknowledgements
The study was supported by the Ministry of education and science of the Russian Federation (unique identification number of the work is RFMEFI58414X0001).

References

[1] Novoselov K, Geim A and Morozov S 2005 *Nature* **438** 197
[2] Chen Z, Lin Y-M and Rooks M 2007 *Physica E* **40** 228
[3] Novoselov K, Geim A and Morozov S 2004 *Science* **306** 666
[4] Novoselov K, Fal'ko V and Colombo L 2012 *Nature* **490** 190
[5] Kawano Y 2013 *Nanotechnology* **24** 1
[6] Lin Y, Dimitrakopoulos C and Jenkins K 2010 *Science* **327** 662
[7] Schwierz F 2013 *Proceedings of the IEEE* **101** 1567
[8] Cheng R 2012 *PNAS* **109** 11588
[9] Han S, Garcia A 2014 *Nature Communications* **5** 3086
[10] Vicarelli L 2012 *Nature Materials* **11** 865
[11] Jouault B 2011 *Phys. Rev.* B **83** 195417
[12] Komissarov I V, Kovalchuk N G, Kolesov E A, Tivanov M S, Korolik O V, Mazanik A V, Shaman Yu P, Basaev A S, Labunov V A, Prischepa S L, Kargin N I, Ryzhuk R V and Shostachenko S A 2015 *Physics Procedia* **72** 450
[13] Shostachenko S A, Vanukhin K D, Ryzhuk R V, Maslov M M, Katin K P, Zakharchenko R V, Minnebaev S V, Pischulina A A and Kargin N I 2015 *Physics Procedia* **72** 419

Prospects for the use of security air flow to prevent ion-molecule reactions in the ionization and drift zone in classical IMS

A V Golovin, N V Makarova, A A Poturuy and V V Beliakov

National Research Nuclear University MEPhI (Moscow Engineering Physics Institute), Kashirskoe shosse 31, 115409 Moscow, Russian Federation

Corresponding author e-mail address: avgolovin@bk.ru

Abstract. The effective transfer of sample problem is relevant in modern analytical equipment. The paper considered a problem in detection trace concentrations of explosives by Ion Mobility Spectrometry (IMS). The investigation deals with sample adsorption on the walls of transport tubes, the ion drift chamber and the chamber of the ion source in ion mobility spectrometer. The sample losses on inlet channel surface and diffusion through penetrable gas channels are comparable with the quantity of sample itself at the sensitivity level of 10-14 g / cm3. The trinitrotoluene (TNT) sorption in different channel materials depending on their sorption properties is analyzed. A new approach preventing sorption of the substance on the chamber walls by security airflow is presented. The study includes gas flow simulation and experiments of protective gas flow setup.

1. Introduction

Ion Mobility Spectrometry is a widely used method for the detection of explosives and narcotics [1-4]. The method is characterized by the ability to separate ions of isomers, high speed of detection and high level of sensitivity, as well as the combination of usability and low cost. However, the construction of spectrometers has vulnerabilities such as "dead space" problem and adsorption of contaminating substances on the internal surfaces. The gas sample passes through the ion mobility spectrometer (Figure 1), can be adsorbed on the walls of transport tubes, the ion drift chamber and the chamber of the ion source [5-7]. The simulation of gas flow in the chamber of the ion source is considered. It is developed the regulation and diagnostic equipment for the formation of gas streams. A new approach to preventing sorption of the substance on the chamber walls by security airflow is presented.

2. Modeling

It is solved an aerodynamic equation for 2D and 3D models of the ion source chamber. The simulation of gas flows is analyzed into the ion source chamber using Comsol Multiphysics package. The equation describes sample transfer probe through a finite volume without mixing gas streams (Figure 2-3). The test substance it is not adsorbed on the walls of the ion chamber due to clean air flow, which ensures the purity of ion drift region and significantly increases the durability of the device and the accuracy of detection of trace explosives and narcotics concentrations. The effect is achieved by using an automated apparatus for experimental testing, including air handling unit, which prevents the gas flow of the substance in contact with the chamber walls.

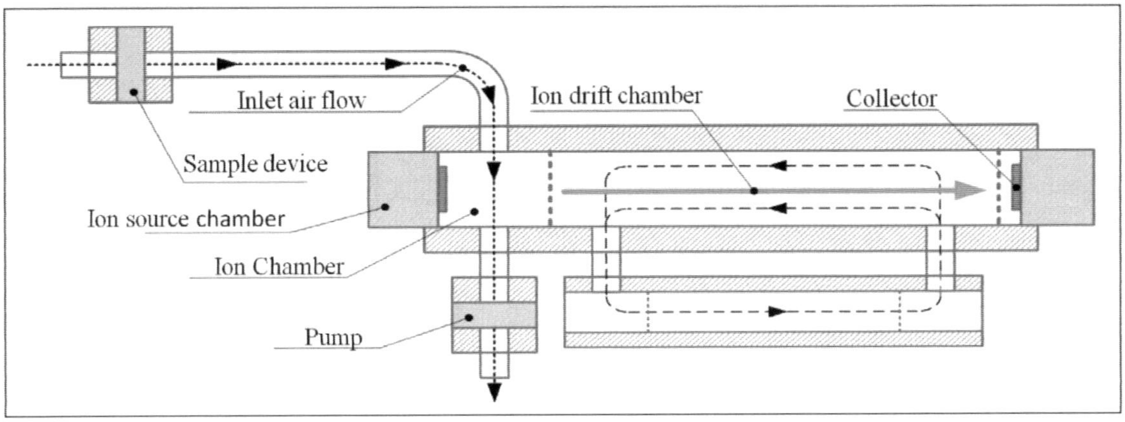

Figure 1. Structural scheme of IMS device with direction of gas flow and ion flux movement in drift tube.

Experimental testing shows that the system of formation of the gas flow runs successfully. The device allows reducing the pollution of ion source chamber and provides the effective transport of substance flow.

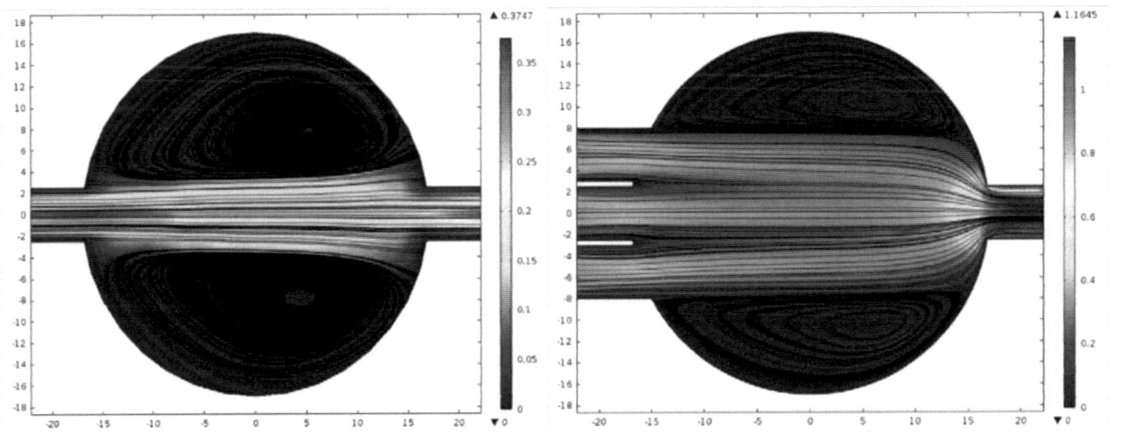

Figure 2. The two-dimensional simulation of gas flow in ion source camera with a complex shape, structure excluding guard stream (left). The two-dimensional simulation of gas flow in ion source camera with complex shape, structure with guard stream (right).

3. Prototyping

The automated device is developed for experimental testing (Fig. 4) including an air flow control unit which does not allow to contact the gas flow of the test substance with the chamber walls.

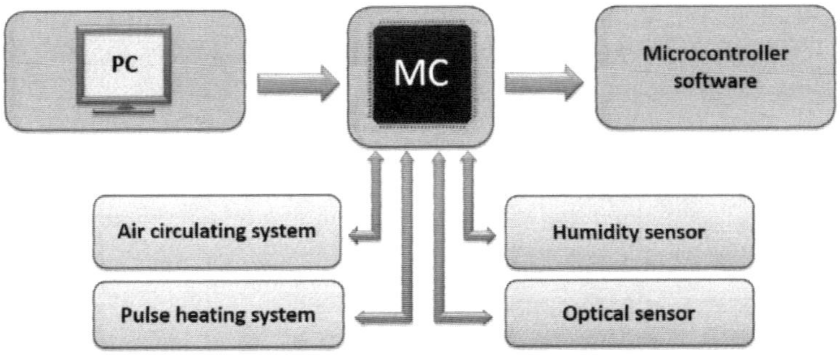

Figure 3. Structural scheme of a test setup for the study of sample tubes. The airflow diagnostic is performed by an optical method.

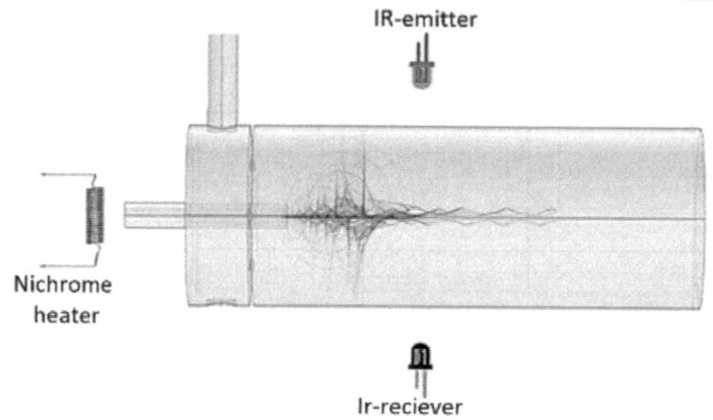

Figure 3. Functional scheme of an optical test setup for the study of sample tubes.

The simulation results confirm the effectiveness of the changes in the sample chamber of the ion source configuration. Original structure airflow test particles are adsorbed on the walls of the ion chamber; the system is contaminated and require cleaning, or reduced accuracy in determining the composition of substances. It is, therefore, necessary to develop a system that improves the accuracy and selectivity of the instrument in the analysis of ultra-small quantities of substances. Such a system is an improved structure of the ion source chamber. Due to protection of the flow of clean air test substance, it is hardly adsorbed on the walls of the ion chamber, which ensures the purity of ion drift region and significantly increases the durability of the device and the accuracy of measurements

4. Experimental data and results

In the first experiments on the creation of a system of formation of the gas flow problem transferring the sample of the test substance over the final volume. The sample did not reach the end of the tube due to the low flow rate. This problem has been eliminated thanks to the selection of the optimum fan speed.

Optimized speed: $V_{in\ put}$ = 7000 rev / min, $V_{out\ put}$ = 11000 rev / min.

At sufficiently low gas flow rates, the flow is always laminar, but with an increase in speed, it is always a transition to turbulent flow, which is time-dependent and spatially inhomogeneous. The speed of the liquid particles, pressure and other characteristics of the medium are changed in time and space randomly even under constant environmental conditions.

In the experiment by asking a different fan speed, installed on the inlet and outlet of ion drift chamber, manages the primary and secondary air flow, resulting in the separation is achieved introduced air samples and walls of the experimental tube. The flow is laminar. The fans are controlled by the PWM signal from the microcontroller. Results of experimental studies conducted in this work are interesting for further optimization of the ion mobility spectrometer.

5. Conclusion

It is necessary to develop a system to prevent leakage of ion-molecule reactions in the drift zone which increases the accuracy and selectivity devices, ensure purity ion drift region and significantly increase the durability of the device and accuracy. The simulation of gas flows in the chamber of the ion source using the package Comsol Multiphysics. The study developed an automated apparatus for experimental testing, including an air-handling unit which does not allow the gas flow of the substance in contact with the chamber walls. Diagnosis airflows provided by the optical method. In the drift, the chamber is created by the counter ions flow of dehydrated and purified air to prevent leakage of ion-molecule reactions in the drift zone. This stream is formed in a closed air circulation system, including the pump and adsorption trap filled with molecular sieves. The task of sample transfer through a finite volume without mixing gas streams has been solved for 2D and 3D models of the ion source chamber. Results of experimental studies are presented in the paper are interesting for further optimization of the block sampling and testing sample ionization.

The effective transfer of sample problem is relevant in modern analytical equipment. The paper considered a problem in detection trace concentrations of explosives by Ion Mobility Spectrometry (IMS). The investigation deals with sample adsorption on the walls of transport tubes, the ion drift chamber and the chamber of the ion source in ion mobility spectrometer. The sample losses on inlet channel surface and diffusion through penetrable gas channels are comparable with the quantity of sample itself at the sensitivity level of 10-14 g / cm3. The study includes gas flow simulation and experiments of protective gas flow setup.

Acknowledgements
This work was supported by the Competitiveness Program of NRNU MEPhI.

References

[1] Eiceman G A, Karpas Z and Hill H H 2013 *Ion Mobility Spectrometry (3rd ed.).* (Taylor & Francis Group) p 45-48
[2] Samotaev N, Golovin A, Vasilyev V, Malkin E, Gromov E, Shaltaeva U, Mironov A and Lipatov D 2014 *Lecture Notes in Electrical Engineering* **268** p 447-451
[3] Laakia J, Kauppila T J, Adamov A, Sysoev A A and Kotiaho T 2015 *Talanta* **132** 889 - 893
[4] Chernyshev D M, Frolov I S, Frolov A S, Mukhanov M S and Sysoev A A 2011 *Journal of analytical chemistry* **66**(13) 5-9
[5] Pershenkov V S, Tremasov A D, Belyakov V V, Razvalyaev A U and Mochkin V S 2006 *Microelectronics Reliability* **46** 641-644
[6] Pershenkov V S, Razvalyaev A U and Tremasov A D 2006 *Proc. 25th Int. Conf. Microelectronics (Belgrade)* (Art. no. 1651051) p 689-692
[7] Samotaev N, Pershenkov V, Belyakov V, Vasilyev V, Golovin A, Ivanov I, Malkin E and Gromov E 2014 *Procedia Engineering* **87** 436-439

Detection heart failures (HF) biomarkers by proton transfer reaction - mass spectrometry and ion mobility spectrometry

Y R Shaltaeva[1], V K Vasilev[1], D Y Yakovlev[1], F Iu Kopylov[2], A L Syrkin[2], P Sh Chomakhidze[2], A A Bykova[2], L K Malinovskaya[2] and A I Skorokhod[3]

[1]National Research Nuclear University MEPhI (Moscow Engineering Physics Institute), Kashirskoe shosse 31, 115409 Moscow, Russian Federation
[2]I.M. Sechenov First Moscow State Medical University, Trubetskaya st. 8-2, 119991 Moscow , Russian Federation
[3]A.M.Obukhov Institute of Atmospheric Physics RAS, Pyzhevskii pereulok 3, 119017, Moscow, Russian Federation

Corresponding author e-mail address: shaltaeva@yandex.ru

Abstract. Exhaled breath contains 1% of volatile organic compounds. The concentration of individual biomarkers in hundreds of volatile organic compounds lies within the range ppm-ppb. In compare with control group the concentrations of acetone, acetic acid, ethanol, propylene biomarkers is significantly higher in HF-PEF group.

1. Introduction

Heart failure (HF) is a preeminent cause of cardiovascular morbidity and mortality. The 5-year survival rate for patients with heart failure, regardless of ejection fraction (EF), is less than 50%. Data from epidemiologic studies of heart failure with preserved EF (HF-PEF) find that the annual mortality is approximately 10%. The diagnosis of HF-PEF is more complicated than the diagnosis of HF with reduced EF (HF-REF). In our previous studies [1-5], we found out that concentration of certain voluntary organic compounds (VOCs) in exhaled breath in patients (pts) with and without HF differs.

2. The experimental scheme

From October 2013 to September 2015 it was enrolled 21 pts with HF-PEF in heart failure group and 16 pts without HF in the control group. HF-PEF is diagnosed according to ESC active guidelines. It is collected fasting exhaled breath samples of all patients in 1L Tedlar bags. Exhaled breath is analyzed using PTR-MS (Compact PTR-MS, Ionicon, Austria). The PTR-MS is gaining popularity modern technique [6-8]. Correlations were evaluated by the method of Pearson and Spearman. All statistical operations were carried out with the help of IBM SPSS Statistic Software, version 22.

The experimental setup is designed and assembled for determination of several volatile organic compounds (VOCs) concentration in the exhaled air samples. The unit consists of the following components: proton transfer reaction - mass spectrometr compact PTR-MS, zero air generator Sonimix 3057, the thermostat, solenoid valve, manometer, flow meter, valves, connecting pipes, computer. The thermostat is taken from chromatograph Shimadzu GC-8A.

The air sample in Tedlar bag, DuPont, USA (bag made of polyvinyl fluoride film) is placed in the thermostat at constant temperature 42 °C. The solenoid valve is mounted inside the thermostat which is also connected with air channels from zero air generator and PTR-MS input. The design allows excluding probe condensation on channels and solenoid surface. The solenoid valve switching makes it possible to supply to PTR-MS an air from zero air generator or exhale probe. The generator air consumption is supported at level 0.5 lpm for optimum operation of generator and avoidance of room air entry. The air consumption is regulated by a valve and controlled by variable area flow meter. The pressure-gauge controlls the air pressure on generator output.

The PTR-MS probe consumption contents 0.05 lpm thus the whole time of 1l probe analysis is 20 min. The experimental study shows that the device consumes 90-95 % of the sample volume during 20 min of analysis. A small amount of air left in the bag. This mode of analysis should be considered as optimal. The analysis of each probe of exhale starts in zero air flow takes 10 min. The sample temperature control mode is maintained during this time. Then the solenoid valve is switched to exhale probe support.

The PTR-MS device has specific feature. The output data of some channels concentration characterized by individual m/z couldn't be unambiguously identified as certain chemical compounds. The output signal is proportional to the number of concentrations of VOCs sum and products of their chemical breakdown in the drift chamber.

Taking into account, the specifics of the exhaled air analysis problems the method of spectrum analysis of m/z values in a row in a certain range is chosen. As follows from literature data [6-7], the substances can not be detected in exhaled air with concentrations greater than the PTR-MS threshold detection unit (0.5 – 1.0 ppb) for the m/z values greater than 141. Consequently, the range of the test spectrum m/z value is 21 - 141 except channels 21, 30, 32 and 37, which correspond to the concentrations of ions generated in the drift chamber. The information from the literature is confirmed by the breath sample analyze experiments. Concentrations do not exceed the device detection threshold for m / z values of 141 to 201.

The specialized software supplied by the manufacturer (the company Ionicon, Austria) controls the instrument PTR-MS. This software uses a proprietary data file format, and in addition, has a number of disadvantages in use. The specialized software tools have been developed for rapid transformation of the data file to format suitable for further analysis. These tools perform the calculation of average concentrations and standard deviations for each m / z values separately for periods of zero flow of air and test sample. The automated exception of transient analysis is carried out at switching the solenoid valve. Duration of the transients is about 1 - 2 minutes. The time has been determined experimentally.

The time of one m / z values spectrum analysis within a selected range is 1 minute. The standard concentration deviations is calculated in the range from 3 to 10% of their average values. The obtained values characterize noise of the device and, apparently, cannot be significantly improved by changing the experimental procedures in existing conditions. Recent results are presented in Table 1.

3. IMS study

It remains an open question of portable HF-PEF diagnostic system. Meanwhile dozens of diseases may be recognized by changes in the chemical composition of volatile organic compounds (VOCs) in the air exhaled by the gas chromatography - ion mobility spectrometry (GC-IMS). For instance some studies show that the mobility spectra from GC-IMS breath analysis of patients with cancer lung damage differs from spectrums of healthy peoples [9-10].

Interest in using Multi Capillary Column MCC IMS for clinical analyses has arisen through concerns regarding the integrity of samples when transported to the laboratory. It is necessary about 10 mL of breath to carry out a full analysis. For investigations of human breath at a high level of humidity a combination of a polar Multi Capillary Column (MCC) (ISC, OV-5, OOO "Siberteh", Novosibirsk, Russia) partly preseparating the analytes is used in combination with a conventional ion mobility spectrometer [9-12]. An IMS coupled to a MCC columns allows the identification and quantification of volatile metabolites occurring in human breath down to the ng/L and pg/L range of analytes within less than 12 minutes and without any preconcentration steps. The analyte passes through the exhalation of 1000 parallel capillaries, each with an internal diameter of 40 micrometers. The overall diameter of 3 mm separation columns, which allows the carrier gas supplied from the optimum for spectrometry at 300 ml/min. The IMS study of acetone, acetic acid control samples shows the sensitivity in ppb range. It is carried out further investigations of exhaled breath samples with improved sample device.

Table 1. Recent results. Adapted from [1] with permission.

Substance name	Patients without CHF The average concentration, ppb (pts = 16)	Patients with CHF The average concentration, ppb (pts = 19)
Acetylene	5.9±1.63	5.68 (4.75-6.79)
HCN	15.16±1.54	15.46±1.8
formaldehyde	1271.17±96.34	1365.93 (1294-1469.5)
methanol	261.31±60.27	256.3 (226.5-330.2)
acetonitrile	17.61 (8.9-46.63)	30.5±27.47
Propylene	107.87±33.64	168.74 (87.5-589.9)
Acetaldehyde	194.21±29.74	197.8 (171.5-275.1)
ethanol	10.43 (9.25-20.94)	15.26 (12.43-32.3)
1,3-butadiene	1.55 (1.25-3.23)	2.57 (1.98-5.1)
acetone	325.72±97.46	683.4 (333.9-3084.7)
acetic acid	22.94±4.1	34.3±12.5
dimethyl sulphide	5.0±2.0	5.92 (3.04-8.36)
isoprene	16.31±5.22	12.8±7.2
methyl vinyl ketone, methacrolein	3.04±1.5	4.3±3.8
methyl ethyl ketone	1.6±0.7	1.93 (1.32-4.08)
peroksiatsetilnitrat	0.66 (0.54-0.8)	0.65 (0.5-0.97)
benzene, dimethylsulfoxide	0.76 (0.68-2.29)	0.98 (0.65-1.5)
terpenes	3.79 (2.84-6.98)	3.22 (1.83-4.8)
2-methyl-3-2-ol	0.89 (0.74-2.57)	1.1 (0.86-2.7)
toluene	1.18 (0.9-3.63)	1.66 (1.3-3.1)
phenol	2.44 (1.7-10.33)	2.93 (2.0-10.9)
dimethylfuran, dimethylpyrazole	0.55 (0.41-1.43)	0.69 (0.48-1.2)
styrene	0.37 (0.29-0.54)	0.42 (0.34-0.7)
xylene, benzene-C8	0.64 (0.59-1.19)	1.82 (0.79-4.9)
propyl	0.6 (0.44-1.07)	0.75 (0.54-1.4)
NO^+	0.71 (0.62-1.07)	0.8±0.1
O_2^+	16.35±1.5	16.9±1.5
$H_3O^+ (H_2O)$	0.23±0.05	0.26±0.08

4. Results

The baseline characteristics of patients are in Table 2.

Table 2. The baseline statistical characteristics of patients.

Characteristic	HF-PEF (pts = 21)	Control group (pts = 16)
Mean age (range), years	72 (60-75)	64 (38-70)
Male sex - n, %	10 (47.6%)	9 (56.3%)
Hypertension - n, %	21 (100%)	14 (87.6%)
Diabetes mellitus - n, %	8 (38.1%)	2 (12.5%)
Previous myocardial infarction - n, %	10 (47.6%)	2 (12.5%)
Atrial fibrillation - n, %	13 (61.9%)	0
Mean EF, %	60±7	69±4

In compare with control group the concentrations of several biomarkers were significantly higher in HF-PEF group. They are acetone, acetic aldehyde, ethanol, propylene. The median (interquartile range) concentration of acetone in HF-PEF group when compared to control group is 934 ppb (324-2432) versus 322 ppb (280-368), p=0,002; acetic aldehyde 297 ppb (213-441) versus 193 ppb (177-231), p=0,004; ethanol 18 ppb (14-28) versus 10,6 ppb (9-21), p=0,018; propylene 343 ppb (122-873) versus 124 ppb (94-141), p=0,04 respectively.

The median (interquartile range) concentration of biomarkers is presented in Table 3.

Table 3. The median (interquartile range) concentration of biomarkers.

Biomarker\ groups	HF-PEF group (n=21) concentration, ppb	Control group (n=16) concentration, ppb
acetone	934	322
acetic acid	297	193
ethanol	18	10.6
propylene	343	124

The study shows applicability of PTR-MS for HF-PEF diagnostics. Nevertheless it is necessary to confirm results by a systematic statistical study.

5. Conclusion

There is a significant difference in exhaled breath PTR-MS spectrums between patients with HF-PEF and without heart failure. In compare with control group the concentrations of acetone, acetic acid, ethanol, propylene biomarkers were significantly higher in HF-PEF group. The IMS study of acetone, acetic acid control samples shows the sensitivity in ppb range. It is carried out further investigations of exhaled breath samples with improved sample device. Further investigation is necessary to determine the correlation between these biomarkers and levels of natriuretic peptides.

Acknowledgements

This work was supported by the Competitiveness Program of NRNU MEPhI.

6. References

[1] Bykova A, Syrkin A, Kopylov F, Chomahidze P, Pershenkov V, Beljakov V, Samotaev N and Shaltaeva Y 2014 *European Journal Of Heart Failure* **16** 177

[2] Samotaev N, Golovin A, Vasilyev V, Malkin E, Gromov E, Shaltaeva Y, Mironov A and Lipatov D 2014 *Lecture Notes in Electrical Engineering* **268** 447-451

[3] Kopylov F, Syrkin A, Chomakhidze P, Bykova A, Shaltaeva Iu, Beliakov V, Pershenkov V, Samotaev N, Golovin A, Vasil'ev V, Malkin E, Gromov E, Ivanov I, Lipatov D and Iakovlev D 2013 *Klinicheskaia meditsina* **91**(10) 16-21 (*in Russian*)

[4] Kopylov P, Syrkin A, Chomakhidze P, Bykova A, Shaltaeva Y, Belyakov V, Pershenkov V Samotaev N, Golovin A, Vasilyev V, Malkin E, Gromov E, Ivanov I, Lipatov D, Yakovlev D and Betelin V 2016 *Cardiology* **5** 37-41

[5] Lipatov D, Shaltaeva Y, Belyakov V, Golovin A, Pershenkov V, Shurenkov V and Yakovlev D 2015 *Proc. 3rd Int. Conf. Nanotechnologies and Biomedical Engineering (Chisinau) (Springer: IFMBE Proceedings)* **55** 404-408

[6] Beauchamp J, Kirsch F and Buettner A 2010 *J. Breath. Res.* **4** 026006 12

[7] Beauchamp J, Herbig J, Gutmann R and Hansel A 2008 *J. Breath Res.* **2** 046001

[8] de Gouw J, Warneke C, Karl T, Eerdekens G, Van Der Veen C and Fall R 2003 *Int. J. Mass Spectrom.* **365** 223-224

[9] Eiceman G A, Karpas Z and Hill H 2013 *Ion Mobility Spectrometry (3rd ed.)* (Taylor & Francis Group) **406**(11) p 2493–2494

[10] Westhoff M, Litterst P, Freitag L, Urfer W, Bader S and Baumbach J-I 2009 *Thorax.* **64** 744-748

[11] Laakia J, Adamov A, Jussila M, Pedersen C S, Sysoev A and Kotiaho T 2010 *J. Am. Soc. Mass. Spectr.*, **21**(9) 1565-1572

[12] Pedersen C S, Lauritsen F R, Sysoev A, Viitanen A K, Makela J, Adamov A, Laakia J, Mauriala T and Kotiaho T 2008 *J. Am. Soc. Mass. Spectr.* **19** 1361-1366

IOP Conf. Series: Materials Science and Engineering **151** (2016) 012018 doi:10.1088/1757-899X/151/1/012018

The investigation of ionization conditions in the trace amounts detection of heterocyclic compounds by ion mobility spectrometry and mass spectrometry

Y R Shaltaeva[1], A A Sysoev[1], S S Poteshin[1], K I Negru[2], S S Grishin[2], V V Trefilova[2], M I Zuev[2] and E P Baberkina[2]

[1]National Research Nuclear University MEPhI (Moscow Engineering Physics Institute), Kashirskoe shosse 31, 115409 Moscow, Russian Federation
[2]D. Mendeleev University of Chemical Technology of Russia, Miusskaya sq. 9, 125047 Moscow, Russian Federation

Corresponding author e-mail address: shaltaeva@yandex.ru

Abstract. The first part of paper is devoted to the detection of New Psychoactive Substances by ion mobility mass spectrometry study. In the second part of the paper presents a promising approach to prevent the spread of narcotic substances, consisting in the use of field-portable ion mobility spectrometers and finding the correlation between the peaks of the spectrograms of ion mobility and the chemical structure of the compound.

1. Introduction

The spread of narcotic substance one the major problems in the modern world. A growing number of New Psychoactive Substances (NPS) are reported every year by a large number of countries and territories throughout the world. Between 2008 and 2014, a total of 569 NPS were reported to be on the market [1]. The formula of synthetic drugs is constantly changing, making it difficult to interdiction "spice", as each time specific formula is entered in the register of prohibited substances [2-4]. As a result, sales of the mixture is, in fact, legal until restriction of its use.

The second major problem is the difficulty of identifying and traffic restraint [5-7]. Most of the narcotic substance are transported by public transport (trains, planes, etc.) and can be hidden under clothing, in hand luggage, in checked baggage. The first part of paper is devoted to the detection of NPS by ion mobility mass spectrometry study. In the second part of the paper offers a promising approach to prevent the spread of narcotic substances, consisting in the use of field-portable ion mobility spectrometer and finding the correlation between the peaks of the spectrograms of ion mobility and the chemical structure of the compound.

2. Detection of synthetic drugs by Ion Mobility Spectrometry

The six synthetic drugs and cocaine is studied by ion mobility time-of-flight mass spectrometry (IMTOFMS) with atmospheric pressure chemical ionization (APCI) and electrospray ionization (ESI) [8]. It were determined characteristic ion mobility mass spectrometry (IMMS) data, reduced mobility, and limits of detection (S/N=3) for the drugs (figure 1).

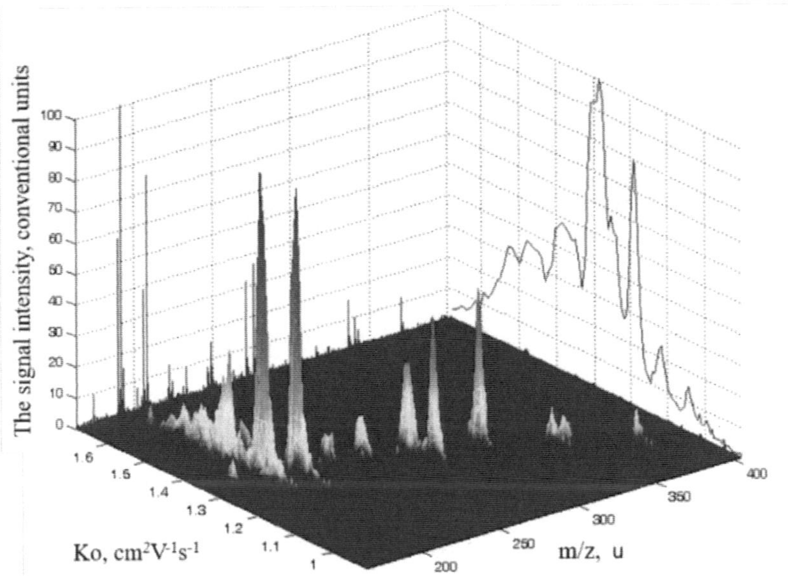

Figure 1. APCI ion mobility mass spectra in a mixture of 8 μM methylone, 5 μM 4-MEC, 8 μM 3,4-MDPV, 6 μM JWH-210, 4 μM JWH-250, 9 μM JWH-203, 7 μM cocaine, and 1 μM 2,6-DtBP in acetonitrile. Adapted from Ref. [8] with permission.

Absolute reduced mobilities in nitrogen were 1.35, 1.28, 1.41, 1.30, 1.18, 0.98, 1.09 and 1.07 $cm^2 \cdot V^{-1} \cdot s^{-1}$, for methylone (M-H)+, methylone (M+H)+, 4-MEC (M-H)+, 4-MEC (M+H)+, 3,4-MDPV (M+H)+, JWH-210 (M+H)+, JWH-250 (M+H)+, and JWH-203 (M+H)+, respectively (Table 1).

Table 1. The values of monoisotopic mass (M) and the reduced mobility of (K_0) of characteristic ions, and the detection limits (S / N = 3) is obtained for the studied drugs (dissolved in acetonitrile) at a data collection duration 100 s. Adapted from Ref. [2] with permission.

Substance name	Monoisotopic mass, M (Da)	The characteristic ions		K_0, $cm^2/(V \cdot s)$	The limit of detection	
		APCI-M, M (Da)	ESI-M, M (Da)		Relative, mol / l	Absolute, g
Metilon	207.09	206.08	208.10	1.35	3×10^{-7}	7×10^{-10}
		208.10		1.28		
4-MEC	191.13	190.12	192.14	1.41	1.5×10^{-7}	4×10^{-10}
		192.14		1.3		
3,4-MDPV	275.15	276.16	276.16	1.18	4×10^{-8}	1.5×10^{-10}
JWH-210	369.21	370.22	370.22	0.98	4×10^{-7}	1.7×10^{-9}
JWH-250	335.19	336.20	336.20	1.09	2.3×10^{-7}	7×10^{-10}
JWH-203	339.14	340.15	340.15	1.07	3×10^{-7}	1.2×10^{-9}
Cocaine	303.15	304.15	304.15	1.2	4×10^{-9}	1.4×10^{-11}

The application allows one to consider IMTOFMS as a good candidate for a method for fast determination of recently appeared designer drugs marketed on the internet as "bath salts," "spice" and "herbal blends." Nevertheless, IMTOFMS stay expensive dimensional laboratory equipment.

The basis of designer drugs is the chemistry of heterocyclic nitrogen compounds, which is the main reason for the paper, presented an investigation of ionization conditions in the trace amounts detection of heterocyclic compounds. The alternative way to massive IMTOFMS consist in the use of field-portable ion mobility spectrometers and finding the correlation between the peaks of the spectrograms of ion mobility and the chemical structure of the compound.

The main reason for the significant interest in the indoles and their derivatives is their wide range of manifestations of the biological activity. Tryptophan is essential amino acid, which included in most proteins. In animal organisms, two hormones are produced from tryptophan, related in chemical structure. One of them is serotonin, which is closely associated with the central nervous system activity, regulates motility and secretion of gastric juice, and the second - melatonin is involved in the control of the change of day and night rhythm of physiological functions. β-indoleacetic acid, which is also produced from tryptophan role hormone belongs flora regulating plant growth. Among the indole derivatives with significant potential physiological activity are hallucinogenic drugs such as lysergic acid diethylamide (LSD), indole alkaloids are widely used in medicine. It should be called reserpine, which has anxiolytic properties, vincristine - drugs used to treat leukemia and cause uterine contractions ergotoxine alkaloids group, including ergotamine, also used as a remedy for a migraine.

One of the devices operating on the method of ion mobility spectrometry is the Kerber IMS, produced by JSC "Yuzhpolymetal-Holding", Russia. This device detects trace amounts of low-volatile and volatile organic compounds of explosives and narcotic substances in the air, on the surface of various objects on the skin and clothing of people. The preliminary exploratory studies were conducted by the Kerber IMS.

3. Results

The dissolved in acetonitrile heterocyclic compounds is detected in low ppb range from evaporated sample (Table 2).

Table 2. The values of monoisotopic mass (M) and the reduced mobility of (K_0) of characteristic ions are obtained for the studied heterocyclic compounds (dissolved in acetonitrile).

Substance name	Monoisotopic mass, M (Da)	K_o,cm^2/(V·s)	K_o,cm^2/(V·s)	K_o,cm^2/(V·s)
imidazole	68	1.715333	1.506	
1-methyl imidazole	82	1.61		
4-methyl imidazole	82	1.569		
2-phenyl imidazole	144	1.674	1.2	
1,2-dimethyl imidazole	96		1.446	
benzimidazole	118	1.793	1.293	
2-methyl benzimidazole	132	1.751	1.313	
2-formyl benzimidazole	146	1.64		
1-methyl-2-formyl benzimidazole	160	1.621		
1,4-diphenyl-5-methylimidazole	234		1.309	0.8625
indazole	118		1.4035	
Benzene	78	1.6435		
Aniline	93	1.79	1.4635	
pyridine	79	1.974		
indole	117	1.7836		
3-formylindole	145	1.6365	1.34325	1.13325
3-acetylindole	159	1.608	1.18	1.076
3 nitrilindol	142		1.427	
Methyl 3- carb. acid	175		1.4585	
1-pentyl-3-formylindol	215	1.649	1.319	0.829333
Grameen	174	1.504	1.3064	1.1284
2-methylindole	131	1.73		
2-phenylindole	193	1.4725	1.33	1.192

Investigated indole derivatives give clear signals on the ion mobility spectrometer Kerber produced by "Yuzhpolymetal-Holding" (YPH) LLC, Russia. The spectrometer has corona discharge ion source. The nature of the spectrogram signal is substantially dependent only on the mass, but the electron density distribution in the molecule under the influence of substituents.

4. Results

The experimental spectrums of studied heterocyclic compounds (dissolved in acetonitrile) shows characteristic peak in reduced mobility range from 1,5 to 1,7 $cm^2/(V{\cdot}s)$. During the experiment and analyzing the results, it is suggested that the particles could be formed, protonated at a specific position of the ring or on a particular atom and complex ions clustered between π- aromatic ring system and reactant ion. It is found that this cluster ion appears on spectrogram more mobile field measurements K_o = 1.6-1.9 cm2/(V·s). The formation and stability of cluster ion are affected by the chemical structure of substances.

5. Conclusion

The article presents retention times of synthetic drugs and their heterocyclic constituents on ion mobility spectrometers and mass spectrometers. The study of the chemical structure of synthetic drugs will improve the means of combating the spread of the illicit substances. It is necessary to conduct a detailed study with machine learning processing of the chemical reactions in the corona discharge ionization chamber for improvement of the methodology and rapid detection of drugs.

Acknowledgements
This work was supported by the Competitiveness Program of NRNU MEPhI.

References

[1] RF Government Decree of 30.06.1998 № 681 2014 "On approval of the list of narcotic drugs, psychotropic substances and their precursors subject to control in the Russian Federation."
[2] Bossong M G, Van Dijk G P and Niesink R J 2005 *Addict. Biol.* **10** 321-323.
[3] Maheux C R, Copeland C R, Pollard M M 2010 *Microgram Journal* **7** 42-49
[4] Kamata H T, Shima N, Zaitsu K, Kamata T, Nishikawa M, Katagi M, Miki A and Tsuchihashi H 2007 *Jpn J. Forensic Sci. Technol.* **12** 97-106
[5] Reitzel L A, Dalsgaard P W, Müller I B and Cornett C 2012 *Drug. Test Anal.* **4** 342-354
[6] Jankovics P, Váradi A, Tölgyesi L, Lohner S, Németh-Palotás J and Kőszegi-Szalai H 2011 *Forensic Sci. Int.* **210** 213-220
[7] Nakajima J, Takahashi M, Seto T, Kanai C, Suzuki J, Yoshida M and Hamano T 2011 *Forensic Toxicol.* **29** 95-110
[8] Sysoev A, Poteshin S, Chernyshev D, Karpov A, Tuzkov Y, Kyzmin V and Sysoev A 2014 *Eur. J. Mass Spectrom.* **20** (2) 185-192

Techniques and instrumental complex for research of influence of microwaves encoded by brain neural signals on biological objects' psycho physiological state

B V Gurkovskiy, B V Zhuravlev, E M Onishchenko, A B Simakov, N Yu Trifonova and Yu A Voronov

National Research Nuclear University MEPhI (Moscow Engineering Physics Institute), Kashirskoe shosse 31, 115409 Moscow, Russian Federation

Corresponding author e-mail address: absimakov@mephi.ru

Abstract. New instrumental technique for research of the psycho-physiological reactions of the bio-objects under the microwave electromagnetic radiation, modulated by interval patterns of neural activity in the brain registered under different biological motivations, are suggested. The preliminary results of these new tool tests in real psycho physiological experiments on rats are presented.

1. Introduction

The Russian Science Foundation announced that for the next decade, one of the Russian science development priorities is the problem of "Neurotechnology and cognitive studies". One of the most important tasks is defined as "the Study of fundamental brain mechanisms and cognitive functions with the use of new methods for directed modulation of neural activity and brain functions. This assumes the creation of new methods and tools for electromagnetic non-invasive brain stimulation, regulating the activity of neurons in the brain using implantable devices. The relevance of these works is determined by the possibility of experimental studies of brain function fundamental mechanisms and a wide range of applications ranging from medicine, pharmacology, and psychiatry to special applications. It should be noted that such studies have started more than half a century ago and widely developed with the use of miniature microelectronic biotelemetry systems that allowed them to influence the biological object and at the same time to record his physiological parameters in free behavior [1,2]. Among the outstanding works on remote effects on the biological object to control its physiological state, motivation, and behaviour, it is necessary to note the work of the Spaniard José Delgado, a pioneer of the electrodes implantation into the brain [3], and studies of V. M. Arbutin, one of the Russian biotechnical systems founders [4]. In these works, the radio channels were used for the transmission of stimulus signals to the implanted electrodes in different brain structures, and information about physiological parameters of biological objects was received for analysis in the opposite direction. Such studies are continuing on the basis of modern micro- and nanoelectronic technology and are essential for the brain mapping, the reconstruction of its functional system, neural interfaces development, including new types of non-invasive human-machine interfaces (EEG, EMG, eye movement, etc.). Another modern method of a non-invasive impact study on the brain is directly related to electromagnetic fields (EMF) exposure on biological objects. Great attention is paid to this problem in recent years, due to continuously growing electromagnetic pollution of the environment. So, mankind concern about its electromagnetic safety. Adequately systematized research work and results in this direction are proposed in the review of the International Commission for the protection from non-ionizing radiation (ICNIRP) [5], and in the following analytical publications [6,7]. An important result of these studies was the person's individual sensitivity detecting even to weak microwave electromagnetic effects [8]. In particular, the relation between the dose and heart rate variability (HRV) [9], and a number of psychological characteristics [10]. This allowed us to formulate the first basic principle of constructing a new methodology and tool, namely, low-intensity microwave electromagnetic radiation absorbed in a thin layer of the skin, causing excitation of the central nervous system - CNS via the receptor apparatus of the skin and peripheral sensory receptors. This stimulation of the CNS is transferred to the basic functional systems of the organism, in particular, changes of HRV and some of the psychophysiological characteristics of the person. The second basic principle is related to the results obtained in [11] and consists in the fact that: the neural activity of different brain structures of experimental animals depends on the prevailing motivation. Temporal patterns of the brain neuron impulse activity reflect the specificity of different motivational states, including hunger, thirst, fear, or aggression, as well as the satisfaction of these motives. The combination of the microwave modulation controlled on the bio - object with a temporary pulse (spike) patterns of brain activity with simultaneous monitoring of several physiological parameters, including behavioral, the foundations of a new research method and the devices proposed in this work.

2. The research methodology

Above were indicated the basic principles of the developed methods and devices to study the effect of the pulsed microwave radiation on biological objects' cognitive function. Let's look at these principles in more details.

1.1. Neurons information codes

In the laboratory of "Functional systems" (P. K. Anokhin Institute of Normal physiology) under the supervision of Professor, Ph. D. Zhuravlev B. V. (one of the authors of this article) was carried out the complex of investigations aimed at the analysis of the inter-spike intervals distribution of neural activity in different brain structures of an animal (rabbit, rat) in his free behaviour and different motivational states. For the experiments, we used both traditional glass microelectrodes [12] and microelectrode matrix in microelectronic application [13]. It was experimentally established on hungry animals that there is a bit of group activity in 12 brain structures (lateral hypothalamus, ventromedial nucleus of the hypothalamus, reticular formation, thalamus, sensorimotor, visual, orbital area of the cortex, hippocampus, etc.). This group of neural activity is characterized by the dominant interspike intervals distribution. As an example, figure 1 shows inter-spike activity and the distribution of the interval pattern neurons of the lateral hypothalamus at the hungry animal.

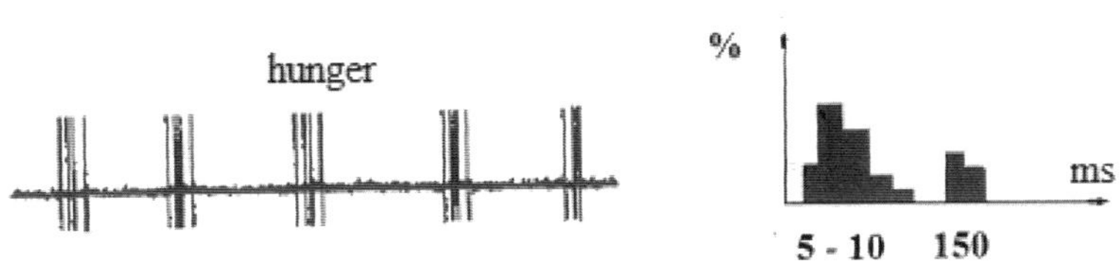

Figure 1. Inter-spike neuron activity and histogram of the inter-spike intervals distribution.

The interval histograms are dominated in the range from 5 to 10 ms and 150 ms. Similar studies in other motivational animal's states (thirst, immobilization, etc.) showed "carrier frequency" corresponding to the 150 ms interval is stored, and the intervals characterizing the motivations change significantly. Thus, all three mentioned specific motivations varied at small intervals between the discharges but had a "carrier" frequency 150 ms, which coincides with the theta - rhythm in the EEG study. This rhythm in the EEG exists in animals and humans during intense, stressful conditions. Further studies have shown that after a satisfaction of all three motivations, there is a reorganization of the interval pattern from group activity to a regular single type, and at the histogram, one interval begins to dominate. For the "hunger" motivation source such a reorganization of neural activity during eating is illustrated in figure 2.

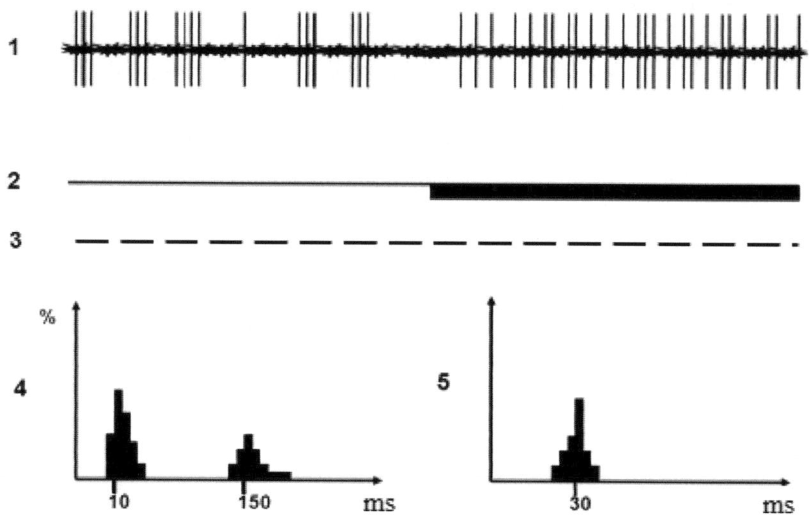

Figure 2. Reorganization of the interval activity pattern of the neuron when animal ate: 1 - the activity of the neuron; 2 - thick line – the beginning of food eating; 3 - timestamp (250 ms); 4 - the neuron activity interval histogram of the hungry animal; 5 - The histogram of the interval activity of the neuron when animal ate.

Thus, as the encoding principle of the stimulating microwave radiation in the developing instrumental complex is proposed to use temporal inter-pulse interval patterns in the neural activity of the brain, during different biological motivations.

1.2. Effect of weak electromagnetic fields on the person's physiological parameters

In recent publications, the authors of this paper [14,15] examined the results of development and application in psychophysiological studies in humans of hardware and software complex for the rapid assessment of the individual effects of weak microwave radiation on heart rate variability of man and a number of psychophysiological parameters. In these works, it was shown that human sensitivity even to weak microwave fields depends on the initial physiological state and health, in particular from the background frequencies and the stability of the heart rate. In addition, subjects with poor health, low activity and mood, high situational and personal anxiety (according to the preliminary psychological tests) were observed more expressed changes in the cardiovascular system. An example of the weak microwave field action (energy flux 50 mW/cm^2; frequency - 930 - 960 MHz, irradiation time was 6 min and the absorbed dose of 110 J/kg) on HRV is shown in figure 3, which shows examples of pulse interval distribution histograms before and after microwave exposure.

Studies of changes in the performance during carrying out complex tests in a group of 25 volunteers in the process the same as the impact of non-modulated EMFs have shown that in 12% of subject's performance indicators did not change and more than 50% of these indicators deteriorated significantly [16]. In [17,18] it was reported about the higher sensitivity of biological objects to modulated fields and resonance phenomena in the functional system of the body, including the CNS, can lead to a significant strengthening of its response to modulated effects, synchronized with biological processes [19].

Thus, in the proposed research methodology affecting microwave electromagnetic field is the physical information carrier encoded with the inherent brain patterning motivation of the biological object, and, even low exposure can lead to synchronous changes of neural activity, and therefore, motivation and behavior of the biological object. Such opportunities are the cornerstone for the further studies. It should be noted that when such studies involving human subjects must ensure a number of moral and ethical standards and safety requirements. Therefore, initial studies under the new methodology were carried out on experimental animals, in particular, on 10 male Wistar rats, in compliance with the principles of humanity set out in European community Directive (2010/63/EU).

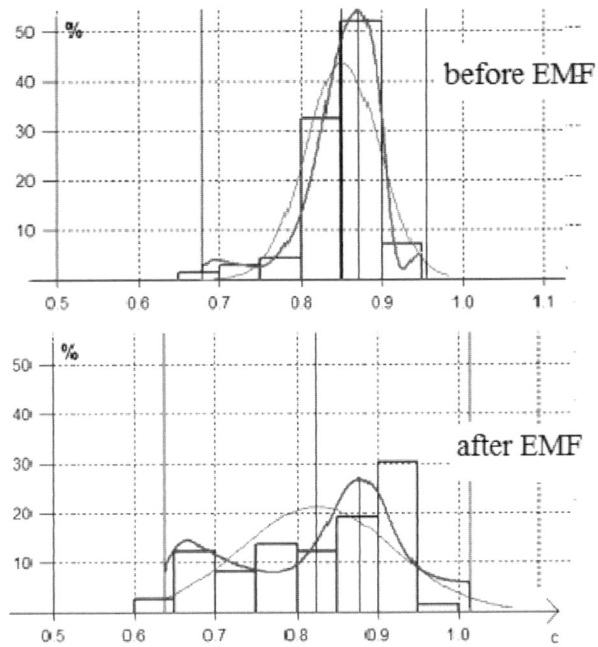

Figure 3. A histogram of pulse intervals distribution before and after a microwave exposure.

3. The instrumental complex

The proposed research methodology determined the structure and composition of the new instrumental hardware and software complex, shown in figure 4.

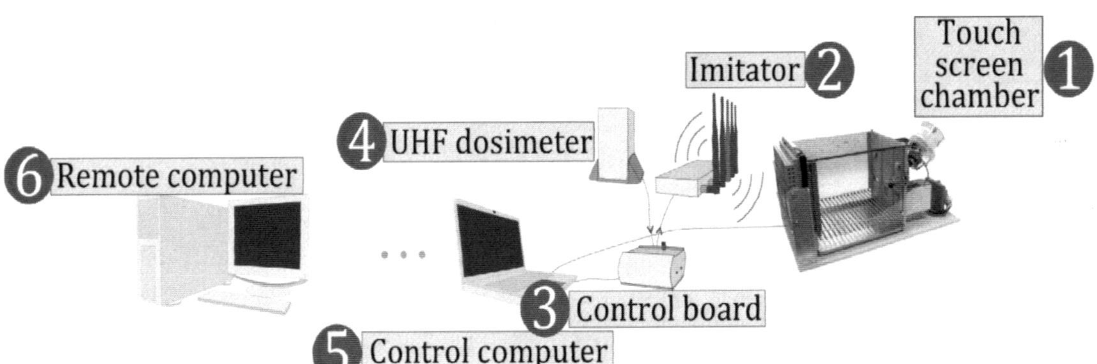

Figure 4. Structure of the research hardware and software complex.

1.3. The main components of a tool set

Unit 1 - Experimental chamber (box) with a lab rat. We used Rat Touch Screen Chamber of the American company Lafayette Instrument where the animals were previously trained to eating and drinking instrumental skills. Photo of the camera is shown in figure 5. The camera is equipped with a video camera that captures moving rats; touch screens, through which the animal gets food and drinking by himself. Implanting into the peritoneal cavity of the rat miniature biotelemetric capsules, it provides a permanent cardio signal recording even under the action of external EMFs.

IOP Conf. Series: Materials Science and Engineering **151** (2016) 012019 doi:10.1088/1757-899X/151/1/012019

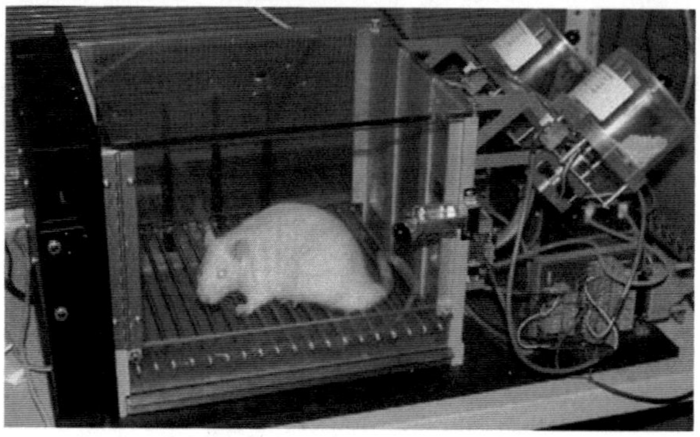

Figure 5. Photo of the experimental chamber with rat animal.

Unit 2 - Imitator. In the background of the figure 5, the antenna of the generator (imitator), providing the irradiation of biological objects with modulated microwave EMF, has been shown. A photograph of this block is demonstrated in figure 6. External digital control of radiating power and microwave signal modulation were provided by using digital potentiometers (type AD5204). The imitator operates in five frequency bands from 0.9 to 2.4 GHz.

Figure 6. Photo of the imitator (generator) of EMF.

Unit 4 - microwave dosimeter. This new device allows controlling some basic dosimetric parameters, such as the flux density of the incident microwave energy (EFD, W/cm^2) and specific absorption rate (W/kg); specific dose absorbed energy (J/kg) [20]. In figure 7: a) photograph; b) the principle of operation of the microwave dosimeter.

a)

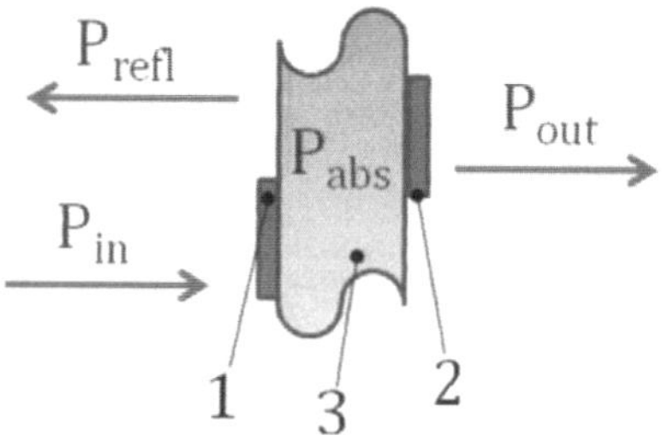

b)

Figure 7. a) The photo of the microwave dosimeter; b) The principle of the device operation.

The principle of operation: 1 and 2 - microwave sensors; 3 - biological tissue phantom; P_{in} - falling energy flux; P_{out} – passed energy flux; P_{refl} - reflected energy flux; P_{abs} - absorbed energy flow. In the absence of microwave radiation, the voltage between sensors 1 and 2 equals zero. When electromagnetic radiation appears, a differential voltage arises, due to absorption in the material layer. This voltage is the quantity that characterizes the absorbed dose of microwave radiation Blocks 3,5 control all units of the complex, collection and preliminary processing of information from all measuring devices. Unit 6 (remote computer) allows remote controlling the blocks.

1.4. Preliminary psychophysiological experiment results

The experiments were performed on 10 Wistar male rats in compliance with the humanity principles, set out in the EU Directive (2010/63/EU). Rats were exposed to microwave radiation with the values of flux density energy, corresponding to SanPiN 2.1.8/2.2.4.1383-03 requirements. The influence of EMF on nondeprived rats' behavior and the behavior of rats after 24 hours food and drinking deprivation was investigated. The rats were placed in the box of hardware-software complex (Rat Touch Screen Chamber, Lafayette Instrument, USA) where they have previously trained feeding and drinking instrumental skills. As a result of experiments, it was shown that, depending on the duration of the systematic GSM band microwave radiation modulated with frequency typical for the motivation of thirst, there was a significant decrease in the drinking acts number (up to 50% of their initial values). The figure 8 illustrates these results. During 13-hour time-modulated GSM radiation, the number of drinking and feeding acts was decreased. The number of food-procuring behavioral acts was not significantly changed when exposed to electromagnetic radiation (EMR) for 1-4 hours. Hence, according to the results of the experiments, electromagnetic radiation, modulated with temporal patterns of the brain neural activity, affects the instrumental behavior of rats. However, it is necessary to increase the number of samples.

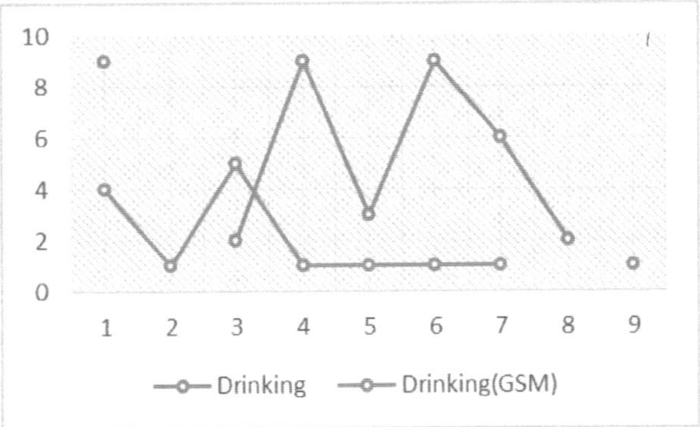

Figure 8. A number of drinking acts of non-deprived rats under the influence of modulated microwave radiation (associated with drinking behavior) for 2 hours. X axis shows different times of the day.

4. Conclusion

The proposed instrumental complex allows investigating the response of the central nerves system (brain) of experimental animals on the microwave electromagnetic radiation, modulated by interval patterns of neural activity in the brain, registered under different biological motivations. The preliminaries tests of this new method and instrumental complex in series of psycho-physiological experiments on rats have shown the changes in their drinking and food-procuring behavioral acts.

The future experiments will be directed to the research of the experimental possibilities for remotely affected by microwave fields on cognitive functions of the brain, which are formed on the basis of the memory mechanisms, the dominant motivational states and evaluated by the results of the behavioral activity. Expected results of the research may have practical importance for medical rehabilitation and psycho-physiological patient's correction.

Acknowledgments

Funding for this study was provided by the Russian Scientific Foundation under Grant #16-19-00167.

References

[1] Levanov V M, Orlov O E and Merekin D V 2013 *The Doctor and Information Technologies* **4** 67-73
[2] Bakalov V P 2001 *Radio and communication* (Moscow) p 27
[3] Jose M R Delgado 1969 *Harper & Row* (New York)
[4] Akhutin V M 1977 *Engineering Psychology* (Moscow) p 149-180 (*in Russian*)
[5] Icnirp.org 2016 *ICNIRP* [online] Available at: http://www.icnirp.org/cms/upload/publications/ICNIRPemfgdl.pdf [Accessed 2 Jul. 2016].
[6] Black D R and Heynick L N 2003 *Bioelectromagnetics* **24** 187-195
[7] Vecchia P, Matthes R, Ziegelberger G, Lin J, Saunders R and Swerdlow A 2009 *Exposure to high frequency electromagnetic fields, biological effects and health consequences (100 kHz-300 GHz)* International Commission on Non-Ionizing Radiation Protection
[8] Electrosensitivity.org.uk. 2016 *Electromagnetic Hypersensitivity (ES)* [online] Available at: http://www.electrosensitivity.org.uk [Accessed 2 Jul. 2016]
[9] Gurkovskiy B V, Murtazina E P, Simakov A B, Zhuravlev B V 2014 *Biomedical Radioelectronics* **4** 23-24
[10] Gurkovskiy B V, Murtazina E P, Zhuravlev B V, Gridneva N A, Trifonova N Yu, Simakov A B 2015 *Biomedical Radioelectronics* **4** 89-91
[11] Sudakov K V, Zhuravlev B V, Kromin A A, Shamaev N N and Timofeeva L V 1987 *Uspekhi fiziologicheskikh nauk* **19** 24-44 (*in Russian*)
[12] Grinchenko Yu V and Shvyrkov V B 1974 *Zh. Vyssh. Nervn. Deyat.* **24** 870
[13] Zhuravlev B V, Simakov A B 1981 *Universities proceedings – Radioelectronics* **12** 56-58
[14] Simakov A B, Onishhenko E M, Zhuravlev B V, Vodokhlebov I N, Gurkovskiy B V and Trifonova N Yu 2016 *Sensors and systems* **4** 9 (*in Russian*)
[15] Simakov A B and Gurkovskiy B V 2015 *Book of Abstracts 8th Dresden Symposium "Hazards – detection and management" (Dresden)* p 30
[16] Samotaev N, Simakov A, Onishchenko E, Gurkovskiy B, Butuzov V, Bocharov Yu, Vodokhlebov I, Zhuravlev B and Murtazina E 2016 *Environmental, Energy and Structural Monitoring Systems (Bari)* (IEEE)
[17] Konovalov V F and Serikov I S 2004 *Radiatsionnaia biologiia, radioecologiia/Rossiiskaia akademiia nauk* **45** 112-115
[18] Pryakhin E A, Tryapitsyna G A and Andreev S S 2007 *Radiation Biology. Radioecology* **3** 339-344
[19] Devyatkov N D, Galant M B and Tager A S 1983 *Biophysics* **28** 895-896
[20] Gurkovskiy B, Onishchenko E and Simakov A 2015 *Proc. of 1st URSI Atlantic Radio Science Conf. (Las Palmas)* p 1

Miniaturized ceramic platform for metal oxide gas sensors array

N N Samotaev

National Research Nuclear University MEPhI (Moscow Engineering Physics Institute), Kashirskoe shosse 31, 115409 Moscow, Russian Federation

Corresponding author e-mail address: nnsamotaev@mephi.ru

Abstract. In work is developing an ultra-fast, low cost and technology flexible process for production array of ceramic MEMS microhotplates for using in semiconductor gas sensors orientated to small series applications, where is sufficient to produce 10-100 samples with a different layout of heaters and membrane per day.

1. Introduction

Currently, gas-sensitive metal oxide sensors are most widely used as discrete microelectronics elements, one type containing gas sensitive material and one microhotplate accordingly for one. This is typical for mobile phones [1], wireless network applications [2-4], in tasks of monitoring air quality [5], and other tasks aimed at large-scale production. On the other side encountered multi gases analysis problems associated with needs to have in same time to activate several types of gas-sensitive materials and respectively independent microhotplates for ones [6].

The multi gases analysis tasks can be divided according to metrological principles for the following:

- The using array of sensors for the analysis of several different types of gases [7];
- Using array of sensors as comparative elements to compensate for external factors [8];
- Building systems such as electronic nose based on neural network algorithm [9].

Technical solving principles of such arrays can be divided into three groups:

- Drawing up of discrete elements microhotplates one body [7];
- Formation on the same chip several separate microhotplates [9, 10];
- Creating one microhotplates large temperature gradient to obtain signals with the same gas sensitive layer located simultaneously at various temperatures [11].

There are microhotplates for gas sensors prepared using standard silicon MEMS technology [12], and to a small extent the combination of ceramics and the classic thick-film technology [13]. This is due in turn to the general trend to reduce the energy consumption of sensor systems using autonomous power sources. On another tern technologically when forming an array of sensors, using both silicon and ceramic MEMS technology absent variation flexibility in the topology by long specificity process occupies a duration from several days (manufacturing photo mask, shadow masks for deposition, sintering processes, and etching, etc.). In our work more attention was focused on the creation of technology, where the time between the design of the array of heaters and final production must take place no more than a few hours.

2. Experimental

Laser micromachining using Minimarker installation [14] with wavelength, 1.064 μm was taken as a basic technological tool. The base material for the microhotplate substrate was monolithic ceramics from aluminum oxide containing 96% alumina because of its cheapness (5 times cheaper than 99.9% alumina) and sufficiently robust for sensor fabrication. As metallization base material, we took platinum-containing composite with additives of organic binder components and borosilicate glass microparticles, which are used for the preparation of finely dispersed paste usable in screen printing process. Microhotplate layout was designed using AutoCAD program and then loaded to the computer of laser engraving machine. According to the previously obtained relationship between the number of laser ablation passes, laser power and depth of engraving, the micromachining process was pre-set. The next steps of fabrication process were the thinning of the whole surface by engraving, cutting of the membrane obtained during the thinning process leading to the formation of beam suspended microhotplate, and dicing the substrate into chips. In the final microhotplate manufacturing step, the platinum ink was deposited on its surface, and then its local laser firing was implemented. The residual, unfired, ink was removed by washing to form microhotplate. Or, alternatively, the ink was fired in an oven, and the heater was formed by local laser ablation.

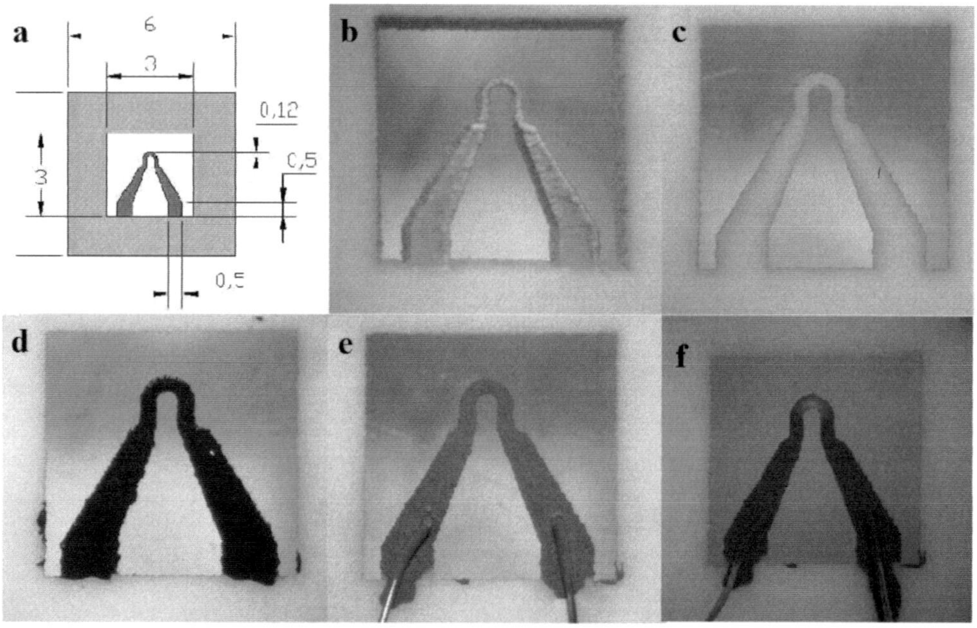

Figure 1. Phase of microhotplate production: (a) design of the microhotplate layout in AutoCAD software, dimensions given in mm (b) engraving by laser Al_2O_3 substrate – view from the bottom side; (c) engraving by laser Al_2O_3 substrate – view from the top side; (d) deposition and annealing of Pt paste on the top of microhotplate; (e) deposition and soldering platinum wire by Pt paste on the top of chip microhotplate (wire diameter is of 100 μm); (f) microhotplate under power load (glowing starts approximately at 600 °C for platinum based material of heater).

Microhotplate obtained within technology tests was made during several hours (photos of the microhotplate on different technological steps are presented in Fig. 1). Microhotplate was packaged in TO-8 still package and studied for the power consumption and heating characteristic time. The following results were obtained of microhotplate with 1,7×1,7 mm chip size: consumption power 150 mW at 450°C (Fig. 2b) and heating characteristic time less than 1 seconds. Subsequent studies showed that for more precise design, the simulation of microhotplate in the Ansys program, which gives more accurate heat and power parameters of the manufactured chip, should be carried out initially.

Using a combination of cheap monolithic ceramic substrate, thick film Platinum paste, laser engraving and AutoCAD software for sensor layout design, microhotplates with good mechanical properties were fabricated. Production speed (laser engraving) of single MEMS structure was the order of 1 minute and 1 hour for group annealing of Platinum paste in a belt furnace. Our experiments show that describing the technology in experience hands gives fantastic speed the design and manufacturing of ready-to-use microhotplate in the order of 1 hour. The cost of materials (alumina substrate and Platinum paste) for the manufacturing single microhotplate size 6×6 mm is about 0.50 Euro. For future decrease, microhotplate power consumption haves two ways. First using ceramic materials with low termoconductivity (As example ZrO_2 [15] or porous Al_2O_3 [16]) or second is related to the application of pulsing heating mode for gas sensitive sensor's layer [17-20].

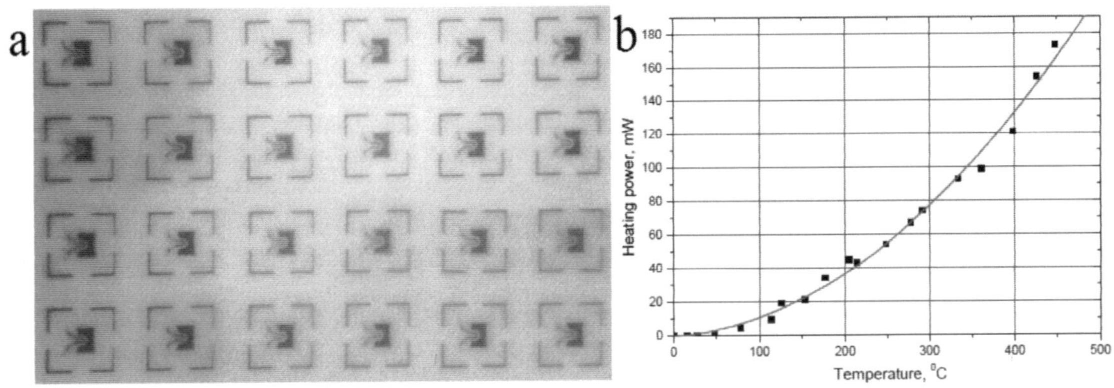

Figure 2. (a) Photo of Al_2O_3 substrate before separation microhotplates on single crystals with 1,7×1,7 mm chip size. (b) Thermal characteristics of microhotplate with 1,7×1,7 mm chip size.

For the manufacture of array of microhotplates by the developed technology is possible to use two lines of action – first is creation by laser micromachining a combine several microheaters on a single chip as shown in Figure 3, or create an array of a set of individual crystals microheaters, as shown in Figure 2 (a).

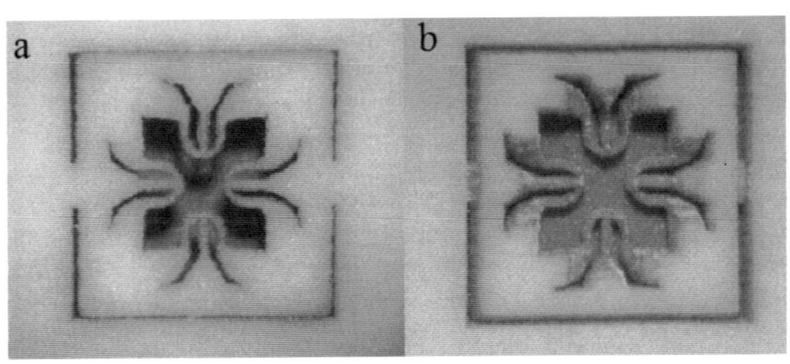

Figure 3. (a) Engraving by laser array of microhotplate on Al_2O_3 substrate – view from the top side; (b) Engraving by laser array of microhotplate on Al_2O_3 substrate – view from the bottom side.

3. Conclusions

During work implementation by laser micromachining of monolithic alumina was made miniature ceramic microheating MEMS platform for an array of metal-oxide gas sensors. MEMS microhotplates is mechanically stable and has lower power consumption in comparison with the classical thick-film technology. The flexible manufacturing process of technology allows integration of gas-sensitive element arrays with different types of sensitive layers and / or operating temperatures on the single ceramic chip.

Acknowledgments
The research was performed within a frame of Russian Federation President Grant for support young scientists №14.Y30.15.7910-MK from 16.02.2015

References

[1] http://www.ccmoss.com/sites/default/files/documents/CC-000015-DS-11-Datasheet%20for%20CCS801.pdf
[2] Baranov A, Spirjakin D, Akbari S and Somov A 2015 *Sensor Actuat. A-Phys.* **233** 279-289
[3] Samotaev N, Ivanova A, Oblov K, Soloviev S and Vasiliev A 2014 *Procedia Engineering* **87** 1294-1297
[4] Aliyu F and Sheltam T 2016 *Sensor Actuat. B-Chem.* **231** 265-275
[5] Blaschke M, Tille T, Robertson P, Mair S, Weimar U and Ulmer H 2006 *IEEE Sens. J* **6** 1298-1307 (art. no. 1703491)
[6] Müller G, Friedberger A, Kreisl P, Ahlers S, Schulz O and Becker T 2003 *Thin Solid Films* **436** 34-45
[7] http://www.sgxsensortech.com/content/uploads/2015/02/1143_Datasheet-MiCS-6814-rev-8.pdf
[8] Karpova E, Mironov S, Suchkov A, Karelin A, Karpov E E and Karpov E F 2014 *Sensor Actuat. B-Chem.* **197** 358-363
[9] Ducso C, Adam M, Furjes P, Hirschfelder M, Kulinyi S and Barsony I 2003 *Sensor Actuat. B-Chem.* **95** 189-194
[10] Vallejos S, Khatko V, Calderer J, Gracia I, Cane C, Llobet E and Correig X 2008 *Sensor Actuat. B-Chem.* **132** 209-215
[11] Haeringer D and Goschnick J. 2008 *Sensor Actuat. B-Chem.* **132** 644-649
[12] Simon I, Barsan N, Bauer M and Weimar U 2001 *Sensor Actuat. B-Chem.***73** 1-26
[13] Vasiliev A A, Pisliakov A V, Sokolov A V, Samotaev N N, Soloviev S A, Oblov K, Guarnieri V, Lorenzelli L, Brunelli J, Maglione A, Lipilin A S, Mozalev A and Legin V A 2016 *Sensor Actuat. B-Chem.* **224** 700-713
[14] http://www.minimarker.ru/tech_data.php
[15] Vasiliev A A, Lipilin A S, Mozalev A M, Lagutin A S, Pisliakov A V, Zaretskiy N P, Samotaev N N and Sokolov A V 2011 *Proc. SPIE - The International Society for Optical Engineering* **8066** art. no. 80660N.
[16] Vasiliev A A, Pavelko R G, Gogish-Klushin S Yu, Kharitonov D Yu, Gogish-Klushina O S, Sokolov A V, Pisliakov A V and Samotaev N N 2008 *Sensor Actuat. B-Chem.***132** 216-223
[17] Samotaev N N, Podlepetsky B I, Vasiliev A A, Pisliakov A V and Sokolov A V 2013 *Automat. Rem. Contr.* **74** 308-312
[18] Samotaev N N, Vasiliev A A, Podlepetsky B I, Sokolov A V and Pisliakov A V 2007 *Sensor Actuat. B-Chem.* **127** 242-247
[19] Veselov D S 2015 *Physics Procedia* **72** 500-503
[20] Veselov D S, Voronov Yu A and Vanyukhin K D 2015 *Physics Procedia* **72** 495-499

Light emitting diode absorption spectroscopy for combustible gas monitoring

S S Fanchenko[1,2], A M Baranov[1], A V Savkin[1] and N N Samotaev[3]

[1]Moscow Aviation Institute (National Research University), Volokolamskoe shosse 4, 125993 Moscow, Russian Federation
[2]National Research Center "Kurchatov Institute", Akademika Kurchatova pl. 1, 123182 Moscow, Russian Federation
[3]National Research Nuclear University MEPhI (Moscow Engineering Physics Institute), Kashirskoe shosse 31, 115409 Moscow, Russian Federation

Corresponding author e-mail address: fanchenko.sergey@kiae.ru

Abstract. A new generation of the infrared light-emitting diodes (LEDs) and photodiodes (PDs) was used to construct an open path non-dispersive infrared (NDIR) methane analyzer. It was shown earlier that its characteristics are suitable for usual alarm systems, but new measurements have shown that cross sensitivity to other alkanes is rather high.

1. Introduction

Infrared light absorption spectroscopy is widely used for about eighty years [1]. The whole infrared region is usually divided into three regions described in Table 1.

Table 1. Infrared usual regions

Region	Wavelength	Vibrations
Far IR	1 mm - 25 μm	Lattice vibrations, rotations
Mid IR (MIR)	25 - 2.5 μm	Fundamentals
Near IR (NIR)	2500 - 800 nm	Overtones

Recently a new generation of semiconductor heterostructure LEDs was developed for MIR and NIR regions [2].

Practically simultaneously with the development of NIR and MIR LEDs the works on their application for gas detection had started [3-8]. IR LEDs are basically double heterostructures grown on a semiconductor substrate by means of liquid phase epitaxy, the narrow-gap band active layer being located between two wide-gap band layers. For NIR region quaternary heterostructures GaInAsSb/AlGaAsSb lattice matched to GaSb substrate are used, the minimal band gap existing in GaSb - 0.726 eV. For MIR region quaternary heterostructures InAsSb/InAsSbP lattice matched to the InAs substrate are used, the minimal band gap existing in InAs - 0.354 eV. Due to the fact that in narrow-gap band semiconductor the non-radiative relaxation processes dominate radiative ones with temperature growth, the MIR LED optical power saturation happens rather quickly. As a result, the optical power of NIR LEDs is by several orders higher than of MIR ones. For NIR LEDs the typical optical power is tens mW, while for MIR ones – tens μW, these numbers being confirmed by LEDs specifications from "Led Microsensor NT"[9].

Fundamental vibrations of main combustive gas molecules are located in MIR region, while in NIR region mainly overtones are located, the absorption intensity being by several orders lower than for fundamental vibrations [10]. So for LEDs the perspectives of using them for NIR region are practically the same as for MIR one, the absorbance measured values being of the same order.

LEDs could be used only in nondispersive infrared (NDIR) absorption spectroscopy due to the fact that their spectral widths are of the same order as the specific absorption bandwidths. The main peculiarity of NDIR scheme is the use of two channels – of the reference channel (for which LED spectra do not intersect absorption bands) and of measurement one (for which LED spectra cover absorption bands). To exclude the effects of additional scattering and absorption, the measurement channel intensity is normalized to reference channel intensity.

Up to now only MIR LEDs [4-7] or halogen lamps with filters [11] were used for combustible gas detection. The reason for it is the fact that the most strong absorption bands are located in MIR region. Despite it, first, trial attempts of using NIR LEDs for methane detection were started at the end of the previous century [3]. Only after NIR LEDs became a commercial product in the second decade of our century, the new research had started [9,12,13]. The obvious advantage of using LEDs is their small dimensions, low energy consumption, and a rather low price. These LEDs could be used firstly in the cases when measurements of rather high concentrations with a rather low accuracy are needed – in such systems as open path gas analyzers, for which the alarm signal is defined on the level of 10% of the low explosion level, its typical values being of the order of several percents for main combustible gases.

2. Experimental

A prototype of an open path LED-based methane analyzer was constructed recently [12]. It consists of two independent units – emitting and detecting blocks mounted on tripods. In emitting block LEDs from "LED Microsensor NT" company are used (at 2.3 μm wavelength for the measurement channel and 1.7 μm for the reference one). LED spectra and methane and water vapor absorption bands are presented in Figure1. These LEDs are characterized by the compact size of 0.35×0.35 mm, low power consumption (<1 mW), short response time – 10-50 ns, operation temperatures up to +150 °C, lifetime 80 000 hours. LEDs are mounted into a package with a thermoelectric module (Peltier element) enabling LED chip temperature stabilization.

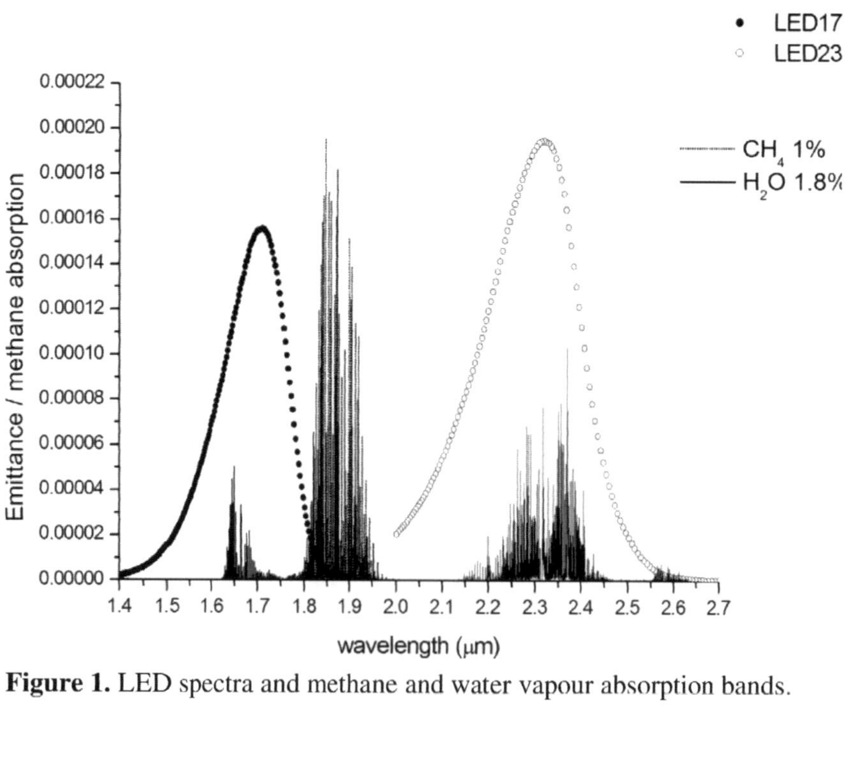

Figure 1. LED spectra and methane and water vapour absorption bands.

In detection bloc "LED Microsensor NT" NIR photodiode is used with following characteristic: cut off at wavelength 2.45 μm, sensitive area diameter 0.3 mm, detectivity $2\text{-}5\cdot10^{10}$ $cm\cdot Hz^{0.5}W^{-1}$. The whole view of the prototype of an open path methane analyzer with the gas cell is given in Figure 2. It was shown in [12], that it could be used for alarm systems.

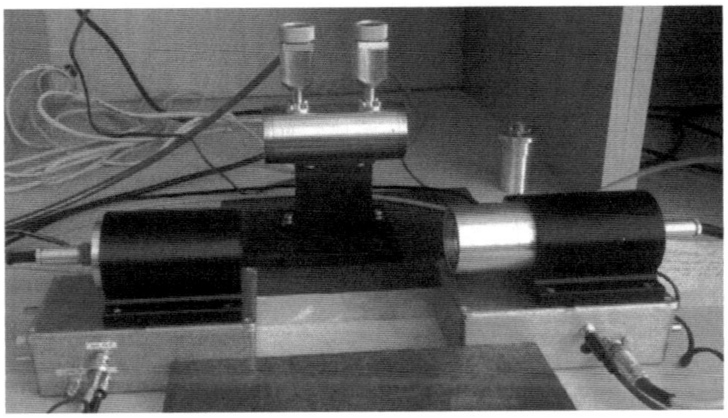

Figure 2. The prototype of an open path LED-based methane analyzer

Besides methane, the developed open path gas analyzer is sensitive to all gases with similar absorption spectra, firstly to the gases from the same homological set, alkanes. Weak dependence on gas content is illustrated in Figure 3, in which the transmittance measurements for both reference and measurement channels are presented for two different alkane mixtures - for 85% isobutane + 10% butane + 5% propane mixture and 70% propane + 30% butane one. The transmittance gets lower after the gas cell is filled with gas. It is seen that after equilibrium is achieved the ratio of the intensities of the measurement and reference channels is quite the same, while the gas content is quite different. For improving the selectivity of the gas analyzer to different gases, one can use LED arrays with a set of LEDs emitting light at wavelengths which are near to specific gas absorption bands. Such LED arrays are also available on the market now [9]. The potentialities of using LED arrays for hydrocarbon detection are discussed in [14].

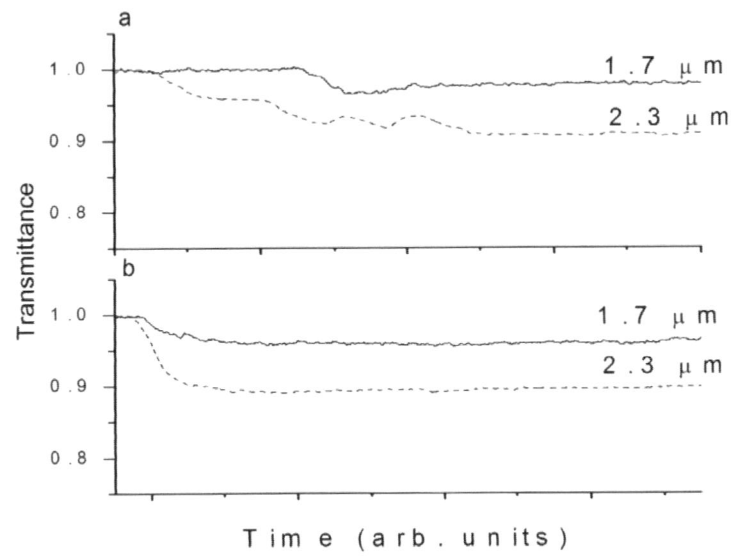

Figure 3. Transmittance for two different gas mixtures for measurement (2.3 μm) and reference (1.7 μm) channels: a - 85% isobutane + 10% butane + 5% propane; b - 70% propane + 30% butane.

3. Conclusions

It is shown that the NIR LED – NIR PD photocouplers permit one to perform NDIR measurements of combustible gases concentrations within the limits needed for standard alarm systems, but with a rather low selectivity to different alkanes.

Acknowledgments
This work was supported by the grant No. RFMEFI57714X0022 from the Ministry of Education and Science of Russian Federation.

References

[1] Hodgkinson J and Tatam R 2013 *Meas. Sci. Technol.* **24** 012004
[2] Stoyanov N D, Salikhov K, Kalinina K, Zhurtanov B and Kizhaev S S 2012 *Proc. SPIE 8257 Optical Components and Materials IX* 82571E
[3] Popov A A, Sherstnev V V and Yakovlev Yu P 1998 *Tech. Phys. Lett.* **24** 73-75
[4] Popov A A, Stepanov M V, Sherstnev V V and Yakovlev Yu P 1997 *Tech. Phys. Lett.* **23** 828-830
[5] Aleksandrov S, Gavrilov G, Kapralov A, Karandashov S, Matveev B, Sothikova G and Stus N 2002 *Proc. SPIE* **4680** p 188-194
[6] Lugovskaya A I, Login S A, Balashov O B, Kuznetsov A A, Vasiliev E V and Chernyak E Y 2003 *Products and Systems. Control, monitoring, diagnostics* **6** 55-59
[7] Matveev B 2014 *Photonics* **48** 80- 91
[8] Kataev M Yu 2015 *Proceedings of TUSUR* **4**(38) p 155-160
[9] LED Microsensor NT (http: ru.lmsnt.com, June 2016).
[10] Workman J and Weyer L 2007 *Practical guide to interpretative near-infrared spectroscopy* (CRC Press)
[11] Makeenkov A, Lapitskiy I, Somov A and Baranov A 2015 *Sensor Actuat. B-Chem.* **209** 1102
[12] Fanchenko S, Baranov A, Savkin A, Petukhov A, Kalinina K, Zhurtanov B and Velikotny M 2015 *Proc. Workshop on Environmental, Energy, and Structural Monitoring Systems (Trento)*(IEEE) p 146-151
[13] Fanchenko S, Baranov A, Savkin A and Sleptsov V 2016 *IOP Conference Series: Materials Science and Engineering* **108** 012036
[14] Fanchenko S, Baranov A, Savkin A, Petukhov A, Kalinina K, Zhurtanov B and Velikotny M 2016 *Proc. Workshop on Environmental, Energy, and Structural Monitoring Systems (Bari) (IEEE)* p 122-125

Investigation of cracks formation in sapphire crystals

S P Malyukov[1], Y V Klunnikova[1], E.V. Kirillova[2]

[1]Southern Federal University, Institute of Nanotechnologies, electronics and Equipment Engineering, Department of Electronic Apparatuses Design, Nekrasovskii per. 44, 347900 Taganrog, Russian Federation
[2]RheinMain University of Applied Sciences, Kurt-Schumacher-Ring 18, D – 65197 Wiesbaden, Germany

Corresponding author e-mail address: spmalyukov@sfedu.ru

Abstract. The analysis of the possible reasons of cracks formation in the sapphire crystals received by the horizontally directed crystallization method on the basis of the received results of numerical simulation of thermoelastic tension in the sapphire crystal is carried out.

1. Introduction

The big size sapphire crystals growth with improved structural characteristics became a priority of modern science and equipment. In the last decades the Czochralsky's method, Kiropoulos's method, the horizontally directed crystallization (HDC) method and Stepanov's method were applied for big size monocrystals growth. These crystals were widely used in electronic technique [1].

The sapphire crystal quality can be characterized by the orientation, density and the nature of dot defects distribution, density of single dislocations, an extension of cracks and blocks borders and their disorientation, the value of residual tension, chemical purity and impurity heterogeneity [1-3].

2. Mathematical simulation of cracks formation in sapphire crystals

There is thermoelastic tension during sapphire crystals growth in the field of temperature gradients. High level of thermoelastic tension in crystals can lead to the crack emergence in sapphire structure. In this case, it is necessary to calculate thermoelastic tension and to develop the software allowing to simulate heat exchange and tension fields at various stages of sapphire crystals growth. Mathematical simulation and calculations should be carried out for all equations which are included in the system, namely: the equations of heat conductivity and thermal elasticity.

These equations [4,5] can be written as:

$$\begin{cases} \mu\Delta u + (\lambda+\mu)\dfrac{\partial\delta}{\partial x} = -\dfrac{\partial(\alpha T)}{\partial x} - F_x, \\ \mu\Delta v + (\lambda+\mu)\dfrac{\partial\delta}{\partial y} = -\dfrac{\partial(\alpha T)}{\partial y} - F_y, \\ \mu\Delta w + (\lambda+\mu)\dfrac{\partial\delta}{\partial z} = \dfrac{\partial(\alpha T)}{\partial z} - F_z, \\ \dfrac{\partial}{\partial x}(a_i\dfrac{\partial T}{\partial x}) + \dfrac{\partial}{\partial y}(a_i\dfrac{\partial T}{\partial y}) + \dfrac{\partial}{\partial z}(a_i\dfrac{\partial T}{\partial z}) = 0, \end{cases} \quad (1)$$

where F_x, F_y, F_z are the components of volume sources; u, v, w are the components of displacements; α is the coefficient of volume heat expansion; λ and μ are the Lame coefficients; Δ is the Laplace operator; i = 1,2,3 are the crystal, melt and crystal powder, respectively; a_i is the thermal diffusion coefficients; T is the temperature.

The boundary conditions for the system (1) are written in the form of relations as previous author's research [6].

The simulation technique and three-dimensional numerical model are developed for calculation of temperatures, displacements, deformations and tension fields at various stages of sapphire growth by the HDC method. It allows carrying out the analysis of the temperature and thermoelastic fields changes in crystal depending on spatial and geometrical characteristics of heaters by means of computing experiment. So on the basis of the analysis, we can redistribute defects in sapphire, thereby having increased quality of the crystals.

The technique of temperatures, displacements, deformations, tensions calculations in system crystal – melt includes the following:

1. The input of basic data: density, heat conductivity and specific thermal capacity of materials, length, height and width of the crucible, Lame coefficients and some other.
2. Calculation of temperatures distribution in crystal – melt system.
3. Calculation of temperatures gradients in crystal – melt system by the method of smallest squares.
4. Displacement calculations as the first approximation in crystal – melt system.
5. Calculation of displacement gradients by the method of the smallest squares and the solution of thermoelasticity equation taking into account gradients of displacements.
6. Calculation of deformations in crystal – melt system on the basis of the received displacements.
7. Calculation of tension in crystal – melt system on the basis of the received deformations.
8. The conclusion of results.

Results of tension calculation are presented in figure 1. In figure 1 it is visible that there is compression tension along lateral faces of the crystal, and at the middle part of the crystal, it can be seen stretching tension. The size of thermoelastic tension depends on the distribution of temperatures.

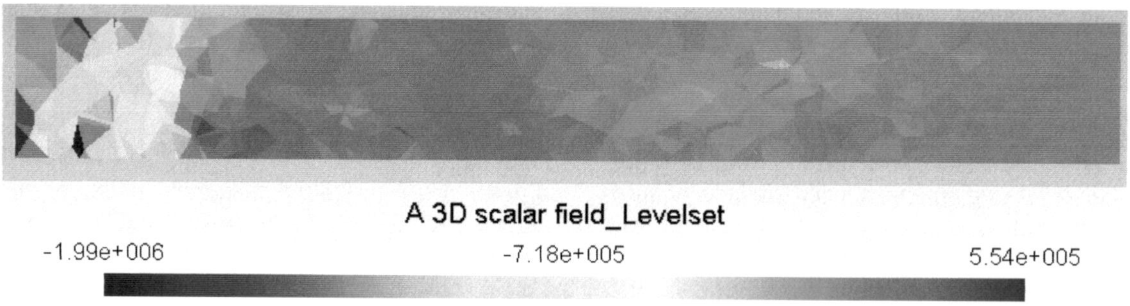

Figure 1. Results of tension calculation in sapphire in crystal – melt system at the vertical section of crystal (the scale is in Pa).

The model for calculation of temperatures, displacements, deformations and tension fields during sapphire growth allows us to apply the numerical experiments and to study the influence of input parameters on the crystals quality. This approach allows to improve the crystal growth process and to receive the crystals with reduced defects level.

The crack in sapphire structure begins to extend when the stress intensity coefficient reaches critical value for this material [1]:

$$K_c = 3.2 \cdot 10^5 \cdot N \cdot m^{-\frac{3}{2}} \tag{2}$$

The results of numerical simulation of thermoelastic stresses in sapphire crystals received by authors earlier were used for the analysis of the possible reasons of cracks formation in the sapphire crystals which are received by the HDC method. The model included calculation of temperatures, displacements, deformations and thermoelastic stresses of the growing sapphire crystal by the method of control volumes on the unstructured grid [6-7].

Critical value of stress intensity coefficient is defined by formula [7]:

$$K_c = \sqrt{2E\gamma}, \quad (3)$$

where E – Young's modulus; γ – the specific superficial energy of sapphire.

We use the following sapphire parameters for simulation [7]: Young's modulusis E = 460 GPa and specific superficial energy of sapphires $\gamma = 1.1\ J \cdot m^{-2}$ when critical stress intensity coefficient is (2).

Cracks growth can lead to sapphire fragile destruction. Cracks can suddenly become unstable in sapphire structure and extend into a material with a certain speed.

Sapphire fragile destruction appears when stress intensity coefficient equals to critical value of stress intensity coefficient [1,7]:

$$K = K_c, \quad (4)$$

when $K = \sigma(\pi a)^{\frac{1}{2}}$ is the stress intensity coefficient; σ is the stress; a is the crack length.

On the basis of stresses calculations during sapphire crystal growth by HDC method, the estimation of formed cracks length in sapphire crystals was made. Fragile destruction of sapphire can appear when the crack in material reaches some critical length at the fixed stress, i.e. for the fixed thermoelastic stress -1.99 MPa the critical crack length will make 0.0082 m.

3. The investigation results

The initial crack emergence is a consequence of high value of internal thermoelastic stresses in sapphire crystal. Further development of crack happens gradually in the sapphire growth process, and it can be connected with a relaxation of thermoelastic stresses.The suggested model adequately describes the influence of thermoelastic stresses on cracks formation in sapphire and can be used for sapphire crystals growth technological process improvement. On the basis of the model the optimum modes of sapphire, crystals growth are chosen. It allows excluding cracking during sapphire growth.

Acknowledgments
This research was supported in part by the Russian Foundation for Basic Research (Grant No. 16-52-12010), Grant of DFG (Kl 1104/3-1).

References

[1] Dobrovinskaya E R, Lytvynov L A and Pishchik V V 2009 *Sapphire. Material, Manufacturing, Applications* (Berlin: Springer)
[2] Shtukenberg A G and Punin Y O 2004 *Optical Anomalies in Crystals*(St. Petersburg: Science)
[3] Hannanov S C, Nikanorov S P and Baholdin S I 2003 *J. Solids Phys.* **45** 1020-1023
[4] Patankar S 2003 *Numerical Solution of Heat Conductivity and Convective Heat Exchange Problems at Flow in Channels* (Moscow: MEI)
[5] Pershin I M 2007 *Analysis and Synthesis of Systems with Distributed Parameters*(Pyatigorsk)
[6] Malyukov S P, Klunnikova Y V and Anikeev M V 2016 *Advanced Materials* (Switzerland: Springer)
[7] Malyukov S P, Klunnikova Y V and Parinov IA 2014 *Advanced Nano- and Piezoelectric Materials and Their Applications* (USA: Nova Science Publishers)
[8] Indenbom V L 1964 *J. Crystallography* **9** 74-83

Effect of mono- and bimetallic nanoparticles Fe, Ni, & Fe/Ni based on carbon nanocomposites on electrocatalytic properties of anodes

K Ranabhat[1], A I Pylinina[1], K S Skripkin[1], E A Sofronova[1], A A Revina[2], V E Kasatkin[2], L N Patrikeev[1,3] and V A Lapshinsky[1,3]

[1]Peoples' Friendship University of Russia, Miklukho-Maklay str. 6, 117198 Moscow, Russian Federation
[2]A.N. Frumkin Institute of Physical Chemistry and Electrochemistry of the Russian Academy of Sciences, Leninsky prospect 31/4, 119071 Moscow, Russian Federation
[3]National Research Nuclear University MEPhI (Moscow Engineering Physics Institute), Kashirskoe shosse 31, 115409 Moscow, Russian Federation

Corresponding author e-mail address: sofronova_ea@mail.ru

Abstract. The optical properties of metallic Fe nanoparticles (NPs), Ni NPs and bimetallic Fe/Ni NPs produced under radiolysis in anaerobic condition based on a chemical reduction in the presence of oxygen and quercetin in reversed micellar solutions, and electrokinetic properties of nanoparticles carbon-based nanocomposites were studied. The possibility of the preparation of coating using different nanocomposites with anomalous electrocatalytic is addressed.

1. Introduction

Nanotechnology belongs to the field of science and engineering dedicated to materials, having dimensions of hundreds of nanometers or less [1]. Metallic and bimetallic nanoparticles have fascinated scientist for over a century and are now heavily utilized in biomedical sciences and engineering. They are a focus of interest because of their huge potential in nanotechnology [2]. Colloidal metal nanoparticles are of interest in both research and technology, due to their specific properties not available in isolated molecules or bulk metals. Because of these properties, nanoparticles have many important applications in catalysis, sensing, and imaging [3]. Recently metal and bimetal nanostructures have emerged as an approach to augment light harvesting in solar cells, giving rise to enhanced optical absorption, improved efficiency and carrier generation [4-7].

2. Experimental setup

Nanoparticles of Fe, Ni and bimetallic nanoparticles Fe/Ni were obtained by the chemical radiation synthesis as described in [8-9]. Optical absorption spectra of the initial solutions and the samples after synthesis of NPs were recorded using spectrophotometer "Hitachi U-3310" with respect to the solution of AOT / isooctane at room temperature in the presence of air. The optical path length quartz cuvette was 1.0 mm.

Comparison of current-voltage characteristics of NP Fe, NP Ni and bimetallic NP Fe/Ni are measured on a disposable "planar" electrodes (Moscow State University) in aqueous 0.5 M solution of H_2SO_4 using a potentiostat IPC-Pro MF.

3.Result and discussion

In figure 1 the optical absorption spectra of mono- and bimetallic nanoparticles Fe / Ni Chem in OMP ($\omega = 5.0$) are illustrated.

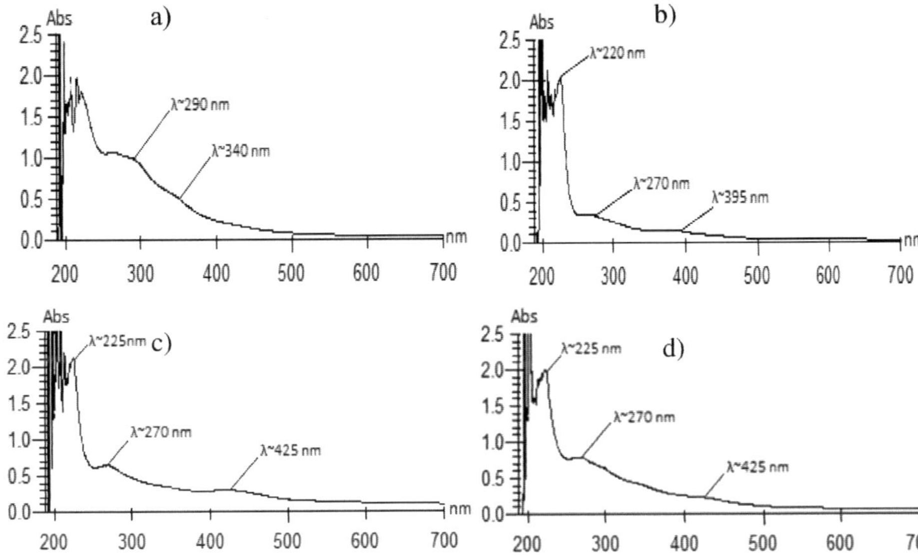

Figure 1. Optical absorption spectra of mono- and bimetallic nanoparticles Chemin RMS: a) NPs Fe; b) NPs Ni; c) NPs Fe/Ni (1:2); d) NPs Fe/Ni (2:1).

On the given absorption spectrum, the maximum intensity is at λ_{max}= 290 nm и λ_{max}= 220 nm, for NPs Fe and NPs Ni respectively [9]. In figure 2(d) absorption spectrum on bimetallic NPs Fe & Ni in the ratio 2:1 is represented, the intensity of the main band is higher than in the NPs Fe/Ni spectrum with lower iron content (1:2). However, about the different functional activity of metallic and bimetallic NPs and nanocomposites based on them, information can be obtained only through the study of electro-catalytic characteristics of nanostructures [4].

Comparison of current-voltage characteristics of NP Fe, NP Ni, and bimetallic NP Fe/Ni is shown in the Figures 2 and 3.

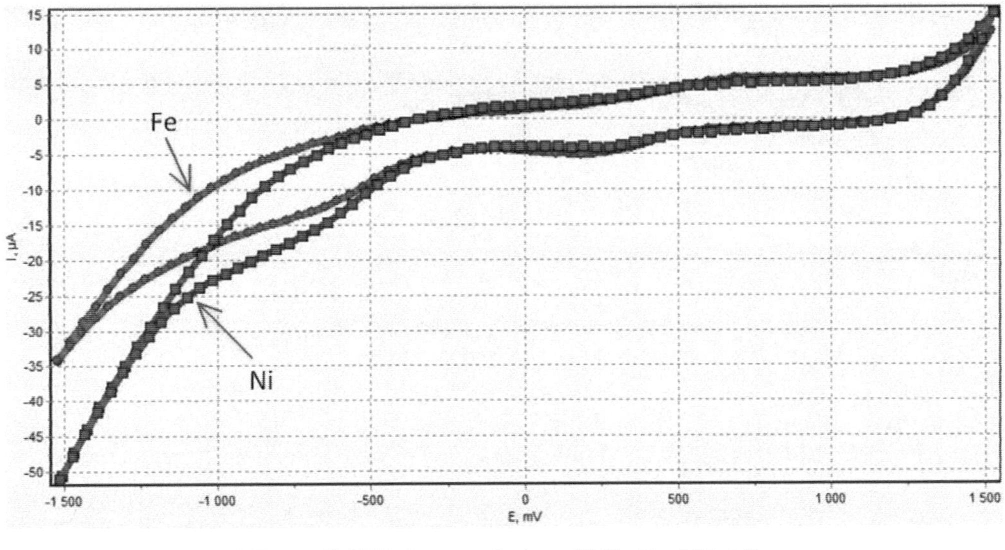

Figure 2. I-V characteristics of NPs Fe, NPs Ni.

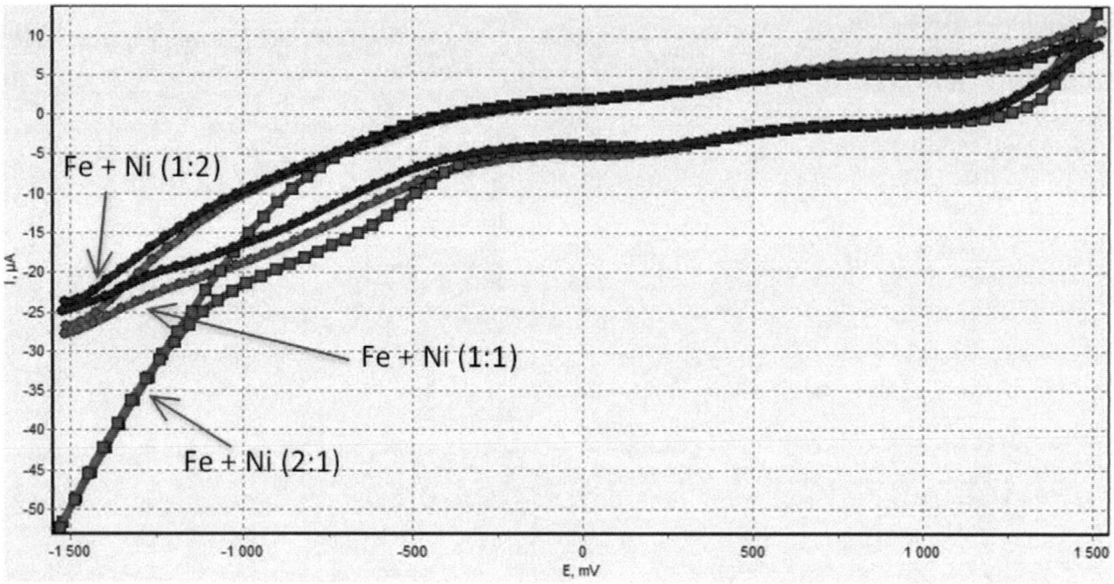

Figure 3. I-V characteristics of bimetallic NPs Fe/Ni with different ratio of iron and nickel.

As one can see, I-V characteristics of NPs of metals depend on the nature of metal, I-V characteristics of bimetallic NPs depends on nature and metal ratio.

I-V characteristics based on activated carbon and NPs Fe, obtained by reverse micellar solutions [4,9] are represented in the Figure 4-7.

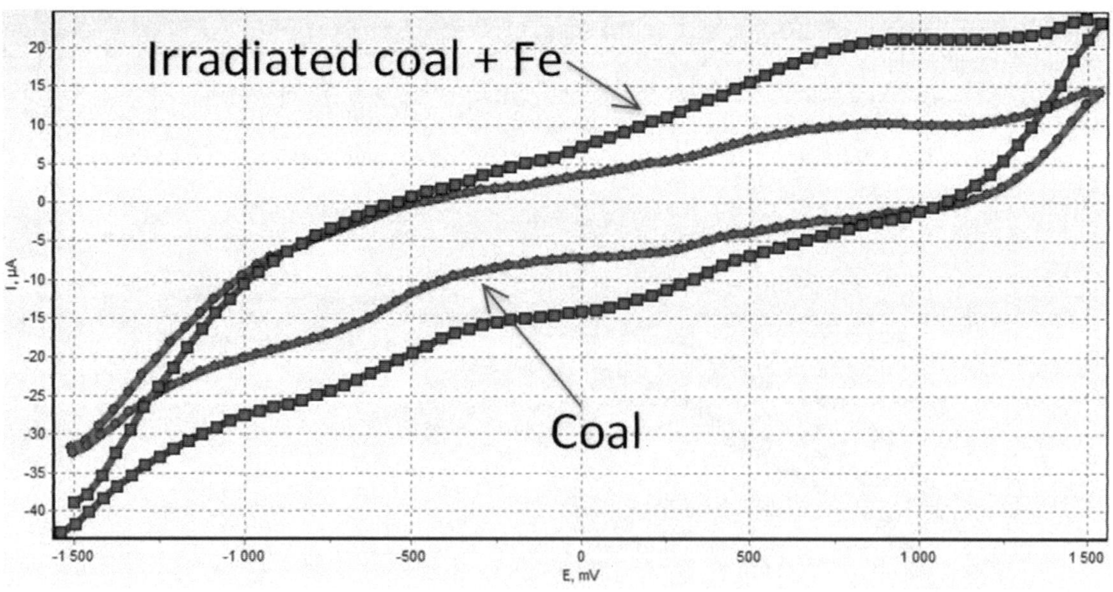

Figure 4. Current-voltage characteristics of the electrodes after irradiation with 20 kHz and coating them with NP Fe of carbon-based material artichoke effect of radiation-chemical modification.

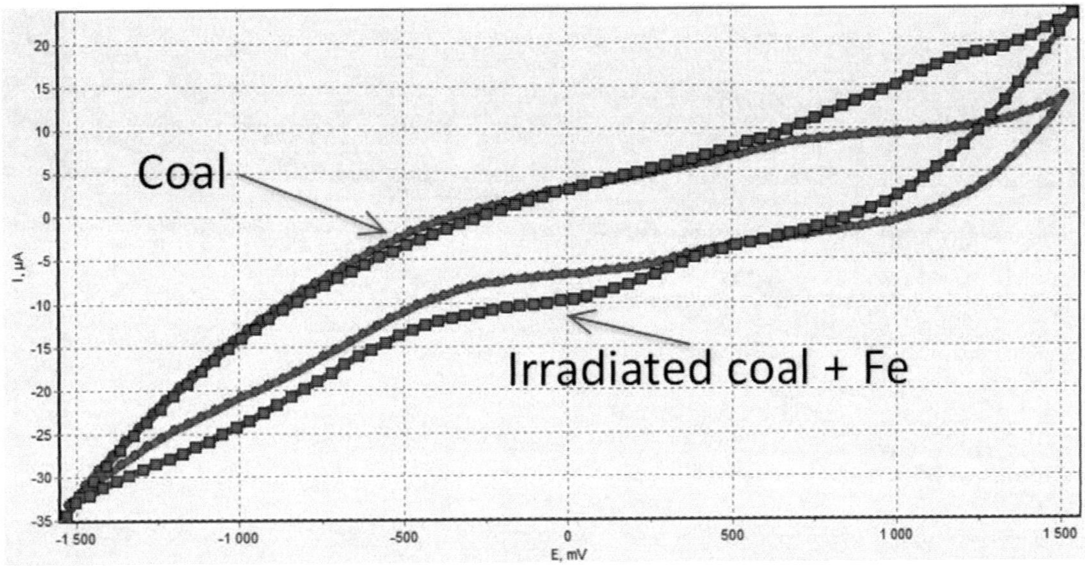

Figure 5. Current-voltage characteristics of the electrodes after irradiation with 20 kHz and coating them with NP Fe effect of the carbon-based material, camelina sativa of radiation-chemical modification.

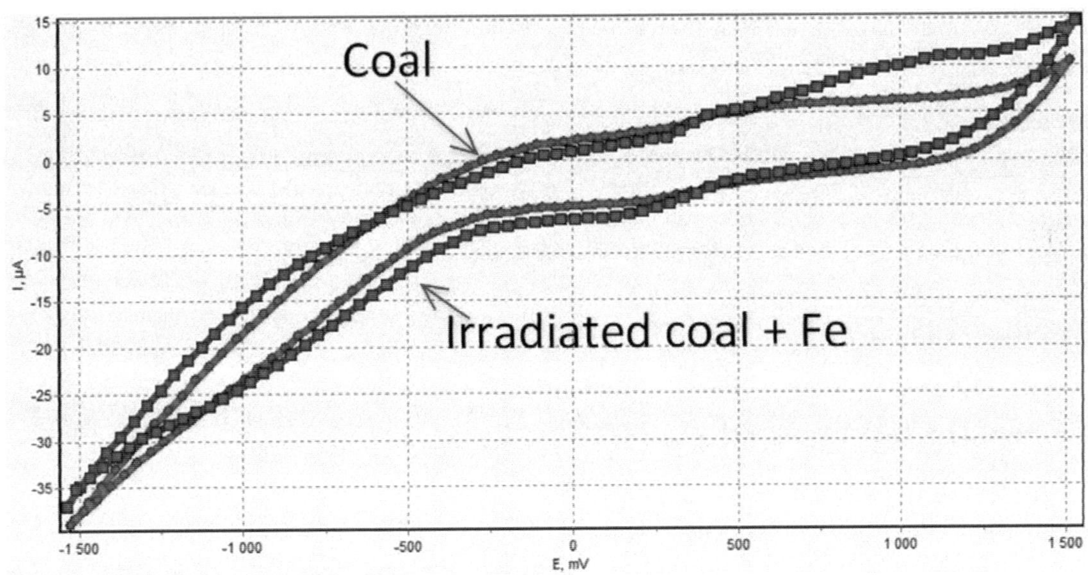

Figure 6. Effect of radiation-chemical modification of the carbon-based material of wheat current-voltage characteristics of the electrodes after irradiation with 20 kHz and coating them with NP Fe.

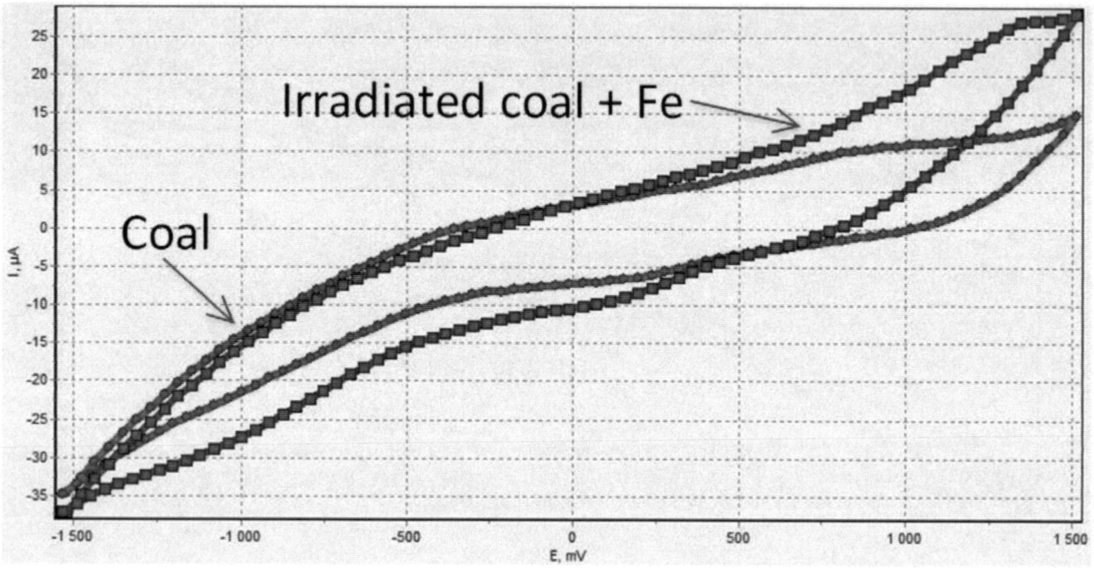

Figure 7. Effect of radiation-chemical modification of the carbon-based material to canola current-voltage characteristics of the electrodes after irradiation with 20 kHz and coating them with NP Fe.

Based on these data it can be concluded that the carbon material based on a preliminary chemical radiation treatment of modified coal from plants (artichoke, camelina sativa, wheat and canola) has a higher electrocatalytic activity.

4.Conclusion

In conclusion, in this paper synthesis and optical property of mono- and bimetallic nanoparticles of Nickel and Iron were studied. The findings that were presented in this paper suggest that the modification of carbon-based nanocomposites coated with nanoparticles of nickel and iron increases the electrocatalytic property of anode for a dye-sensitized solar cell (DSSC) also known as the Gratzel cell.

Acknowledgements

This paper was done in the project №011615-0-000.

References

[1] Salata O V 2004 *Nanobiotechnology* **2** 3
[2] Yen Y C, Chen P H, Chen J Z, Chen J A and Lin K J 2015 *ACS Applied Materials & Interfaces* **7** 1892–1898
[3] http://shodhganga.inflibnet.ac.in/bitstream/10603/11458/6/06_chapter%201.pdf
[4] Revina A A, Lebedeva M V, Shvetsov A A and Rakov E G 2013 *Nanotechnology, science and production* **1**(22) 62-69 (*in Russian*)
[5] Winans J D, Hungerford C, Shome K, Rothberg L J and Fauchet P M 2015 *Optics Express* **23** A92–A105
[6] Shah A, Rahman L, Qureshi R and Rehman Z 2012 *Reviews on advanced materials science* **30** 133-149
[7] Dzhafarov T D, Pashaev A M, Tagiev B G, Aslanov S S, Ragimov S H and Aliev A A 2015 *Advances in Nano Research* **3**(3) 133-141
[8] Dokuchaev A G, Myasoedova T G and Revina A A 1997 *High energy chemistry* **5** 353-356
[9] Shcherbatova E V, Chekmar D V and RevinaA A 2015 *Advances in chemistry and chemical technology* **29** (6) 30-32

IOP Conf. Series: Materials Science and Engineering **151** (2016) 012024 doi:10.1088/1757-899X/151/1/012024

Automotive MEMS sensors based on additive technologies

A A Vasiliev[1,4], A V Sokolov[4], A V Pisliakov[4], K Yu Oblov[2], N N Samotaev[2], V P Kim[3], S V Tkachev[3], S P Gubin[3], G N Potapov[4], Yu V Kokhtina[4] and A V Nisan[4]

[1]National Research Center "Kurchatov Institute", Akademika Kurchatova pl. 1, 123182 Moscow, Russian Federation
[2]National Research Nuclear University MEPhI (Moscow Engineering Physics Institute), Kashirskoe shosse 31, 115409 Moscow, Russian Federation
[3]AkKo Lab LLC, Gilyarovsky str. 65/1, 129110 Moscow, Russian Federation
[4]Ostec Enterprise Ltd., Moldavskaya str. 5/2, 121467 Moscow, Russian Federation

Corresponding author e-mail address: A-A-Vasiliev@yandex.ru

Abstract. The application of MEMS devices is one of the recent trends in sensor technology. However, traditional silicon MEMS have some intrinsic limitations, when applied to the monitoring of high temperature/high humidity processes. Thin ceramic films of alumina, zirconia or LTCC fixed on rigid frame made of the same ceramic material in combination with ink and aerosol jet printing of functional materials (heaters, temperature, pressure, gas sensitive elements) provides a cheap, flexible, and high-performance alternative for silicon MEMS devices used as gas sensors, gas flowmeters, lambda probes, bolometric matrices for automotive and general application.

1. Introduction

Recently, the main part of semiconductor and thermocatalytic gas sensors used in gas analytic instruments in the world is fabricated using thick film technology [1]. The application of this technology enables the fabrication of rather stable gas sensors, the elements of the sensor being fabricated by printing of conductive (Pt-based) or resistive (RuO_2 based) ink (paste) for the heater of the sensor and metal oxide semiconductor (doped SnO_2, ZnO, TiO_2 nanopowder) or thermocatalytic layer (Pt-Pd loaded Al_2O_3) ink for the sensing layer. The sensors produced by this technology are characterized by relatively high power consumption at working temperature. For example, the sensor described in [1] consumes ~ 280 mW at working temperature of 450°C. This power consumption is relatedto thelarge size of the sensor made using thick film technology; the optimization of sensor layout permits to decrease power consumption down to ~ 100 – 120 mW, but a further decrease in power consumption seems to impossible within this approach.

Last years, this limitation was overcome by the application of silicon MEMS technology [2] used for the fabrication of sensor micro-hotplates. Using this technology, not only gas sensors were fabricated, it was appliedfora large number of different devices, like flow meters, microbolometers, etc. At the moment, the main disadvantages of this technology are related with limited working and technological (necessary for the formation of sensing layers) temperature of the sensor and with the application of expensive and sophisticated equipment for the fabrication of sensors [3-8].

A possible solution of these problems was suggested in [9]. This solution consists in the application of ceramic MEMS substrates made of thin films of alumina, zirconia, or LTCC (thickness 10 – 20 μm) stretched on arigid frame made of the same material as the material of themembrane. This approach permits us to avoid problems related to thedifference in expansion coefficients typical of silicon based MEMS. The advantages of this device layout become more pronounced if the functional elements of MEMS device are produced by aerosol or ink get technologies providing flexibility of the technological process and low cost of the production.Indeed, the modification of sensor layout is related only to the modification of program controlling the printing process, and the ink, in contrast to sputtering process, is deposited only where it is necessary, but not everywhere in the sputtering chamber. In addition, the application of printing process enables the avoidance of photolithographic process and platinum etching.

These ceramic MEMS can operate at high ambient temperature and, in general, under harsh environmental conditions, and, therefore are very promising for automotive gas concentration and gas flow sensors. We consider the possibility to use the ceramic MEMS for most type of sensors used in thecar industry: gas, flow, pressure, acceleration, IR, and other sensors.

2. Experiment

The basis of these MEMS devices is 12 μm thick alumina films made by anodic oxidation of aluminum. This possibility was discussed in publications [9,10]. Large area (>48×60 mm) films remain flat after annealing at 1000°C. Therefore, the application of thin alumina films enables the deposition and further annealing of almost any functional materials: platinum and ruthenium dioxide based heaters requiring annealing at ~ 850°C, sensing layers based on SnO_2, ZnO, TiO_2, etc., requiring formation temperature of ~ 700 - 750°C, high TCR materials for microbolometers.

Other possibilities for the fabrication of ceramic MEMS is given by thin slip cast HTCC based on yttria-stabilized zirconia and LTCC. These ceramic films can be 10-20 μm thick. An example of ceramic MEMS made of thin LTCC is presented in Fig. 1. This element was fabricated by slip casting of LTCC film (20 μm thick) followed by the co-pressing of this film with thick LTCC ceramics with ahole. In the next generation of the sensor,substrates,we will fabricate thewhole substrate with such elements containing several hundreds of individual chips.

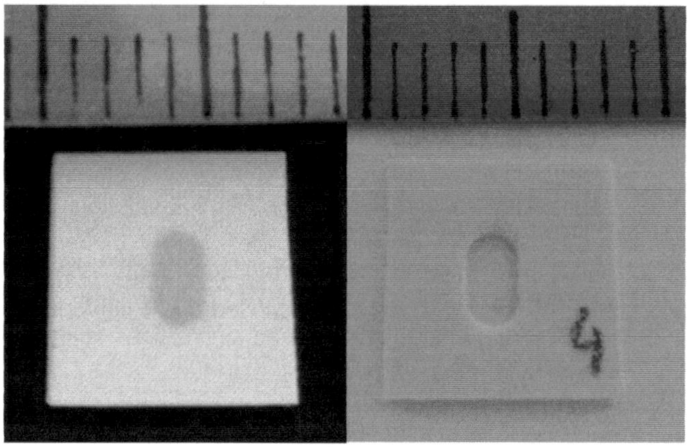

Figure 1. Top and bottom view of sensor chip made of 20 μm thick LTCC ceramics fixed on 0.5 mm LTCC frame.

Heating elements are fabricated by aerosol and ink jet printing using ink containing 3-5 nm particles of Pt as described in [10]. After drying and annealing, these prints give 20-40 μm wide lines stable at working temperature > 500°C.

An example of the microheater deposited on 12-micron thick aluminum oxide film is presented in Fig. 2. The size of the microcantilever triangle heater is of about 1 (base of thetriangle) to 2 mm, the width of platinum lines is about 35 μm.

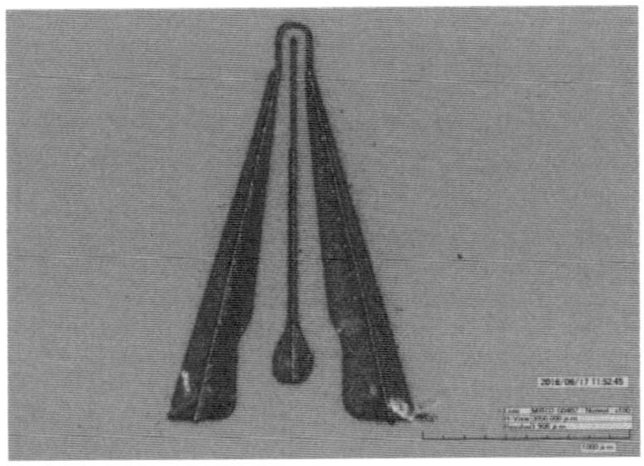

Figure 2. Platinum microheater fabricated using platinum aerosol jet printing over alumina membrane with athickness of 12 μm. Heater size is of ~ 1 (base) × 2 mm. Hot area is located in the vertex of the triangle, where the line width is 35 μm.

After deposition, the ceramic membrane was cut by alaser beam (optic fibre laser with a wavelength of 1.06 μm and maximum power of 50 W normally used as amarker) to form sharp triangle cantilever with thehot area located in the vertex of thetriangle. This sensor layout permits us to minimize heat losses from the hotplate due to minimization of thickness and area of heat conducting element. In addition, the application of cantilever type micro-hotplate improves the stability of microheater, because, in contrast to thewhole membrane, heating of cantilever does not lead to bending of the hot element.

A sensing layer of the gas sensor was deposited by ink jet or using drop coating technique. Sensing material was prepared using tin (II) sulfate as aprecursor, tin was oxidized in solution by hydrogen peroxide. The application of this procedure enables direct synthesis of powder with aspecific area of about 100 m^2/g (particle size of about 10 nm). The printable ink was prepared by mixing this powder with asolution of ethylcellulose in terpineol.

3. Results and discussion

We tested thermal properties of micro-hotplates based on thin alumina with Pt aerosol jet printed microheater.

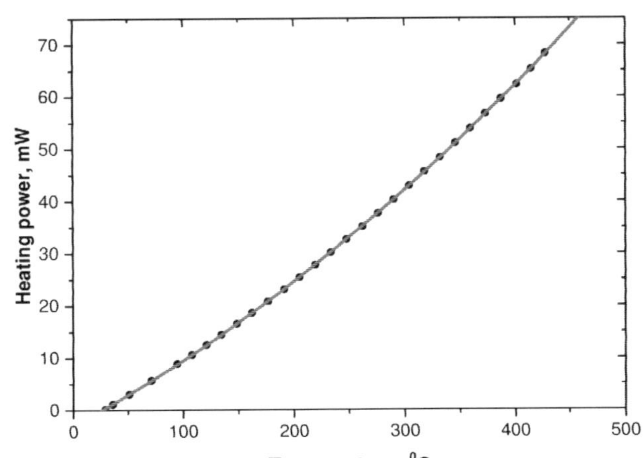

Figure 3. Heating power of cantilever type gas sensor (Fig. 2) as a function of working temperature. The micro-hotplate consumes approximately 70 mW at working temperature of 450°C necessary for methane detection.

The heating power consumption of alumina MEMS is about 70 mW at permanent heating to 450 °C. The plot illustrating the power consumption as a function of working temperature is presented in Fig. 3. The temperature of the microheaterwas calculated using the value of TCR measured preliminarily in the furnace.

This value is the same as the power consumption of micro-hotplate fabricated using silicon MEMS technology [11] working at about 350 °C. At the same working temperature our ceramic MEMS micro-hotplate consumes ~ 55 mW, therefore thethermal efficiency of ceramic MEMS is at least not worse compared to silicon MEMS. Short thermal response time (<100 ms) of the microheater allows the fabrication of fast temperature sensors and enables the operation of gas sensors in pulse heating mode permitting, in turn, selective measurement of gases (CO and H_2, for example) and application of ceramic MEMS in mobile and autonomous instruments, because average power consumption of gas sensor in this case is below 1 mW. It was demonstrated that the detection limit of CO and hydrogen using this pulsing heating mode is about 1 ppm.

4. Conclusion

The application of ceramic MEMS with thin membrane element in combination with aerosol and ink jet technology of functional element printing enables flexible and cheap fabrication of various elements of electronics for harsh environment and automotive application.

The advantages of ceramic MEMS technology is most pronounced when these sensors are used for high temperature/harsh environment application. Indeed, the ceramic substrate used for such micro-hotplateis annealed at 1000 °C. Therefore any reasonable operation and technological temperature can not damage it.

Among these elements, which could be fabricated using a combination of ceramic MEMS and aerosol/ink jet technique are air quality sensors, lambda probes, gas flow sensors and other devices working at high temperature. This approach gives a possibility to fabricate sensors with characteristics exceeding the characteristics of silicon MEMS devices, but free of disadvantages of the last: ceramic MEMS is characterized by full compatibility of materials, high working temperature, and high-cost efficiency at medium scale production (up to several million units per year). The application of platinum ink jet printed conductive lines are characterized by good adhesion of platinum to alumina after annealing at a temperature of ~ 800 °C. Any other material but ceramics can not withstand this technological temperature. Decrease in production cost of printed micro-hotplates is related to a decrease in permanent production expenses: silicon MEMS technology needs not only expensive equipment but is also characterized but high service cost, which can not be attractive if the production scale is not extremely large.

Acknowledgments

The research was performed within a frame of Russian Federation President Grant for support young scientists №14.Y30.15.7910-MK from 16.02.2015

References

[1] http://www.figaro.co.jp/en/product/docs/tgs2611_product_information_rev02.pdf
[2] http://www.sgxsensortech.com/content/uploads/2014/07/MICS-SMD-Devices.pdf
[3] Veselov D S 2015 *Physics Procedia* **72** 500-503
[4] Veselov D S, Bakun A D, Voronov Yu A, Kireev V Yu and Vasileva O V 2016 *J. Phys. Conf.* **748** 012015
[5] Veselov D S and Voronov Yu A 2015 *Physics Procedia* **71** 423-427
[6] Veselov D S and Voronov Yu A 2016 *J. Phys. Conf.* **747** 012022
[7] Veselov D S, Voronov Yu A and Vanyukhin K D 2015 *Physics Procedia* **72** 495-499
[8] Veselov D S, Bakun A D and Voronov Yu A 2016 *J. Phys. Conf.* **748** 012017
[9] Vasiliev A A, Pisliakov A V, Sokolov A V, Samotaev N N, Soloviev S A, Oblov K, Guarnieri V, Lorenzelli L, Brunelli J, Maglione A, Lipilin A S, Mozalev A and Legin V A 2016 *Sensor Actuat. B-Chem.* **224** 700-713
[10] Vasiliev A A, Sokolov A V, Legin A V, Samotaev N N, Oblov K Yu, Tkachev S V, Gubin S P, Potapov G N, Kokhtina Yu V and Nisan A V 2015 *Procedia Engineering* **120** 1087-1090
[11] http://www.sgxsensortech.com/content/uploads/2014/07/1084_Datasheet-MiCS-5524-rev-8.pdf

IOP Conf. Series: Materials Science and Engineering **151** (2016) 012025 doi:10.1088/1757-899X/151/1/012025

Artifacts caused by microwave radiation in cardiac radio telemetry

A B Simakov, B V Gurkovskiy, E M Onishchenko, N Y Trifonova and D S Veselov

National Research Nuclear University MEPhI (Moscow Engineering Physics Institute), Kashirskoe shosse 31, 115409 Moscow, Russian Federation

Corresponding author e-mail address: absimakov@mephi.ru

Abstract. The effect of the cardiac radio biotelemetry on human heart rate variability has been investigated. The burst transmission mode with a high duty cycle minimizes the influence of microwave artifacts on the reliability of transmitted data.

1. Introduction

Radio telemetry is one of the main techniques of humans' physiological parameters registration at their free behaviour and a cardiac signal is the most important factor in the assessment of their psycho-physiological state. Usually, the cardiovascular reactions to microwave exposure by the heart rate variability (HRV) analysis are evaluated [1,2]. Well-developed algorithms and software for HRV analysis, corresponding to the today's international standards [3,4] were developed. Recently assumed, that low power telemetry transmitters had no effect on the human cardiovascular system (CVS). However, in the last decade shown, that even low-intensity electromagnetic fields (EMF) might change cardiogram parameters [5]. Therefore, this work focuses on the study of artifacts arising when measuring the physiological parameters, in particular, HRV, due to exposure to microwave radiation from the cardiac telemetry data channel.

2. Research procedure and experimental installation

Figure 1 illustrates the details of suggested research procedure and experimental installation.

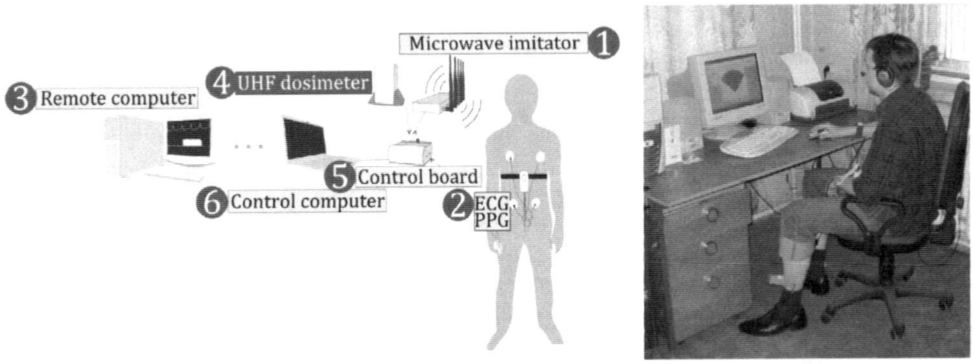

(a) (b)

Figure 1. (a) Block-diagram of the instrumental set and (b) photo of the person under study.

The experiment involved 25 volunteers (15 men and 10 women) aged 19 to 63 years. They participated in the investigation in compliance with all ethical standards and safety regulations.

During the experiment, the volunteer was located in a well-shielded room as shown in figure 1(b). At the beginning of the study, before EMF exposure, the volunteer was in a free state and his heart rate was recorded by pair of optical photoplethysmography (PPG) sensors marked by 2 in figure 1(a). Then he has been irradiated in a blind mode by externally controlled microwave source within the frequency and power range of mobile GSM Standard (frequency band of 930 – 960 MHz and energy flow less than 50 $\mu W/cm^2$). Simultaneously, we were recording a cardio-intervals-gram (CIG).

The microwave generator indicated by 1 in figure 1(a), simulates the GSM based radio telemetry transmitter used in cardiac monitoring. Figure 2(a) shows its photo.

To measure the absorbed dose and to monitor microwave radiation level in safety control system we used a new portable bio-dosimeter [6] marked by 4 in figure 1(a). Figure 2(b) shows a photo of the bio-dosimeter.

A control board indicated by 5 in Figure 1(a) manages all devices of the research installation and performs the preliminary data processing. Control computer labeled by 6 in figure 1(a) records data and transmits it to a remote server labeled by 3 in figure 1(a) for final processing and storage the measured information.

(a)

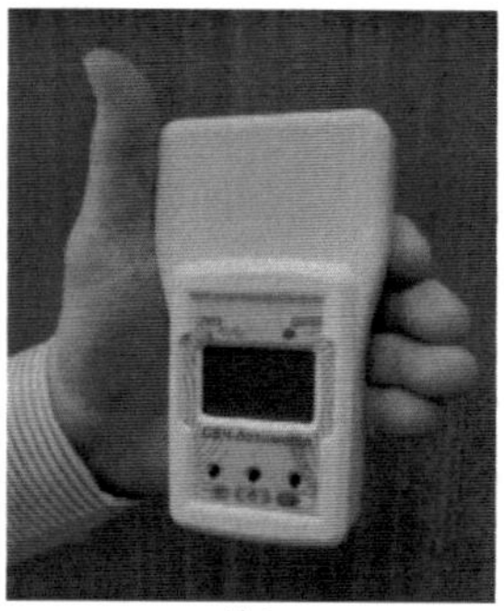

(b)

Figure 2. The photos of (a) simulator of radio telemetry transmitter and (b) portable bio-dosimeter.

3. Details of the study

The features of radio telemetry system operation were simulated by three timing protocols. All protocols started with five-minute initial testing before microwave exposure and ended with the same back-end testing after microwave exposure.

The first protocol simulated a rapid data radio transmission. The duration of one microwave exposure session was 30 seconds with a one-minute break. It was followed by a five-fold repetition of this session. The second protocol involved the following sequence of operations: three minutes of radiation exposure and a three-minute break with a three-time repetition of the session. The third protocol was a model of the six-minute long session without repetition.

We used the following methods for HRV visualization, analysis and parameters evaluation: the distribution of heart rate intervals; histograms of CIG; power spectral density (PSD) of CIG; autocorrelation of CIG and scattered charts analysis. Figure 3 shows an example of experimental heart rate intervals distribution: before and after the microwave radiation when a volunteer in accordance with a second protocol was tested.

The results demonstrate very large differences in individual sensitivity of people to low-intensity microwave radiation. They show a high individual sensitivity of the specific persons to weak microwave signals simulating the signals of GSM based real radio telemetry systems used in cardiac monitoring. This suggests the possibility of significant side effects i.e. artifacts with commonly used methods of data measurement and transmission in the cardiac monitoring.

The statistical estimations of these effects for 25 persons and three timing protocols have been performed, which simulate the effects of remote monitoring systems in cardiology shows the following:

— under microwave irradiation in accordance with the first protocol no significant changes of the cardiac rhythm were detected;
— under exposure in accordance with the second protocol, 70% of the persons had an increase in a heart rate of 3 – 4 beats per second;
— under exposure in accordance with the third protocol, almost 86% of the subjects had the individual drift of HRV and for 28% of them, the drift did not disappear in five minutes after exposure.

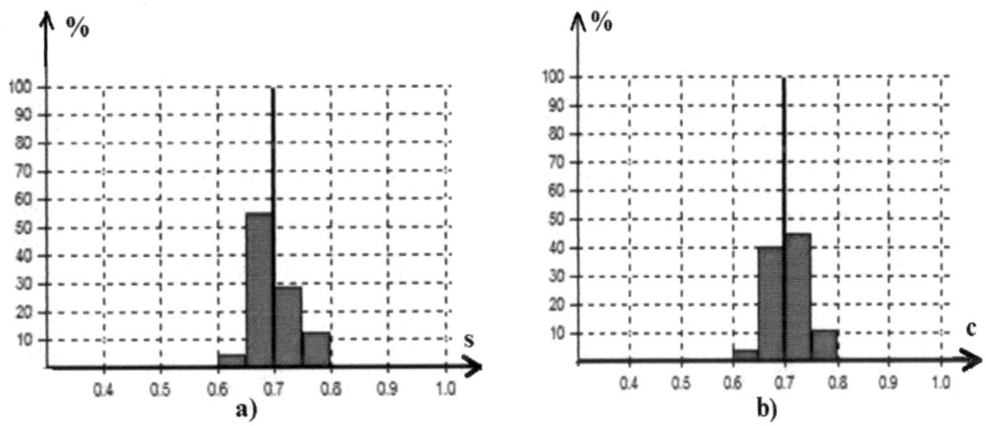

Figure 3. The histograms of experimental heart rate intervals distribution according to the second-time protocol: a) before EMF action; b) after EMF exposure.

4. Conclusion

Experimental modeling of the side effect of radio telemetric systems on human cardiac signals demonstrated their peculiar sensitivity to low-intensity transmitter irradiation in the microwave range. The transmitting signals can change cardiac activity and produce, so-called biological artifacts, providing the data corruption. At the first time protocol (30 sec of EMF), no significant changes of the cardiac rhythm were detected. The increasing of exposure duration leads to artifacts. Therefore, the reduction of bio radio telemetry artifacts is possible by accumulation data and transmission them in a burst mode with a large duty cycle.

Acknowledgments
The Russian Scientific Foundation supported this work (grant 16-19-00167).

References

[1] Andrzejak R, Poreba R, Poreba M, Derkacz A, Skalik R, Gac P, Beck B, Steinmetz-Beck R and Pilecki W 2008 *Ind. Health* **46** 409–17
[2] Rajendra Acharya U, Paul Joseph K, Kannathal N, Lim Ch M and Suri J S 2006 *Med. Biol. Eng. Comput.* **44** 1031–51
[3] Heart rate variability: standards of measurement, physiological interpretation and clinical use 1996 *Eur. Heart J* **17** 354–81
[4] Malik M, Cripps T, Farrell T and Camm A J 1989 *Med. Biol. Eng. Comput.* **27** 603–11
[5] Marko S. Markov 2015 *Electromagnetic Fields in Biology and Medicine* (Boca Raton, FL: CRC Press)
[6] Gurkovskiy B, Vodohlebov I., Onishchenko E and Simakov A 2015 *1st URSI Atlantic Radio Science Conference (Electronic Materials)*

Radiation hardness improvement of analog front-end microelectronic devices for particle accelerator

A G Miroshnichenko, A S Rodin, A S Bakerenkov and V A Felitsyn

National Research Nuclear University MEPhI (Moscow Engineering Physics Institute), Kashirskoe shosse 31, 115409 Moscow, Russian Federation

Corresponding author e-mail address: ASBakerenkov@mephi.ru

Abstract. Series of schematic techniques for increasing radiation hardness of the current mirrors is developed. These techniques can be used for the design of analog front-end microelectronic devices based on the operational amplifiers. The circuit simulation of radiation degradation of current transmission coefficients was performed for various circuit solutions in LTSpice software.

1. Introduction

The analog front-end microelectronic devices are widely used in modern accelerators technology [1]. For experimental physics research in nuclear physics, it is necessary to use particle accelerators and detectors for particle flow registration. The analog front-end microelectronic devices are a part of the ionizing radiation detector which forms the current signal. It enables to measure the energy and pulse of charged particle. Primary signal gain, usually, are carried out using the operational amplifier. Location of the electronic functional units based on operational amplifiers in the zone of ionizing radiation impact requires the use of radiation-hardened electronic components.

Radiation degradation of integrated circuits (ICs) is the result of the accumulation of positive charge in silicon dioxide and interface trap buildup at the Si/SiO2 interface over the passive base region of the semiconductor structure during irradiation. It leads to a change of threshold voltages of MOS-structures and to decrease of the current gain of bipolar transistors [2]. As the result, there are arise functional and parametric failures of electronic equipment under ionizing radiation impact.

One of the basics elements in operational amplifier schematic is a current mirror, which usually consist of bipolar transistors susceptible to radiation degradation of the current gain [3]. Therefore, there is the dependence of the current transmission coefficient on the total absorbed dose. It leads to radiation drift of input electrical parameters of operational amplifiers, such as input current and offset voltage. The increase of radiation hardness of the current mirrors will improve the reliability of electronic equipment, which includes the operational amplifiers. Currently, there is series of schematic and topological techniques for radiation hardness increase of bipolar transistors, such as additional implantation of dopant in subsurface areas of the semiconductor structure and the use of protective rings. For the implementation of these methods, it is necessary to complicate the technological process additionally. This complication leads to degradation of electrical parameters of semiconductor devices. In this work the schematic techniques, which will increase the radiation hardness of current mirrors of bipolar transistors without technology complication and deterioration of ICs parameters, are presented. The proposed methods are based on the use of radiation drift compensation circuit in current mirrors outputs.

2. Bipolar transistor radiation degradation model

For the schematic techniques development to radiation hardness increase of current mirror circuits, it is necessary to understand the physical mechanism of radiation degradation of bipolar transistors under irradiation. At the Si/SiO_2 interface over the passive base region, occurs buildup of interface traps, which acts as recombination centers. As the result, the surface recombination rate increases. It leads to an increase of recombination component of the base current and to rising of charge carrier injection from the lateral emitter surface to passive base region. The above process does not affect the collector current. This process is controlled by the gradient of concentration of the minority charge carriers in the base active areas along the way from the emitter to collector. This mechanism shows that ionizing radiation leads to increase of base and emitter currents, without affecting the collector current. The irradiation of bipolar p-n-p transistors 2N2907 was performed by Co-60 isotope to validate this conclusion [4]. Transistors samples were irradiated at room temperature (25 ± 2 °C) to total dose 72 krad (SiO_2), the dose rate was 10 rad(SiO_2)/s. The dependences of the collector current, base current and static transistor gain on the emitter-base voltage are presented in Fig.1 and Fig.2.

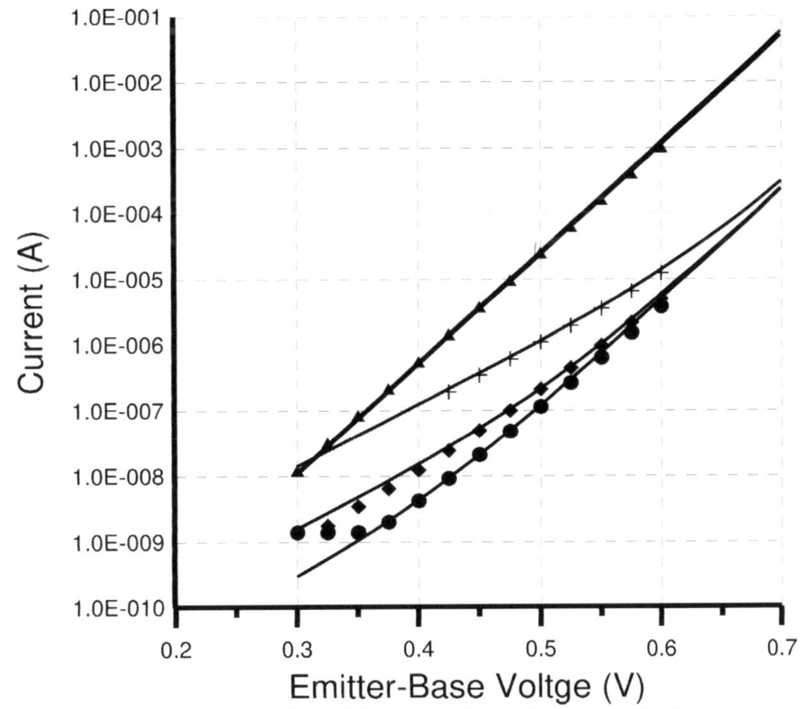

Figure 1. Experimental and theoretical (solid curve) dependences of the base current on the emitter-base voltage for 2N2907 transistor before irradiation (●) and for various doses (12 krad(Si) – (♦), 72 krad(Si) – (+)); the dependence of the collector current on the emitter-base voltage before and after irradiation (▲).

From the data, it is clear, that collector current remains practically constant during irradiation, while the gain degradation, which is determined by a base current increase. Analysis of experimental data also shows that the base current depends on total ionizing dose approximately linearly at a fixed emitter-base voltage. Since the base current does not change significantly, the reciprocal value of static gain $1/B = I_b/I_c$ is directly proportional to the total ionizing dose and can be used in modeling for comparative evaluation of various schematic techniques radiation hardness.

3. Comparative analysis of radiation hardness of the current mirrors on bipolar transistors

The circuit of the current mirror is presented in Fig. 3a. This circuit is a simple current source on bipolar transistors. This circuit has low output resistance. It connected with the Early effect in the transistor (T2). Cascode circuit in the current mirror presented in Fig. 3b. It enables to increase the output resistance and to suppress Early effect.

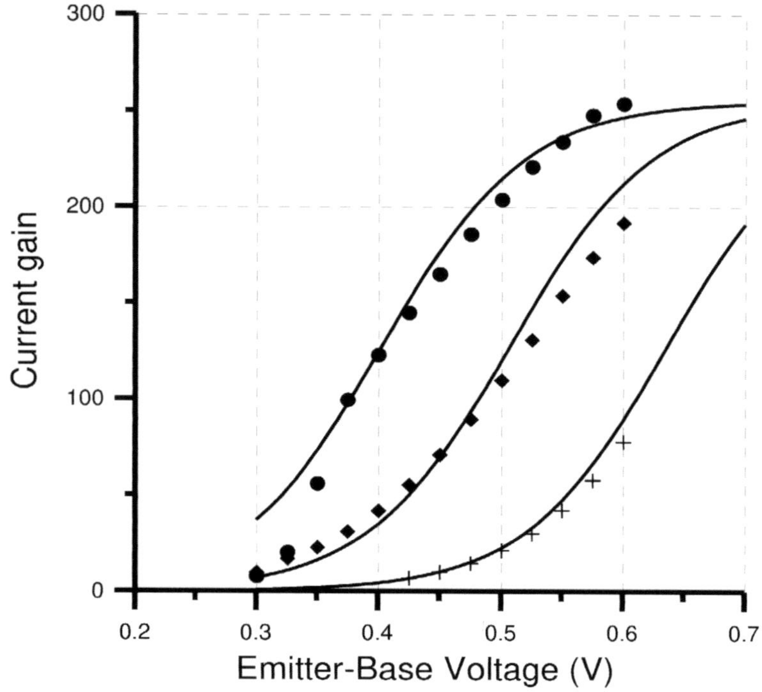

Figure 2. The dependence of the static gain of the transistor on the emitter-base voltage before irradiation (●) and various doses (12 krad(Si) – (♦), 72 krad(Si) – (+)) [4].

Circuit simulation of radiation degradation of the current transmission coefficients of simple and cascode circuits was performed in LTSpice IV, assuming that the reciprocal value of the static current gain of bipolar transistors is proportional directly with total ionizing dose during irradiation. The simulation results are presented in Fig.4.

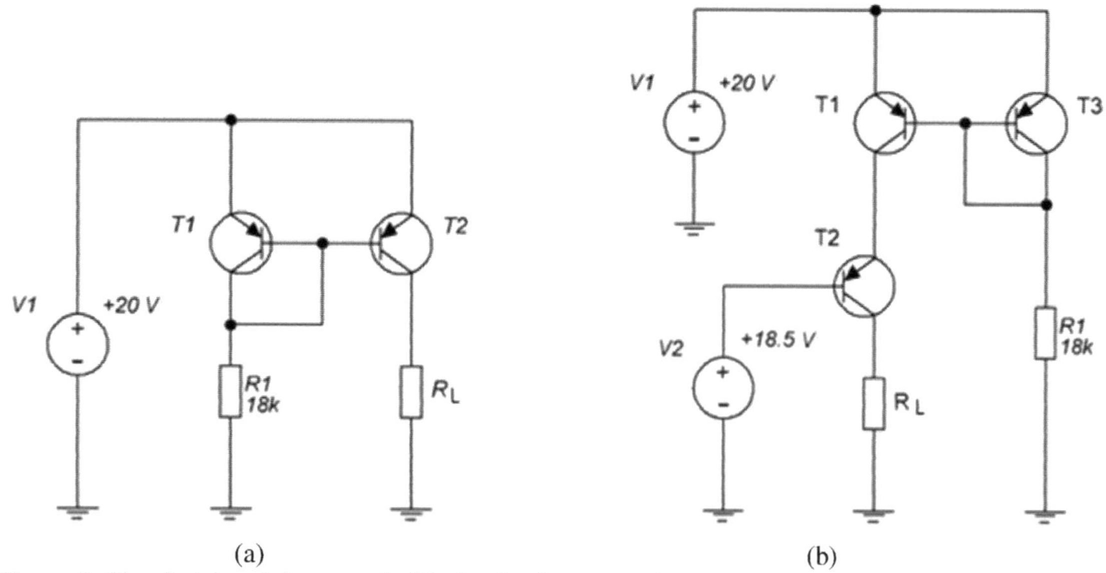

Figure 3. Simple (a) and the cascode (b) circuit of current mirror.

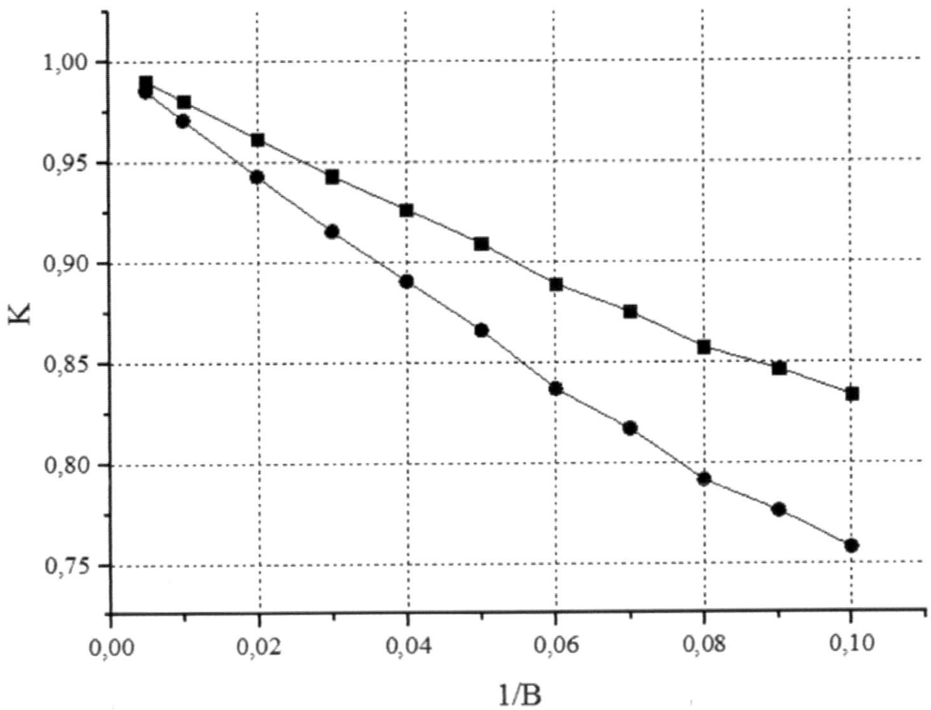

Figure 4. The dependence of the current transmission coefficients of simple (—■—) and cascode (—●—) circuits of current mirror on the reciprocal value of static gain of bipolar transistors.

From the data, it is clear, that the degradation of input current in cascode circuit is faster than in simple circuit. Assuming that a parametric failure of the devices occurs if the current transmission coefficients differ from 1.0 by more than 10% (K = 0.9), we can obtain that the parametric failure of cascode circuit occurs at total dose level which is lower by 1.54 times that in simple current mirror circuit. The decreasing of failure total dose level is connected with the degradation of the emitter current transmission coefficient of transistor T2 (Fig. 3b).

To increase the radiation hardness of cascode circuit without losing its advantages, including the high output resistance, the circuit of a current mirror which presented in Fig.5, can be used. The circuit simulation results of the dependence of the current transmission coefficient on the reciprocal value of the static current gain of bipolar transistors for the radiation hardened circuit, and traditional cascod circuits are presented in Fig.6. From the data, it is clear, that the improved circuit of the current mirror enables to increase the radiation hardness by 1.54 times. The radiation hardness is increased by the compensation of the collector current decrease of transistor T2 using a base current of transistor T4. The emitter current of transistor T4 is fixed by the circuit of a current source which includes transistors T1, T3 and resistor R1.

The values of output resistance for all the schematic techniques of current mirrors, which were described above, were obtained by the circuit simulation. The values of output resistance and critical reciprocal value of static gain, which corresponds to parametric failures of bipolar transistors, are presented in Table 1. From the data, it is clear, that the improved circuit of cascode current mirror is more radiation-hardened than the traditional cascode circuit of the current mirror. Nevertheless, the improved circuit of cascode current mirror has significantly less output resistance than traditional cascode circuit of the current mirror.

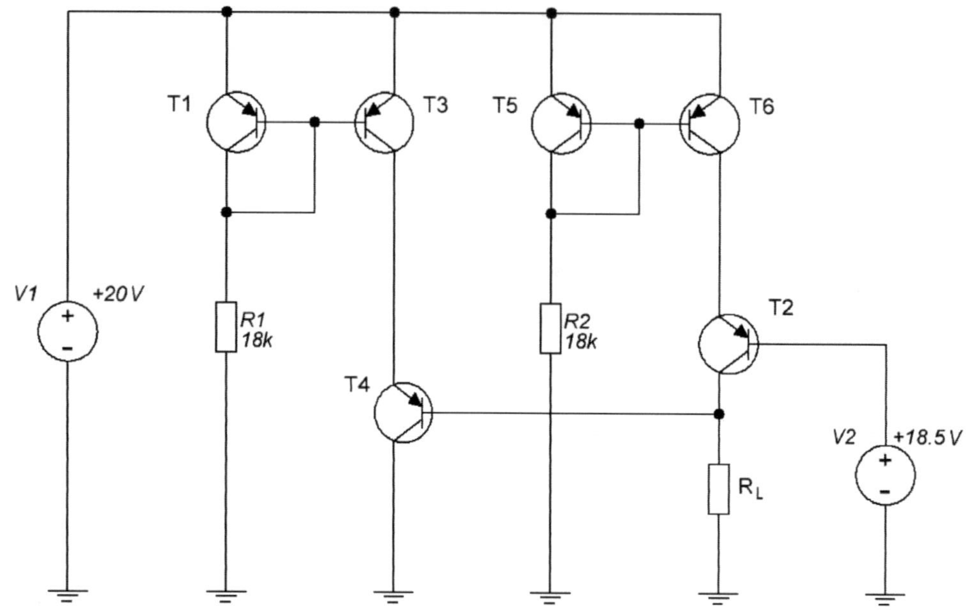

Figure 5. The radiation-hardened cascode circuit of current mirror.

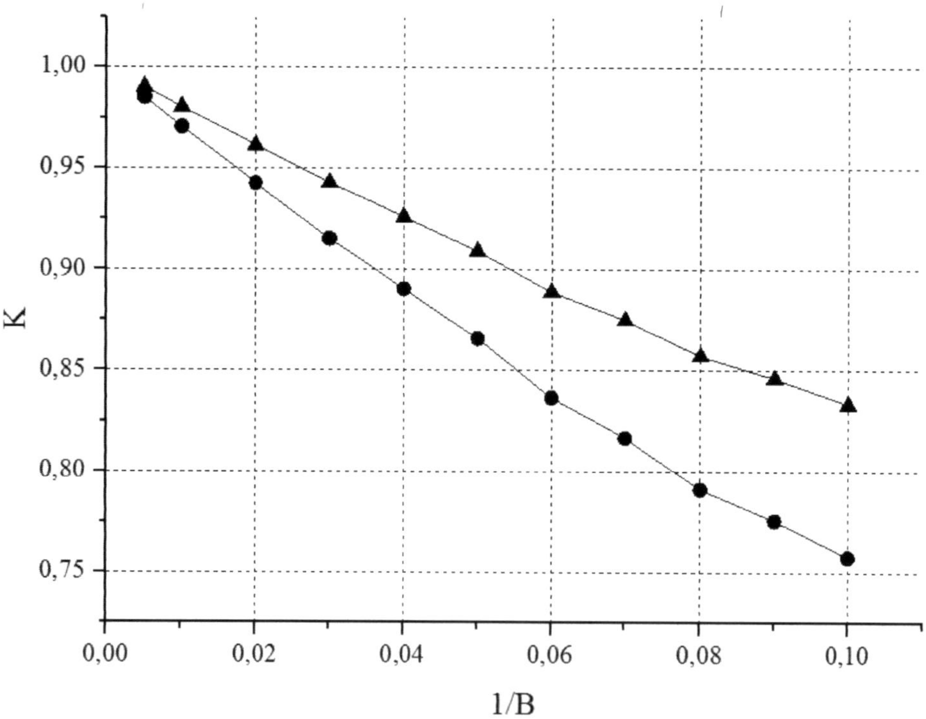

Figure 6. The dependence of current transmission coefficient of the radiation-hardened circuit (—▲—) and the traditional cascode circuit (—●—) on the reciprocal value of static gain of bipolar transistors (circuit modeling results)

IOP Conf. Series: Materials Science and Engineering **151** (2016) 012026 doi:10.1088/1757-899X/151/1/012026

Table 1. The characteristics of simple, traditional cascode and radiation-hardened circuits of current mirrors

	Simple circuit of current mirror	Traditional cascode circuit of current mirror	Radiation-hardened circuit of current mirror
Output resistance	100kOhm	5 MOhm	2 MOhm
$(1/B)_{critical}$	0,054	0,035	0,054

4. Conclusion

Based on the physical model of radiation degradation of the bipolar transistor and experimental data showing that the reciprocal value of static current gain is proportional directly to the total ionizing dose, a comparative analysis was performed for the various schematic techniques of current mirrors by the use of this statement during circuit simulation in LTSpice.

The circuit simulation enables us to estimate the value of the output resistance for each of the considered circuits. The improved circuit of cascode current mirror has less output resistance than traditional cascode circuit of the current mirror. The improved circuit of cascode current mirror is more radiation-hardened than the traditional cascode circuit of the current mirror.

Therefore, these techniques will potentially enable to increase the operation time of electronics in conditions of ionizing radiation impact.

Acknowledgements
This work was supported by the Competitiveness Program of NRNU MEPhI.

References

[1] Ullan M, Wilder M, Spieler H, Spencer E, Rescia S, Newcomer F M, Martinez-McKinney F, Kononenko W, Glillo A A and Diez S 2013 *Nucl. Instrum. Meth.* A **724** 41
[2] Rashkeev S N, Cirba C R, Fleetwood D M, Schrimpf R D, Witczak S C, Michez A and Pantelides S T 2002 *IEEE T. Nucl. Sci.* **49** 2650
[3] Vakhneko M S, Komleva V A and Lebedev A A 2015 *Sensors & Systems* **1** 28 (*in Russian*)
[4] Bakerenkov A S, Pershenkov V S, Solomatin A V, Belyakov V V and Shurenkov V V 2014 *Appl. Mech. Mater.* **565** 138

IOP Conf. Series: Materials Science and Engineering **151** (2016) 012027 doi:10.1088/1757-899X/151/1/012027

Simulating nanosecond voltage comparator with improved radiation hardness compatible with PECL logic

A A Lebedev, A E Nazarenko, M S Vakhnenko, V A Komleva and A Y Nikolaev

National Research Nuclear University MEPhI (Moscow Engineering Physics Institute), Kashirskoe shosse 31, 115409 Moscow, Russian Federation

Corresponding author e-mail address: Lebedev741@mail.ru

Abstract. The typical voltage comparator circuits are considered. Shown that the use of series-parallel connection of cascades improves the accuracy, speed and hardness to single event transient effects.

1. Introduction

Ultrafast voltage comparators (VC) are widely used to construct parallel analog-to-digital converters (ADC) and also for recording high-frequency signals in nuclear electronics. To improve performance, these VC are compatible with PECL output. To reduce power consumption while maintaining high performance, generally used gated VC with trigger-latch. However, such VC comparisons accuracy is about of 5-10 mV. Two architectures are used to improve the accuracy of comparing the analog signal with a reference:

- with horizontal build-up stages;
- with vertical build-up stages.

The first case increases the switching delay by narrowing the bandwidth. The dynamic range of the input signal is reduced in the second. Commercially available CN compatible with the PECL, typically operate at a supply voltage from 3.3 V to 5 V. In the VC with vertical build-up stages specified supply voltage range is a challenge.

The main purpose of this work is to create a VC with a vertical build-up stages, operating at supply voltages from 3,3 V to 5 V and having an order of magnitude higher than the comparison accuracy with the VC given in Figure 1 and Figure 2.

2. Single stage voltage comparators with trigger-latch

First, [1] of the comparator circuit (Figure 1) is made on a differential stage with a trigger-latch. Using this architecture, most suitable for the ADC having 4-6 digits. However, the use of an emitter follower at the entrance reduces the performance of the VC. The second [2], [3] circuit (Figure 2) is made without the input of the emitter follower, with the bases of trigger-latches are connected to the low impedance emitter follower output bus to enhance the performance. The transient analysis in this configuration allows for a given technology clock frequencies up to 1 GHz. When using emitter follower clock speed is reduced to 500 MHz. The accuracy of the above VC static mode determined VC gain be approximately equal to 20.

IOP Conf. Series: Materials Science and Engineering **151** (2016) 012027 doi:10.1088/1757-899X/151/1/012027

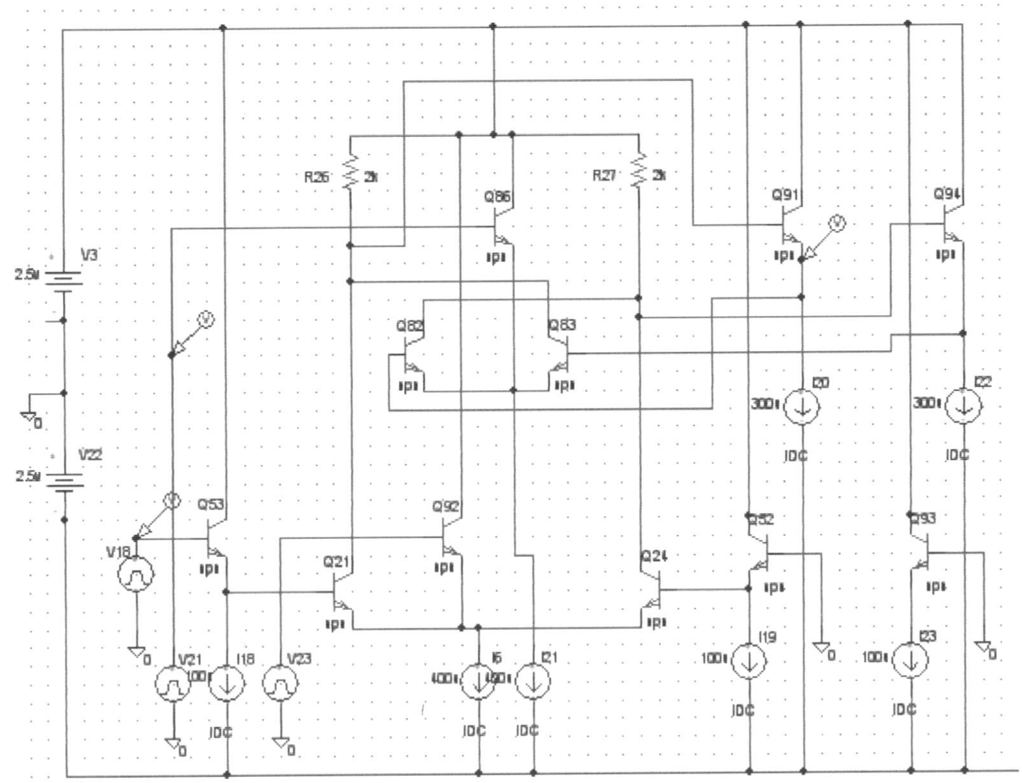

Figure 1. Schematic diagram of the single-stage VC with emitter follower output.

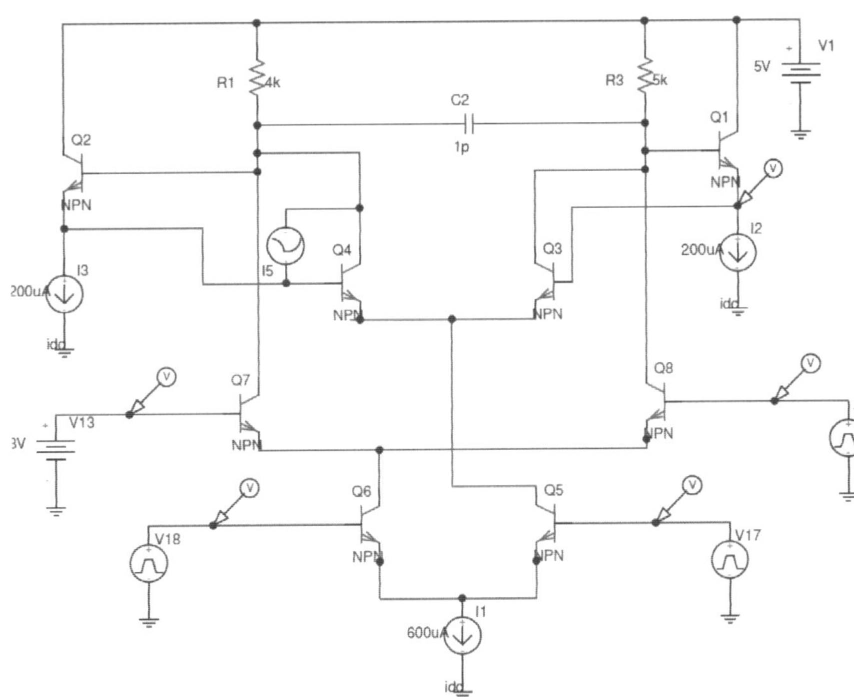

Figure 2. Schematic diagram of the VC with improved speed.

IOP Conf. Series: Materials Science and Engineering **151** (2016) 012027 doi:10.1088/1757-899X/151/1/012027

3. Voltage comparator with series-parallel switching stages

To ensure the functioning of the VC at the supply voltage from 3,3 V to 5 V a series-parallel cascode including the first and the second stage is used (Fig. 3). Static gain is determined by the following equation:

$$K_u = \frac{R_{26}}{r_{eQ21}} \times \left(1 + \frac{R_{37} \times \gamma}{r_{eQ103}}\right) \tag{1}$$

where:

$$\gamma = \frac{R_{inQ103}}{R_{inQ103} + r_{eQ107}} < 1 \tag{2}$$

It can be shown that in such VC gain factor is about 5-10 times greater than in previous CN. Reducing the switching delay, in this case, is performed by increasing modal current of the second stage in the dynamic mode.

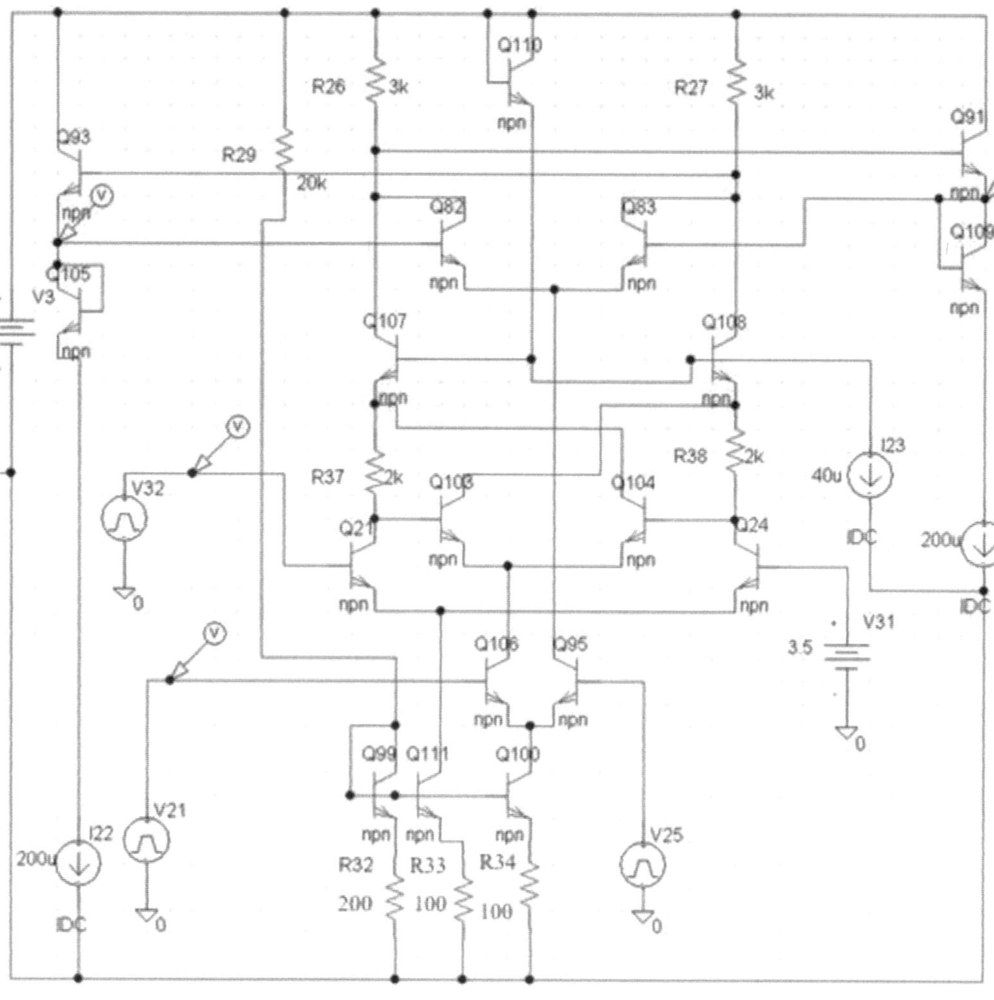

Figure 3. Schematic diagram of the VC with series-parallel switching stages.

Basic time characteristics of the two types of VC are shown in Fig. 4.

IOP Conf. Series: Materials Science and Engineering **151** (2016) 012027 doi:10.1088/1757-899X/151/1/012027

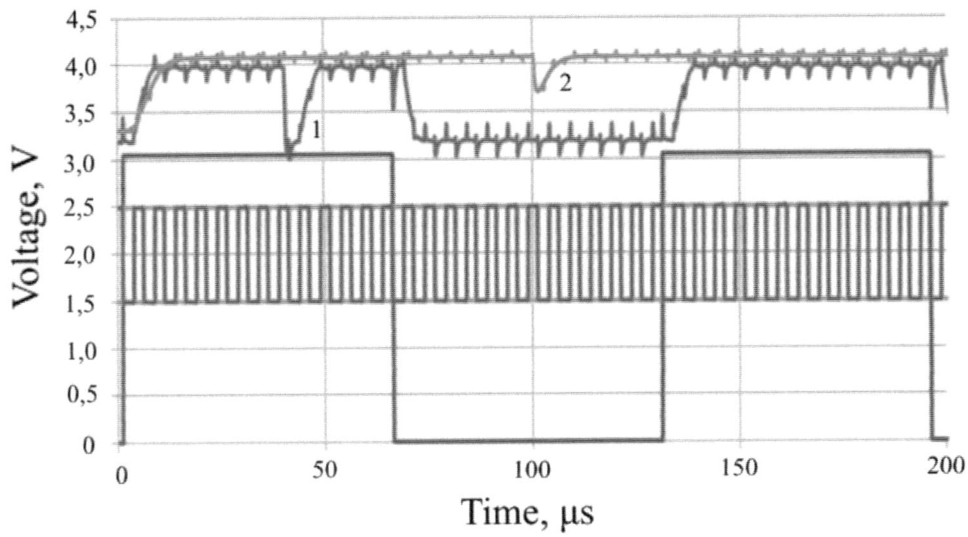

Figure 4. Time characteristics of the proposed VC (1 – single-stage VC with emitter follower output , 2 – VC with improved speed).

The sampling frequency is from 200 to 500 MHz and is mainly determined by the frequency of unity gain transistor (Ft = 5 GHz). Switching Time - 2 ns. Power dissipation ~ 5-10 mW.

4. Single event transient impact on the transition process.

Fig. 4 shows the reaction to the impact of SET of considered VC [4]. For the first comparator (index 1 in Figure 4) the transition process is 3 ns at an amplitude of 0.7 V. For the second - 1.7 ns, and 0.4 V, respectively. Transition process time of SET is reduced due to the increasing modal current of the second stage in the dynamic mode. Increasing the voltage gain influence the degradation of the gain transistor β slightly.

5. Conclusions

VC with a series-parallel switching stages can increase the sensitivity of the circuit is almost an order of magnitude in comparison with the single-stage circuits CL and reduce the time of transient after SET.

Acknowledgements
This work was supported by the Competitiveness Program of NRNU MEPhI.

References

[1] U.S. Patent 5510734, Apr. 23,1996
[2] Peetz B, Hamilton B D and Kang J 1986 *IEEE J. Solid-state circuits* SC-21 **6** 997
[3] Kacprzak T 1988 *IEEE J. Solid-state circuits* **23** 260
[4] Adell P, Schrimpf R D, Barnaby H J, Marec R, Chatry C, Calvel P, Barillot C and Mion O 2000 *IEEE T. Nucl. Sci.* **47** 2616

Aid system for visually impaired people at spatial orientation

K A Palaguta, A I Krukov, S M Troikov and I S Shubnikova

Moscow state university of mechanical engineering (MAMI), Avtozavodskaya st. 16, 115280 Moscow, Russian Federation

Corresponding author e-mail address: ka129@yandex.ru

Abstract. The paper is devoted to the analysis of possibilities of infrared lasers and ultrasonic sensors of distance measurement in assisting visually impaired people.

1. Introduction

According to the world health organization worldwide, there are 37 million blind people and 124 million with low vision [1]. At the same time, according to UN projections, by 2020 the number of blind people worldwide could grow to 75 million people. According to some origins, the number of the counted blind and visually impaired in Russia is 218 thousand people, including wholly blind – 103 thousand. Modern technology allows visually impaired people and people with damage to the visual system to improve object recognition and orientation in space, which leads to improving their quality of life.

2. Statement of the problem

Visually impaired people are usually well oriented in space, but do not notice small obstacles in its path, such as curbs, steps up or down, small holes. To use the white cane they don't have, but not noticed in time these obstacles can injure the person [2-6]. To date has developed and offers such tools to help visually impaired people, a cane as Mygo [7], cane SmartCane [8], LaserCane walking stick [9].These canes or using laser sensors and then do not require contact with the surface, or they use ultrasonic sensors, but then at the end of the stick is a wheel or a sphere moving on a surface on which the moving people. Each stick is placed a pair of sensors: one to detect obstacles like a step up, and another to detect obstacles like a step-down. A significant drawback of all the proposed devices is the fact that they emphasize vision problems from its users. For people with weak eyesight is usually undesirable. Therefore, we consider the choice of sensors for the system to aid visually impaired people. These converters must be such that its dimensions could be integrated into a conventional walking stick. The outer diameter of this cane is typically 22-30 mm. an Additional requirement is that walking man did not need to touch the surface on which movement occurs, the end of the stick. The distance at which it detects the obstacle shall be not less than 1.5 – 2.0 m. Usually, depending on the used in the manufacture of materials, a cane may take weight from 0.1 to 0.4 kg but if you use canes made of titanium with adjustable length weight may be up to 0.8 kg. This fact also needs to be considered when choosing a sensor.

3. Experimental

For measuring small distances apply infrared (IR) laser and ultrasonic distance sensors. Infrared distance sensors contain an IR transmitter led and IR receiver photodetector. Infrared led emitting an infrared signal, and an infrared photo-detector detects the reflected signal. The amplitude of this reflected signal is usually inversely proportional to the distance from the object to the infrared transceiver. The maximum distance is the threshold for determining the presence of the object. An example of such a sensor is the sensor range Sharp GP2Y0A02YK0F (figure 1) is an analog infrared sensor to determine the presence of an object at a distance of from 20 to 150 cm [9].

Figure 1. Infrared sensor Sharp GP2Y0A02YK0F.

The main disadvantage of this sensor is that the distance measurement is used triangulation method. For this reason, the measurement results depend on the angle of incidence of IR radiation on a controlled surface, which is unacceptable for the considered problem.

A significant drawback of optical, including laser, sensor is bad resistance to weather conditions. For example, contamination of the detector leads to the impossibility of detecting objects, while ultrasonic sensors with the same degree of contamination continue to work, but with less sensitivity.

The system to help people with impaired vision is currently developed. The main purpose of such systems is to provide information on the nature of the obstacles. As measurement sensors in such a system typically uses a pair of ultrasonic sensors. This seems the most reasonable, the more that existing ultrasonic sensors, such as used in Parking Systems (Parktronic), have small dimensions, allowing to place them inside a regular orthopedic cane. Ultrasonic sensors can have separate transmitters and receivers, and combined.

The first of these is the sensor HC-SR04 (figure 2). This ultrasonic sensor can measure the distance to obstacles in the range from 2 cm to 4 m. It is the most preferable in terms of price/quality, but its dimensions, components 40×20×15 mm, not allow you to place it inside the cane.

Figure 2. Ultrasonic sensor HC-SR04.

Parking sensors use a combined emitter and receiver. The most effective parking sensors at a distance from 25 cm to 1.8 meters are the sensors which correspond to a desired range of distances.

There are separate modules ultrasonic sensors (sonar), such as LV-MAXSONAR-EZ1 (figure 3), allowing to detect objects at distances up to 6.45 m. This sonar is one of the typical decisions on the basis of ultrasonic sensors MaxSonar-UT firms Maxbotix. Module output signals on the distance of the object present in three forms: analog signal, PWM and serial data R_x and T_x. There is also a sealed design given module ultrasonic sensor LV-MAXSONAR-WRC1 (figure 3 b).

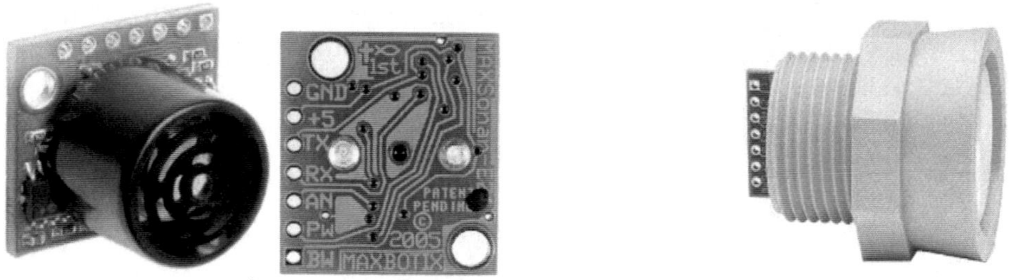

a) b)

Figure 3. Modules ultrasonic sensor LV-MAXSONAR:
a) LV-MAXSONAR-EZ1; b) LV-MAXSONAR-WRC1

Examples of ultrasonic sensors are shown in Table 1, where added the most accessible sensor ParkMaster A [10]. The diameter of the sensors themselves is almost the same and ranges from 16.4 mm for the sensor used with the LV-MAXSONAR-EZ1 to 18.8 mm sensor ParkMaster A. the Problem is, the Board size of the module LV-MAXSONAR-EZ1 is 19.9×22.1 mm, which eliminates the placement of the Board of the stick. As a way out of this situation, you can offer accommodation module LV-MAXSONAR-EZ1 in rubber cane tip diameter reaches 50 mm, and use this module to detect obstacles like a step or a curb. Sensor to detect the steps down it is advisable to place the handle of the cane and used as a sensor ParkMaster or A sensor selected from the module LV-MAXSONAR-EZ1.

Table1. Examples of ultrasonic sensors.

The name of the sensor	Measured distance, m	Resolution, sm	Temperature measuring range, C°
MA40E7S-1	0,2 – 3	1	–40 – +80
MB1100	0,2 – 6,5	1	–40 – +65
DYP-ME007Y	0,3 – 3,5	1	–45 – +70
ParkMaster A	0,3 – 2,59	1	–40 – +80

Surface mount the sensor on the body of the cane implies the contact with the dust particles, water and maybe some reagents, which are applied to the road surface, resulting in a preferred use of the sensor is totally enclosed. Most of the sensors that are built in sealed design, protection of its internal structure correspond to the degree of IP67, which allows it to be completely closed against the ingress of dust and moisture. You should also use a sensor with a broad temperature range (-40÷70°C). Another option to consider is the beam width, which is usually contained in the technical documentation for the sensor or can be determined experimentally.

After assembly of the device, load the program memory of the microcontroller and layout of all elements in the cane, it was necessary to test the sample.

To do this, use normal laboratory conditions with obstacles of the type of steps down and steps up.

The program was configured so that when approaching certain obstacles used the same algorithm like parking sensors on the cars that is, the closer the obstacle, the higher the pitch of the sound, and the farther away from obstacles so that it will be lower (figure 4).

The fundamental difference from the rest of the competing devices is to scan all the objects not only in the horizontal plane, but also below it, i.e. the presence of ledges, stairs leading down, and wells and hole.

IOP Conf. Series: Materials Science and Engineering **151** (2016) 012028 doi:10.1088/1757-899X/151/1/012028

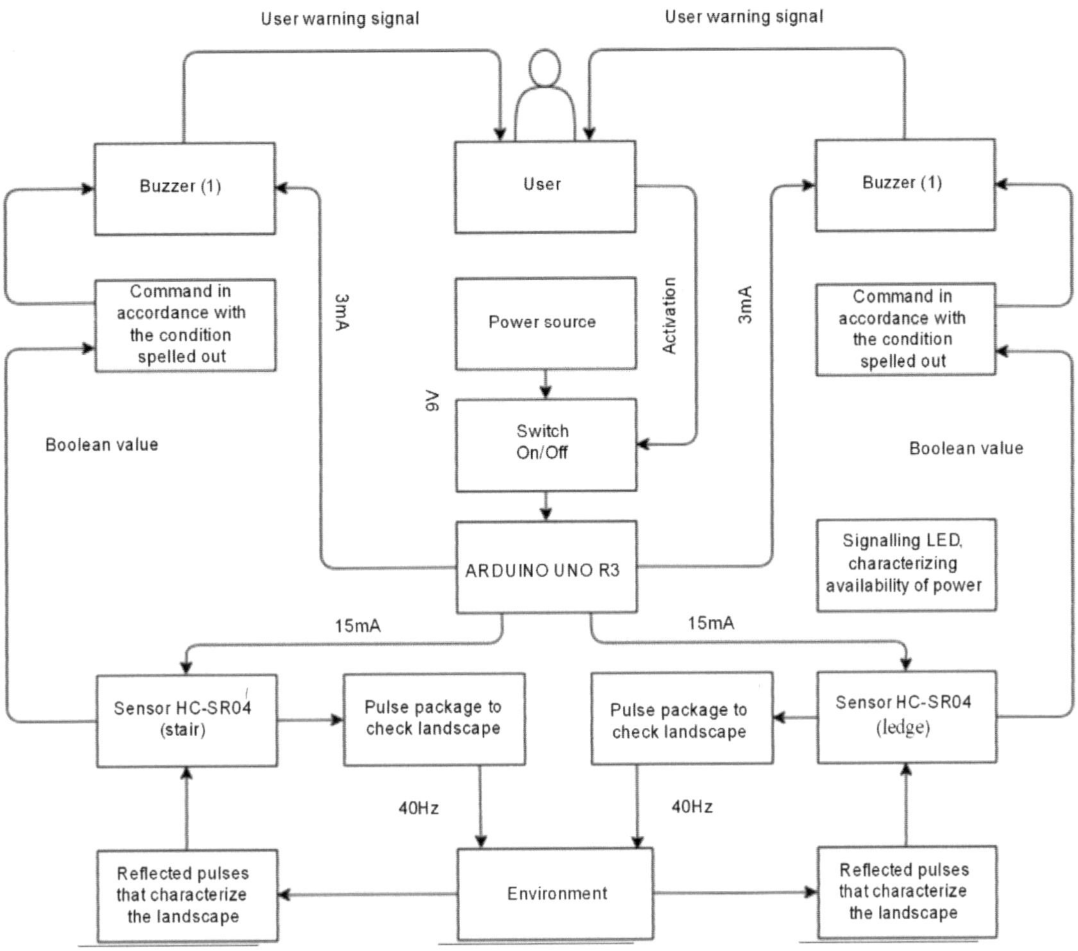

Figure 4. Block diagram of the device of functioning.

4. Conclusion

The device showed good results when scanning a space in motion, in laboratory conditions. At the exit to the street, it was immediately revealed that one of obvious drawback is the device output. In a typical street environment due to noise sound with conventional piezoelectric sensors is not audible when using headphones. The user can also experience some discomfort because of people with impaired vision, a person receives compensation for a lack of vision in enhanced auditory senses, and a person cannot prevent them from functioning.

It can be concluded that the most effective would be to use not only one type of output signal (signaling of danger) but several at the same time. Additional information channel for a person can be a source of vibration in the handle of the cane. In this case, when using the vibration signal in the form of periodic bursts of pulses can be frequency of the vibration signal coding type of obstacle (step up or step down), and the frequency of the pulse train to be used for coding distance to the obstacle, i.e. the closer the obstacle, the higher the frequency of the pulse train. In order to be able to use the cane as a regular cane for walking, it is necessary to highlight the moments when it takes a fixed position in space. It can be implemented that including the system processing of the measuring information two-axis accelerometers. In the layout system aid visually impaired people was used 2 ultrasonic sensors HC-SR04, which provided the performance of the system if an obstacle is detected and the type of the step up, and step down.

IOP Conf. Series: Materials Science and Engineering **151** (2016) 012028 doi:10.1088/1757-899X/151/1/012028

References

[1] *Blindness and Visual Impairment.* Springer Science + Available from: http://dx.doi.org/10.1007/springerreference_223576
[2] Dakopoulos D and Bourbakis N 2010 *IEEE T. Syst. Man.* **40** 25-35
[3] Oktem R, Aydin E and Cagiltay N 2008 *Proc. 34th Annual Conf. Industrial Electronics Society (Orlando)* (IEEE) p 2982-2987
[4] Peasley B and Birchfield S 2013 *Proc. Workshop on Robot Vision (Clearwater Beach)* (IEEE) p 197-202
[5] Ray A, Behera L and Jamshidi M 2008 *IEEE Syst. J.* **2** 258-272
[6] Adams R and Bischof L 1994 IEEE *T. Pattern Anal. Mach. Intell.* **16** 641-647
[7] Hirsch U, Kuntze H B, Brecht R and Kilthau A 1992 *IEEE Int. Conf. Robotics and Automation (Nice)*(IEEE) p 670-675
[8] Wu W, Au L, Jordan B, Stathopoulos T, Batalin M, Kaiser W, Vahdatpour A, Sarrafzadeh M, Fang M and Chodosh J 2008 *Proc. 3rd Int. Conf. Body Area Networks (Brussels)*(ICST)
[9] Donovan S, Lim C, Diaz N, Browner N, Rose P, Sudarsky L R, Tarsy D, Fahn S and Simon D K 2011 *Parkinsonism & Related Disorders* **17**(4) 240-245
[10] Mutiara G A, Agung A A G and Handayani R 2015 *Proc. 1st Int. Conf. Wireless and Telematics (Monado)* (IEEE)

High accuracy magnetic field sensors with wide operation temperature range

I S Vasil'evskii[1], A N Vinichenko[1], D I Rubakin[1], I A Bolshakova[2] and N I Kargin[1]

[1]National Research Nuclear University MEPhI, (Moscow Engineering Physics Institute), Kashirskoe shosse 31, 115409 Moscow, Russian Federation
[2]Lviv Polytechnic National University, Bandera street 12, 79013 Lviv, Ukraine

Corresponding author e-mail address: ivasilevskii@mail.ru

Abstract. n+InAs(Si) epitaxial thin films heavily doped by silicon and Hall effect magnetic field sensors based on this structures have been fabricated and studied. We have demonstrated the successful formation of highly doped InAs thin films (~100 nm) with the different intermediate layer arrangement and appropriate electron mobility values. Hall sensors performance parameters have been measured in wide temperature range. Obtained sensitivity varied from 1 to 40 Ω/T, while the best linearity and lower temperature coefficient have been found in the higher doped samples with lower electron mobility. We attribute this to the electron system degeneracy and decreased phonon contribution to electron mobility and resistance.

1. Introduction

Magnetic field sensors are widely used in MegaScience research projects as the magnetic field is used as a key tool for the control of charged particle beams. In various magnetic field control systems, the high precision magnetic field sensors must have good linearity over a wide range of magnetic field (1 ÷ 8 T), the low drift of sensitivity and resistance over a wide temperature range - from cryogenic to high (4,2 ÷ 600 K). Furthermore, the sensors should be compact in size. For a number of experimental systems, an additional requirement is a long-term stable operation of such sensors in conditions of charged particles, neutrons and gamma rays radiation [1,2]. Traditional silicon sensors cannot withstand such operating conditions.

A3B5 compound semiconductors offer an advanced performance due to high electron mobility, high electron density and thin film technology. Among them InAs and InSb thin film heterostructures are most attractive due to high doping level, providing the electron degeneracy and thus, low temperature drift [3,4]. However, the mechanical deformations due to the mismatch of the lattice constants of GaAs substrate and InAs or InSb make difficult to obtain high-quality InAs layers during epitaxial growth. Direct overgrowth of InAs causes propagation of the dislocations from nucleation layer to the overlying heterostructure active layers, enhancing the scattering of electrons, thus increasing the device resistivity. In the direct epitaxial growth of the relaxed InAs films, the dislocations' density decreases quickly with an increase of film thickness, and the quality improvement is observed with a layer thickness of > 1.5 ÷ 4 microns [5,6]. At the standard speeds and temperatures above 500 °C, the growth of InAs on GaAs leads to the formation of a rough surface and stacking faults. The electron mobility in the films with the thickness less than 100 nm is about 400÷1000 $cm^2/(V \cdot s)$ that is far below the mobility of the defect-free InAs.

The purpose of the present work is the development of epitaxial thin films n+InAs(Si) heavily doped with silicon and engineering of Hall effect magnetic field sensors based on this structures.

2. Material and device fabrication

2.1. Sensor material growth and properties

The structures were grown by solid-source molecular beam epitaxy Riber Compact 21T on GaAs and sapphire substrates. The samples differ by nucleation/transition layer construction (samples of the 1st, 2nd and 3rd type) located between GaAs and the doped layer of ~ 100 nm n+InAs(Si). Transition layers were introduced to optimize the epitaxial film quality before the growth of n+InAs films. For the samples on GaAs substrate GaAs buffer and superlattice AlAs/GaAs were followed by the AlGaAs layer, then, in the samples of 1st type i-InAs relaxation layers, for the 2nd type - $In_xAl_{1-x}As$ sublayer with InAs content of 75÷90 %, for the 3rd type - the metamorphic transition $In_xAl_{1-x}As$ buffer with the change of the composition from 20 to 90 % at a thickness of 1.1÷1.2 μm. 4th type sample was highly doped InAs layer grown directly on the sapphire substrate.

Since the gradual increase of InAs mole fraction up to 100% is difficult to implement due to the inertia of the cells effusion, the final part of the transition buffer is implemented as a superlattice $InAs/In_xAl_{1-x}As$. In such a structure, the sharpest drop of In composition and the maximum strain at the beginning of the InAs film growth were implemented in the structures of the 1st type. The 2nd type provided a more seamless transition from GaAs to InAs lattice. The samples of the 3rd type provided the most smooth distribution of misfit dislocations at the relaxation of $In_xAl_{1-x}As$ layer. The parameters of the samples (mobility μ and Hall electron concentration n_H) are listed in the following Table.1. The density of dislocations and stacking faults in the InAs layer affect the electron transport properties. To analyze them, the Hall concentration and the electron mobility were measured in the wide temperature range (5÷295 K).

The samples of the first type have low electron mobility and low surface roughness. The introduction of the $In_xAl_{1-x}As$ sublayer with the thickness of 25÷30 nm in the samples of the 2nd type increases the mobility of electrons by 5÷15 %. The highest electron mobility was observed in the samples of the 3rd type with the metamorphic buffer with the thickness of 1.0÷1.2 μm, as well as the largest increase in mobility at the nitrogen temperature.

The surface morphology of the films is associated with the growth conditions, temperature and pressure of arsenic. If the temperature was increased, the large defects were observed, probably due to the clusterization of the metallic In. Excess of As leads to the formation of the stacking faults and appearance of the rough surface. The optimal conditions were observed at T of about 460 °C and the pressure of arsenic of $4\div5\cdot10^{-6}$ Torr. The morphology of striated relief was not observed in the samples of the 3rd type with a thick metamorphic buffer. This is probably caused by the increase of the true temperature of the substrate due to an increase of absorption of infrared heating at the growth of layers with a high content of InAs.

Sample 346 was grown on a sapphire substrate without a transition layer. X-ray diffraction revealed the oriented polycrystalline InAs structure of this film. Due to ultra-high Si doping level and poor crystalline quality it demonstrated very low electron mobility values and negligible temperature dependence of the mobility. In turn this can be used to suppress the temperature coefficients of resistance and sensitivity.

2.2. Sensor fabrication

The sensors have a form of Greek symmetric cross performed by contact photolithography and mesa wet etching in $H_3PO_4:H_2O_2:H_2O$ etchant. The active channel width was 200 μm. Contact metallization formed by lift-off lithography and Ti/Au (100 nm / 800 nm) vacuum deposition. The wafers have been passivated by 100 nm Si_3N_4 PECVD deposition to eliminate a surface long thermal oxidation and the damage during the followed processing. For the contact pads opening ICP dry etching of Si_3N_4 have been consequently performed. Hall sensor view is shown in Figure 1. Dies were cutted by disc saw. Au wires have been bonded by splitted electrode technique for the electrical measurements.

Figure 1. Hall sensor photomicrography.

3. Experimental performance

The sensors have been measured in the DC constant current regime at the different temperatures within the range 5÷295 K. Current values were 10 or 20 mA. As the Hall voltage U_H in common case is dependent on the various physical factors – temperature, magnetic field, heterostructure properties etc. Hall sensors performance is described by the series of parameters, defined as follows:

- sensitivity $S = \frac{1}{I_0}\frac{\partial U_H}{\partial B}$, where I_0 is sensor current, shown in Figure 2;
- moreover, temperature sensitivity coefficient $STC = \frac{1}{S_0}\frac{\partial S}{\partial T}$;
- resistance temperature coefficient $RTC = \frac{1}{R_0}\frac{\partial R}{\partial T}$;
- nonlinearity factor $\delta(B) = \frac{U_H(B) - U_{lin}(B)}{U_H(B)}$, where $U_{lin}(B)$ is the linear fit of experimental $U_H(B)$ dependence for the magnetic field interval 0.2-4 T. S_0 and R_0 refer to the room temperature. Low field data were excluded due to the divergence of $U_H(B)$ at $B \to 0$, accompanied by noise signals and systematic magnetic field creep embedded in the superconducting solenoid. Thus δ(B) presented the maximal nonlinearity. The dependence of δ on doping density is shown in Figure 3.

The device properties have been measured by DC current technique in CryoFree closed cycle helium cryostat with the superconductive solenoid in a magnetic field up to 6 T.

Hall sensor and material parameters measured at T=295 K are listed in Table 1.

Table 1. Material and Hall sensor properties for the different structure types measured at room temperature.

Sample	Type	n_H, 10^{13} cm^{-2}	δ, %	S, Ω/T	STC, 10^{-3} %/K	RTC, 10^{-4} %/K	μ_H, cm^2/(V·s)
162	1	1.93	0.19	32.7	3.5	23.0	5520
163	3	1.60	0.79	39.8	4.4	200	6440
252	2	3.29	0.34	19.1	2.5	19.7	3100
246	2	6.19	0.14	10.1	1.4	18.9	3620
119	1	1.05	0.03	6.0	1.8	13.1	1980
346	4	4.95	0.11	1.3	1.2	1.5	190

It can be seen from Table 1 data that the higher doping and higher electron concentration devices demonstrated lower RTC and STC. The reason is high Fermi energy due to high doping level, providing low-temperature variation of electron concentration. Resistance variation also depends on electron mobility temperature dependence. Low electron mobility caused by high scattering rate on ionized impurities and structural defects. The highest RTC were observed in the samples with a combination of lower electron concentration and high electron mobility. In the polycrystalline sample 346 very low mobility resulted in reduced temperature variations of sensitivity and resistance.

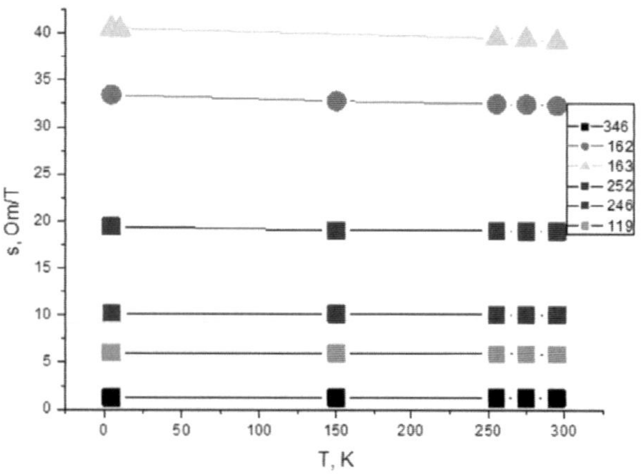

Figure 2. Temperature dependence of sensitivity for the different samples.

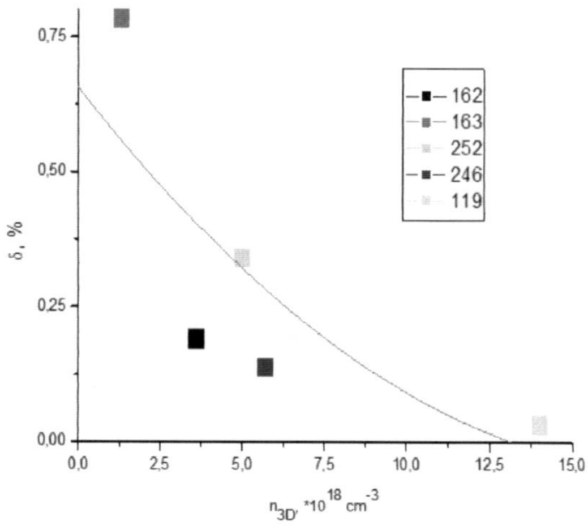

Figure 3. Correlation between nonlinearity and Si doping density.

It is worth notice that at low temperature (5÷6 K) the Hall signal and magnetoresistance hadn't demonstrated any quantum oscillations with increasing magnetic field, that usually observed in quantum well heterostructures with high electron mobility two-dimensional electron gas [7]. We account it to the relatively low mobility values due to direct InAs film doping and three-dimensional character of electron system in conductive thin films.

4. Conclusion

We have demonstrated the successful formation of highly doped InAs thin films (~100 nm) with the different intermediate layer arrangement and appropriate electron mobility values. Hall sensors have been fabricated on these structures and their performance parameters have been measured in wide temperature range. Obtained sensitivity varied from 1 to 40 Ω/T, while the best linearity and lower temperature coefficient have been found in the higher doped samples with lower electron mobility. We attribute this to the electron system degeneracy and decreased phonon contribution to electron mobility and resistance.

Acknowledgements
This work was supported by the Competitiveness Program of NRNU MEPhI.

References

[1] Bolshakova I, Vasilevskii I, Viererbl L, Duran I, Kovalyova N, Kovarik K, Kost Y, Makido O, Sentkerestiova J, Shtabalyuk A and Shurygin F 2013 *IEEE T. Magn.* **49** 50-53
[2] Bolshakova I, Belyaev S, Bulavin M, Brudnyi V, Chekanov V, Coccorese V, Duran I, Gerasimov S, Holyaka R, Kargin N, Konopleva R, Kost Y, Kuech T, Kulikov S, Makido O, Moreau P, Murari A, Quercia A, Shurygin F, Timoshyn S and Vinichenko A 2015 *Nucl. Fusion* **55**(8) 083006
[3] Okamoto A and Shibasaki I 2003 *J. Cryst. Growth* **251** 560-564
[4] Geka H, Okamoto A and Shibasaki I 2005 *J. Cryst. Growth* **278** 614-618
[5] Wang P D, Holmes S N, Le T, Stradling R A, Ferguson I T and de Oliveira A G 1992 *Semicond. Sci. Tech.* **7** 767-786
[6] Nee T E, Lin R M, Hsieh L Z and Chang L B 2002 *J. Vac. Sci. Technol. A* **20** 1128-1131
[7] Suga K, Kindo K, Ishida S, Okamoto A and Shibasaki I 2004 *Physica B* **346-347** 470-475

Thin coatings and films hardness evaluation

V M Matyunin, A Yu Marchenkov, A N Demidov and M A Karimbekov

Moscow Power Engineering Institute, Krasnokazarmennaya st. 14, 111250 Moscow, Russian Federation

Corresponding author e-mail address: mpei_tm@mail.ru

Abstract. The existing thin coatings and films hardness evaluation methods based on indentation with pyramidal indenter on various scale levels are expounded. The impact of scale factor on hardness values is performed. The experimental verification of several existing hardness evaluation methods regarding the substrate hardness value and the "coating – substrate" composite hardness value is made.

1. Introduction

Surface coatings and films are widely used in micro- and nanoelectronics. Usually, the indentation with pyramidal indenter is used for evaluation of coatings' hardness, because it is known as one of the most simple and high efficient methods. Frequently, the indentation is the only possible method of thin coatings mechanical properties control, because thin films and coatings are not suitable for preparing the proper samples for other types of mechanical tests. However, the hardness determination of thin items, including the thin coatings, requires the observance of the condition, concerning the relation φ between item thickness H and indent depth h ($\varphi = H/h$). According to the international standard documents, regulating the hardness evaluation procedure, at least $H/h = 10$ should be provided for correct hardness estimation. The violation of this condition inevitably leads to the substrate properties influence on the obtained hardness value. Moreover, for example, for 1 μm-thickness coating the indent depth h should not exceed 0,1 μm or 100 nm which corresponds to nanoscale level (in accordance with ISO 14577-1:2002). However, it is known that the hardness values sufficiently increase with the indentation load (and along with that, the indent depth) reduction due to scale factor impact [1,2]. Therefore, hardness values, determined on the nanoscale level, may be greatly higher, than even on micro scale level and even more on macro scale level.

If it is not possible to provide the required value of φ, (that is φ sufficiently less than 10), the substrate will exert the influence on coating hardness test result. In this case, it is necessary to evaluate the hardness of substrate thereafter the hardness of "coating - substrate" composite system and after which calculate the coating hardness using the obtained results. The equations for calculation of coating hardness for such cases were proposed in several papers, for example [3,4].

2. Experimental section

This paper is focused on the research of scale factor impact on thin coatings hardness, evaluated with pyramidal indenter on macro and micro scale levels with the observance of $\varphi \geq 10$ conditions. Hardness tests were made at TiN-based 23…25 μm thickness coatings with carbon steel (0.45 %C) substrate. Hardness was evaluated using the Instron Tukon 2500 and Buehler MicroMet 5124 Vickers hardness testers with the various indentation loads F (from 0,049 N (5 gf) up to 2,94 N (300 gf)).

The results of tests are performed in Table 1. Indents depths h, diagonals d, volumes V_i, strained volumes beneath the indents V_{eps} as well as hardness values for different indentation loads F are submitted.

Table 1. Indents depths h, diagonals d, volumes V_i, strained volumes beneath indents V_{eps} as well as hardness values of the TiN-based 25 μm coating for indentation with different loads F.

No.	F		d, μm	h, μm	φ	V_i, μm³	V_{eps}, μm³	HV		Indentation scale range
	gf	N						N/mm²	kgf/mm²	
1	5	0.049	1.8	0.26	96	$1.43\cdot10^{-1}$	$1.37\cdot10^{1}$	28 057	2 860	micro
2	10	0.098	2.6	0.37	68	$4.14\cdot10^{-1}$	$3.97\cdot10^{1}$	25 722	2 622	micro
3	15	0.147	3.4	0.49	51	$9.61\cdot10^{-1}$	$9.23\cdot10^{1}$	24 064	2 453	micro
4	20	0.196	4.0	0.57	44	$1.51\cdot10^{0}$	$1.45\cdot10^{2}$	22 740	2 318	micro
5	25	0.245	4.5	0.65	38	$2.24\cdot10^{0}$	$2.15\cdot10^{2}$	22 563	2 300	micro
6	50	0.490	6.6	0.95	26	$7.00\cdot10^{0}$	$6.72\cdot10^{2}$	20 758	2 116	micro
7	100	0.980	9.8	1.40	18	$2.24\cdot10^{1}$	$2.15\cdot10^{3}$	19 159	1 953	micro
8	200	1.960	14.3	2.05	12	$7.04\cdot10^{1}$	$6.76\cdot10^{3}$	17 952	1 830	micro
9	300	2.940	18.4	2.63	10	$1.49\cdot10^{2}$	$1.43\cdot10^{4}$	16 128	1 644	macro

Elastic-plastic strained volumes under the indent V_{eps} were determined using the equation, introduced by paper authors [5], according to which strained volume under the indent is directly proportional to indent volume V_i:

$$V_{eps} \approx 96 \cdot V_i \,, \tag{1}$$

where V_i clearly depends on the indent depth h:

$$V_i = 8{,}167h^3 . \tag{2}$$

As it follows from Table 1, the indentation was made on macro and micro scale levels, and the hardness values increase sufficiently with the indentation load reduction. More clearly the scale factor impact on hardness tests results can be performed in figure 1, where the relations between Vickers hardness HV and indentation load F (a) and between Vickers hardness HV and indent depth h (b) are shown.

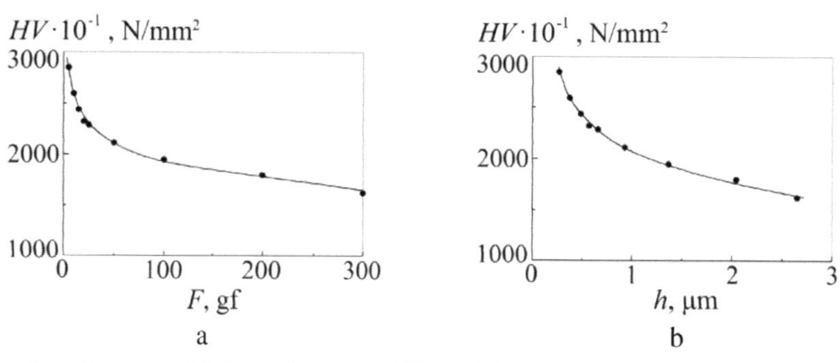

Figure 1. Relations between Vickers hardness HV and indentation load F (a) and between Vickers hardness HV and indent depth h (b) obtained for TiN-based 25 μm coating.

The reasons for scale factor impact on hardness tests results are commonly divided into physical, mechanical and instrumental [6]. One of the main reasons is the multiple reducing the strained volume of material under the indent with the transition from macro to micro scale level. In the experiment, presented above, the strained volume reduced approximately by 3 orders of magnitude (see the table 1).

Among others, the additional experiments for coating hardness evaluation providing the $\varphi < 10$ conditions were made. These tests were made at TiN-based 6 μm coating with carbon steel (0.45 %C) substrate. The results of these tests are represented in Table 2.

Coating hardness values were evaluated with the regard to substrate influence. The experiments were shown that equation

$$H_f = H_s + \frac{H_c - H_s}{2C_2 \frac{H}{h} - C_2^2 \left(\frac{H}{h}\right)^2}, \tag{3}$$

where H_f is the coating hardness; H_s is the substrate hardness; H_c is the "substrate - coating" composite system hardness; $C_2 = 2\sin^2 11°$, proposed by [4], has the close approach of calculated hardness value to the hardness value, evaluated $\varphi \approx 10$. Hardness values, calculated using the equation (3) are performed in Table 2.

Table 2. Indentation tests results of TiN-based 6 μm coating in case of observance and violation of the condition, regulating the φ relation.

No.	*F*		"Substrate - coating" composite system test					HV_s		HV_f calculation using the equation (3)		Error *Δ*, %
	gf	N	*d*, μm	*h*, μm	φ	HV_c N/mm²	HV_c kgf/mm²	N/mm²	kgf/mm²	N/mm²	kgf/mm²	
1	25	0.245	3.9	0.56	10.5	32 373	3 300	3 120	318	32 373	3 300	0
2	50	0.490	5.6	0.80	7.5	28 940	2 950	3 002	306	30 195	3078	6.7
3	100	0.981	10.7	1.53	3.9	15 804	1 611	2 708	276	29 597	3017	8.6
4	150	1.472	16.9	2.41	2.5	9 614	980	2 560	261	27 044	2451	26.0

3. Conclusions

The following conclusions can be made by the results of experiments. The comparison of several coatings hardness values should be made not with the equal indentation loads but with the equal indent depths. It provides the equality of indent volumes, strained volumes beneath the indenter and, as a result, the similar scale factor impact. The substrate impacts on coating hardness value if $\varphi < 10$, and the less the φ, the more the impact. The experimental verification of different equations, suitable for coatings hardness evaluation at $\varphi < 10$ with regard to the substrate hardness and the "substrate - coating" composite system hardness, performed that equation (3) can be successfully used if $\varphi \geq 4$. In the case of $\varphi < 4$ the error Δ starts to increase sharply (see the table 2), what can be explained by increasing the role of substrate and the impact of the scale factor.

Acknowledgements
The work was carried out in Moscow Power Engineering Institute (Moscow, Russia) with financial support from Russian Science Foundation (project code 15-19-00166)

References

[1] Atkinson M 1998 *J. Mater. Sci.* **33** 2937-47
[2] Matyunin V M, Marchenkov A Yu and Demidov A N 2013 *Metals Tech.* **6** 53-54
[3] Pozdeeva T I, Khlyntsev V P, Kozlov V M and Sidelnikov V K 1979 *Electr. Mater. Treat.* **5** 51-53
[4] Jonnson B, Hogmark S 1984 *Thin films.* **114** 257-69
[5] Matyunin V M, Dragunov V K and Marchenkov A Yu 2015 *Piezoelectrics and Nanomaterials: Fundamentals, Developments and Applications* ed I A Parinov (New York: Nova publishers) chapter 6 p 133-68
[6] Fedosov S A, Peshek L 2004 *Determining mechanical properties of materials by microindentation: Modern foreign methods* (Moscow: Moscow State University Press) p 26-34

Carbon dioxide gas sensor based on optical control of color in liquid indicator

K Yu Oblov, A V Ivanova, S A Soloviev, S V Zhdanov, Yu A Voronov and V V Florentsev

National Research Nuclear University MEPhI (Moscow Engineering Physics Institute), Kashirskoe shosse 31, 115409 Moscow, Russian Federation

Corresponding author e-mail address: avivanova@mephi.ru

Abstract. A new optical carbon dioxide sensor based on the change in glow intensity of the Europium-III complex, caused by CO_2 absorption to various pH-indicators (thymol blue, phenol red and cresol red) of carbon dioxide was developed, and its sensitive properties were studied.

1. Introduction

CO_2 is significant used in industry gas for various applications, which include the production of chemicals, it also acts as an inert agent for food packaging, used for the manufacture of beverages, refrigeration systems, welding systems, fire extinguishers, water purification systems, etc. [1,2].

In the agro-industry, CO_2 is widely used for packing in a modified atmosphere, where its purpose is to inhibit the growth of spoilage bacteria [3,4]. For example, in active packaging technologies, the CO_2-generating system could be considered as a supplementing method for oxygen cleaning [5]. High CO_2 level (10-80 vol.%) is desired for food products such as meat and poultry, in order to inhibit the growth of microorganisms on the surface and to extend shelf life. Maintenance of the CO_2 concentration in the packages with food [6], for example using nested sachets [7,8] should be carefully monitored, since CO_2 permeability is 3-5 times higher than the oxygen one for most plastic films, so CO_2 is absorbed by many products, such as meat and poultry.

Real-time processes monitoring is essential for effective process control. The rapid development of the biological processes use together with the agro-industry has led to an intensive search for new sensors, able to provide real-time information about the status of processes.

Conventional methods for determining the CO_2 include, among others, infrared spectrometry [9] or electrochemical sensors based on liquid or solid electrolytes [10,11]. The most popular sensors used for CO_2 gas sensing in biotechnological applications are based on electrochemical measurements that may be obtained from a variety of materials. Optical CO_2 sensors with low detection limits are used in various industrial processes, such as environmental monitoring, pollution control, biotechnology and food industries [12]. They are based on CO_2 Lewis acid properties; namely, by reaction with an aqueous Lewis donor, the pH change occurs and could be monitored using absorbance or the fluorescence of the pH values. Generally, acid-base indicators are immobilized in so-called membranes of the solids sensors [13], made of gas-permeable polymers such as ethyl cellulose [14], the sol-gel [15], silicone [16], composite materials [17], or they are directly attached to the tip at the end of the optical fiber or in its core [18]. In addition to the indicator, the membrane often comprises a quaternary ammonium hydroxide [19] and/or ionic liquids at room temperature [20] for an ion pair with the basic form of the pH indicator, as well as to supply water required to form a hydrated ion pair for the absorption of CO_2 from the atmosphere by forming lipophilic carbonate buffer.

Gaseous CO_2 is measured by infrared absorptiometry as well as electrochemically, using the Severinghaus electrode [21]. Since the infrared absorption band at the wave carbon dioxide is in the range of 4200-4400 nm, infrared absorptiometry is used in optical sensors. On the other hand, the CO_2 sensor with the Severinghauz electrode consists of a glass electrode that detects changes in the pH-level in surrounding bicarbonate ionic solution after it enters and subsequent hydration of CO_2 through the thin film. However, a very sensitive IR absorptiometry-based sensor is influenced by water vapor, and by itself is a very expensive system. To overcome these issues, the study and improvement of the CO_2 sensor was made. In recent years the CO_2-sensors operating by colorimetric and fluorimetric pH-sensitive indicators changes or sensitive to CO_2 colored indicators were developed.

Optical CO_2 sensors come in two types. One of them is based on the change in color of pH-indicator: timolsulfonftalein (thymol blue), fenolsulfonftalein (phenol red), cresol red, etc. Another sensor based on the change of color CO_2-sensitive luminescent colored indicator, for example, 1-hydroxypyrene trisulfonate and ruthenium complexes. This method is simple and convenient. In the case of the fluorescent phenol-based CO_2 sensor the fluorescence intensity change with a shift of the fluorescent flow color from fluorescent (phenol anion) to not fluorescent (a neutral form of phenol) is used. However, the number of substances for which fluorescence level changes under the influence of CO_2, severely limited. To solve this problem, the CO_2 sensor that runs on a combination of color change technology and fluorescent flow was developed. The own glow should not be changed under the influence of CO_2. The principle of this sensor is based on the overlap of the wavelengths of the own inner glow range CO_2 absorption range when using the pH-indicator.

The study of the CO_2 sensor with two dyes, one of which is basic, was published. Ruthenium fluorescent complex is used as a basic indicator.

A new optical carbon dioxide sensor was developed, and its sensitive properties were studied. The sensor is based on the change in glow intensity of the Europium-III complex, caused by CO_2 absorption to various pH-indicators (thymol blue, phenol red and cresol red) of carbon dioxide. For all CO_2 sensors operating on the pH-indicators the glow intensity at 612 nm was registered and increased with an increase in carbon dioxide concentration. The best results were obtained using the thymol blue indicator. In its case, the sensitive film response time to the carbon dioxide was 4 seconds, at the transition from 100% nitrogen atmosphere to 100% CO_2. The recovery time of the sensitive film in the reverse transition to a 100% nitrogen atmosphere was 36 seconds. The response time when using the indicator phenol red and cresol red was 4.4 seconds and 8.8 seconds respectively. The recovery times were 39.2 seconds and 56.6 seconds respectively. Currently, the research of the sensor behaviour in the air is conducted to improve the composition of the sensitive film [22].

2. Experiment

2.1. Reagents

As a chemical indicator responsive to the concentration of carbon dioxide, bromothymol blue indicator was chosen. The indicator allows determining a change in pH values in the range of 6.0 to 7.6 while changing its color from yellow to blue through green colors (Fig. 1) [23].

Figure 1. pH-scale of the bromothymol blue indicator.

2.2. Design of the reaction chamber

The reaction chamber of the installation is a camera for the reaction between the liquid indicator and the gas tested for the presence of carbon dioxide. Upon contact with carbon dioxide, the liquid in the chamber changes its optical properties, namely the colors. This change must be captured electronically by optical sensors of the measuring system, and after the subsequent transformations are presented to the user in the digital form.

The reaction chamber is a container, which is made of an optically transparent in the visible range of light frequencies organic polymer (plexiglass). It consists of two parts and has inside a cuvette for placing the indicator fluid and channels for supplying the tested gas (Figure 2). The main problem of the chamber is the evaporation through the supply channels of the indicator fluid. In this regard, necessary to increase the hydrodynamic resistance of the channel. The balance between the free access of the tested gas into the chamber and minimal evaporation of the test liquid should be maintained. The dynamics of this indicator depends on the characteristics of the environment: atmospheric pressure, air temperature and the temperature of the cuvette. Necessary to consider that while increasing the temperature of the cuvette, the temperature of the indicator fluid increases as well, which in turn reduces the coefficients of its viscosity and increase volatility coefficient. This leads to an increase in fluid velocity in the channels and further to its evaporation, and to the throttled gas in the channels. The hydrophobic membrane structures were installed to reduce the flow rate (Figure 3). Their aim is to pass gas into the chamber and not let go of the camera the indicator fluid. At the ends of the camera the optocoupler (LED and photodiode) is set. The LED is a light source of constant intensity and a narrow frequency range [24]. The photodiode allows investigating changes in the intensity of light passing through the liquid indicator when painting indicator to a different color under the influence of carbon dioxide.

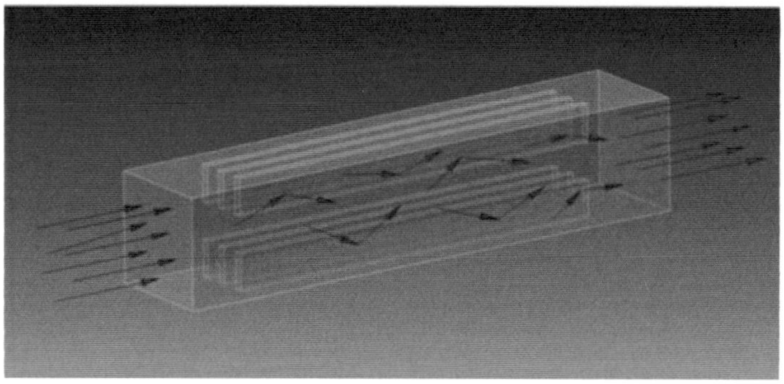

Figure 2. 3D-model of the reaction chamber.

Figure 3. The composite camera with hydrophobic membranes for retaining the indicator liquid (the side view).

2.3. Measuring system design

To analyze the changes in the CO_2 concentration and liquid color, the electro-optical method of inspection and measurement is applied. The color of the indicator fluid is directly dependent on the level of carbon dioxide in the atmosphere that comes with this liquid contact. At low CO_2concentrations liquid indicator changes its color, and the light radiation passing therethrough does not change in intensity. With an increase in CO_2 concentration in the atmosphere contacting with the liquid, the fluid color change from light blue to yellow. In consequence of this changes the intensity of light radiation passing through the liquid. It falls on the photodiode, which causes further signal enhancement, which is received in the system of measurement and digitization (Figure 4).

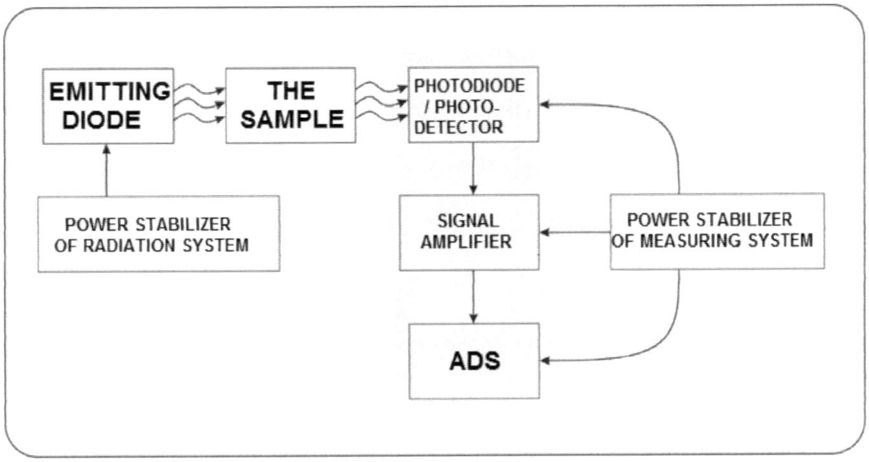

Figure 4. The scheme of the carbon dioxide concentration in the gas measuring apparatus.

As seen in Figure 4, the device scheme comprises seven constituents [23]. The emitting semiconductor diodes are used as a source of monochromatic radiation. They irradiate the sample with light of the desired frequency spectrum with specified intensity, which varies during the passage of the light beam through the test sample. The voltage regulator of the radiation system power supply is applied in the system. Its presence in the system is determined by the need to reduce current fluctuations and, as a result, fluctuations of the LED light intensity. DC stabilizer is built on the integrated high-precision linear regulators. The features of such linear regulators are: high load capacity; small change - subsidence of the output voltage of the regulator changes the load current; low noise; small temporal and thermal drift. The sample is placed between the emitting diode and photodiode. Outgoing from the emitting diode light passes the sample and reaches the photodiode or photodetector. The photodiode detects changes in the intensity of the radiation coming from it.

3. Conclusion

The paper describes the optical carbon dioxide sensor based on the change in light intensity Europium-III complex, caused by CO_2 absorption to various pH-indicators of carbon dioxide. To analyze the changes in the CO2 concentration and liquid color, the electro-optical method of inspection and measurement was applied. The best results were obtained using the thymol blue indicator. In its case, the sensitive film response time to the carbon dioxide was 4 seconds, at the transition from 100% nitrogen atmosphere to 100% CO_2.

Acknowledgments

The research was performed within a frame of Russian Federation President Grant for support young scientists №14.Y30.15.7910-MK from 16.02.2015

References

[1] Jones D 1998 *Introduction to Fiber Optics* (Naval Education and Training Professional Development and Technology Center)

[2] Geib D 2003 *Multiplexing of Extrinsic Fabry-Perot Optical Fiber Sensors for Strain Measurements* ed. by Thesis M S (Virginia Polytechnic Institute and State University)
[3] Culshaw B and Dakin J 1989 *Optical Fiber Sensors: Systems and Applications* (Boston: Artech House)
[4] Giallorenzi T, Bucaro J, Dandridge A, Sigel G, Cole J, Rashleigh S and Priest R 1982 *IEEE J. Quantum. Elect.* **18** 626-665
[5] Krohn D A 1988 *Fiber Optic Sensors: Fundamental and Applications* (Instrument Society of America, Research Triangle Park, North Carolina)
[6] Udd E 1991 *Fiber Optic Sensors: An Introduction for Engineers and Scientists* (Wiley, New York)
[7] Udd E 1992 Fiber Optic Sensors, *Proceedings of SPIE*, CR-44
[8] Udd E 1996 *Proceedings of IEEE*, **84** 884−894
[9] Tracey P M 1991 *IEEE T. Ind. Appl.* **27** 1
[10] Yu F T S and Shizhuo Y 2002 *Fiber Optic Sensors* (Marcel Decker Inc. New York)
[11] El-Sherif M A 2005 *ABBI* **2** 161-170
[12] Inaudi D and Glisic B 2008 *Proc.13th Int. Symposium on Deformation Measurement and Analysis (Lisbon)* p 1-10
[13] Mendez A 2007 *Proc. 4th Pan-American Conference for Non Destructive Testing (Buenos Aires)* p 1-11
[14] Jenny R 2000 *Fundamentals of Fiber Optics: AnIntroduction for Beginners* (Volpi Manufacturing USA Co New York)
[15] Casas J R and Paulo J S 2003 J. *Bridge Eng.* **8** 362-373
[16] Berthold J W 1995 *J. Lightwave Technol.* **13** 1193-1199
[17] Connelly M C 2005 *Fiber Sensors* (Elsevier Ltd Limerick)
[18] Udd E, Schulz W, Seim J, Corones J and Laylor H M 1998 *Fiber Optic Sensors for Infrastructure Applications* (Oregon Department of Transportation,Washington D.C)
[19] Wang Z 2003 *Intrinsic Fabry-Perot Interferometric FiberSensor Based on Ultra-Short Bragg Gratings forQuasi-Distributed Strain and TemperatureMeasurements* ed. by Thesis M S (Virginia Tech)
[20] Fidanboylu K and Efendioglu H S 2009 *Fiber Optic Sensors and Their Applications* (Karabuk, Turkey)
[21] Severinghaus J W and Bradley A F 1958 *J. Appl. Physiol.* **13** 515-520.
[22] Nakamura N and Amao Y 2003 *Analytical and Bioanalytical Chemistry* **376** 642–646
[23] Sabnis R W 2007 *Handbook of Acid-Base Indicators* (CRCPress)
[24] http://www.vishay.com/docs/81519/bpw21r.pd

IOP Conf. Series: Materials Science and Engineering **151** (2016) 012032 doi:10.1088/1757-899X/151/1/012032

Maximum power point tracking for optimizing energy harvesting process

S Akbari[1], P C Thang[1] and D S Veselov[2]

[1]Moscow Aviation Institute (National Research University), Volokolamskoe shosse 4, A-80, GSP-3, 125993 Moscow, Russian Federation
[2]National Research Nuclear University MEPhI (Moscow Engineering Physics Institute), Kashirskoe shosse 31, 115409 Moscow, Russian Federation

Corresponding author e-mail address: akbarisaba@gmail.com

Abstract. There has been a growing interest in using energy harvesting techniques for powering wireless sensor networks. The reason for utilizing this technology can be explained by the sensors limited amount of operation time which results from the finite capacity of batteries and the need for having a stable power supply in some applications. Energy can be harvested from the sun, wind, vibration, heat, etc. It is reasonable to develop multisource energy harvesting platforms for increasing the amount of harvesting energy and to mitigate the issue concerning the intermittent nature of ambient sources. In the context of solar energy harvesting, it is possible to develop algorithms for finding the optimal operation point of solar panels at which maximum power is generated. These algorithms are known as maximum power point tracking techniques. In this article, we review the concept of maximum power point tracking and provide an overview of the research conducted in this area for wireless sensor networks applications.

1. Introduction

There has been a growing interest in using energy harvesting techniques for powering wireless sensor networks [1-6]. Wireless sensor nodes are usually powered by batteries [7-8]. Therefore, the operation time of wireless sensors is usually limited by the finite capacity of onboard batteries which can be a crucial factor in those applications requiring the use of stable power supplies. It is necessary to note that battery replacement in some networks can become a complicated procedure due to the large number of sensors deployed [9] or the harsh environment where the monitoring network is located [10]. Energy can be harvested from the sun, wind, vibration, heat, etc. Ambient source selection can be carried out based on the availability of energy sources as well as the type of application considered. As ambient energy sources have an intermittent nature, it is reasonable to develop multisource energy harvesting platforms [4,11]. Moreover, energy harvesting process can be optimized. In the context of solar energy harvesting, it is possible to develop algorithms for finding the optimal operation point of solar panels at which maximum power is generated. These algorithms are known as maximum power point tracking techniques [12,13] and for instance, can be used to increase the efficiency of the storage medium charging process [14]. The structure of this paper is as follows. In Section II, we overview three common methods for maximum power point tracking and in Section III the results of some research in this area is presented. In Section IV, a conclusion is made based on our survey.

2. Maximum power point tracking methods

Maximum power point tracking has been widely used in solar energy harvesting systems [12,13,15]. Solar panels generate the maximum amount of power at an optimal point which is a function of solar irradiance and temperature. An MPPT device can be realized by utilizing several techniques. In order for the MPPT circuit to be used for wireless sensor networks applications, it is necessary that the energy consumption of the MPPT circuit be as low as possible.

Fractional open circuit voltage (FOCV) is one of the methods used for MPPT techniques according to which the maximum power point is a constant fraction of the solar panel open circuit voltage [16]. Based on this method, there is a linear relationship between the maximum power point voltage V_{mpp} and the open circuit voltage V_{oc}. The relationship is mathematically expressed as $V_{mpp} = K_1 V_{oc}$. The open circuit voltage is measured periodically. Maximum power point can be also tracked by using the fractional short circuit current (FSCC) method in which case the solar cell maximum power point current I_{mpp} changes linearly with respect to the short circuit current I_{sc} [17] and in terms of equations is described as $I_{mpp} = K_2 I_{sc}$. The constants K_1 and K_2 are used to calculate the values of maximum power point voltage and current from open circuit voltage and short circuit current respectively.

Another method called perturb, and observation (P&O) involves "perturbing" the solar panel operating point periodically by changing its voltage in order to find the maximum power point (MPP).The solar panel power is controlled before and after each perturbation. An increase in the amount of power after a voltage perturbation means that the solar panel operating point has moved closer to the MPP, and the perturbation should be held in the same direction. A decrease in the amount of power after a voltage perturbation means that the solar panel operating point has moved away from the MPP and the perturbation needs to be reversed. The implementation of this method requires a more complex circuit with respect to the fractional open circuit approach [18,19].

3. Related work

The authors in [20] developed an MPPT algorithm for wireless sensor networks applications which searches for the optimal point by perturbing the solar panel operating point. If an optimal point is reached the control algorithm evaluates the power gradient in the neighborhood of the working point as the changes in weather conditions may alter the maximum power point. The platform performance was compared with the one which used a diode. The maximum power point tracking algorithm generated more power with respect to the system based on the diode. The MPPT system delivered power as low as 1 mW while the other system could not extract power at levels below 10 mW.

The research presented in [21-23] apply the FOCV method for tracking the maximum power point. According to reference [21], the authors use a pilot cell for measuring the open circuit voltage which allows for performing continuous maximum power point tracking.

The Everlast platform [14] is another example which used the FOCV method for tracking the maximum power point of a solar panel. One of the reasons for choosing this method reason was based on measuring the values of K_1 and K_2. K_1 was chosen due to its narrower range of change (which was between 0.8 and 0.68) in comparison with the short circuit current method (varying between 0.25 and 0.68). The authors developed a sample and hold circuit for realizing the MPPT method. The circuit is illustrated in Figure 1. The timer 555 was used to signal the circuit to sample the solar panel voltage and turns the pulse frequency modulation regulator on and off. The voltage from the solar panel is divided since V_{mpp} is a fraction of V_{oc}.

The research demonstrated in [24] analyzed the design issues concerning indoor light energy harvesting for powering wireless sensors nodes. The authors simulated the performance of a solar panel under a 500 lux illuminance. The simulated operation voltage of the solar panel and its power without using an MPPT circuit were 3.14 V and 129 μW. The solar panel yielded its maximum power point at 151.59 μW. An attempt was made to apply a microcontroller based MPPT circuit. According to the results of their analysis, the mentioned MPPT method was not considered to be effective since the power gain of the MPPT circuit was 22.6 μW which was less than the power consumption of the

MPPT microcontroller having a value of 300 μW. As the authors have mentioned, even the use of four solar cells cannot compensate for the microcontroller power consumption. The paper also addressed the use of the FOCV method. The authors considered a circuit consisting of an ultra-low power comparator, MOSFET switch and a low drop regulator implementing the MPPT method which according to their calculations could result in a 30 μW power consumption.

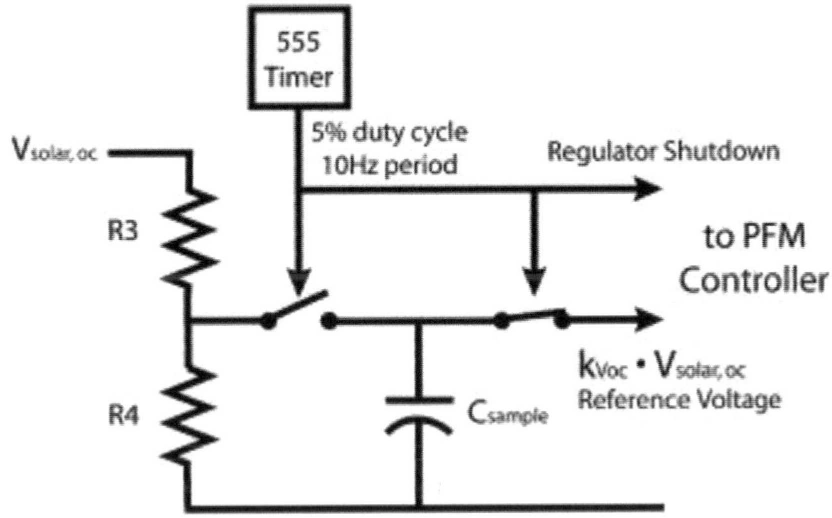

Figure 1. The sample and hold circuit developed in [14]

The measured values for MPP voltage and power for an illuminance of 500 lux were 2.59 V and 139.2 μW which indicate a power gain of 10.2 μW. As the authors have stated, if four solar cells have been used the power gain of the MPPT controller is equal to 40.8 μW and by subtracting the mentioned value from the power consumption of the MPPT circuit, the power gain applied to the load is almost 10.8 μW. The authors proposed to use a directly coupled system with a diode to prevent the reverse current flow to the solar panel. Based on their method, the average V_{mpp} for an irradiance range 100-1000 lux was calculated and, the values of capacitance and the equivalent series resistance of the capacitor were determined such that V_{mpp} could be achieved.

Reference [25] presented an MPPT platform designed to operate in indoor environments. The platform is based on utilizing the FOCV principle for tracking the MPPT. A sample and hold circuit is developed in order to realize the MPPT method. The authors estimated the power loss which can occur during the sampling process. Sampling efficiency was evaluated for a time interval 0 to 7200 seconds. As a result of the experiment, the maximum efficiency can be achieved at the time interval 60 to 300 seconds. A period of 60 seconds was chosen as the duration of the sampling process which according to the authors of the paper is considered appropriate taking into account the rapid changes in the light level. The method developed in [18] finds the MPP by nulling the difference of the solar panel average power for two consecutive subintervals of the charging time.

4. Conclusion

We have overviewed the FOCV, FSCC and P&O methods which are considered as three common MPPT techniques. Also, a survey concerning the application of MPPT for wireless sensor networks was presented. Regardless of the type of the MPPT method, it is necessary that the power consumption of the MPPT circuits be as low as possible, an issue which has been addressed in the current research trend to broaden the perspectives of using these circuits for wireless sensor networks applications.

References

[1] Winston K G Seah, Zhi Ang Eu and Hwee-Pink Tan 2009 *Proc. 1st Int. Conf. on Wireless Communication, Vehicular Technology, Information Theory and Aerospace and Electronic Systems Technology (Aalborg)*(IEEE) p 1
[2] Baranov A, Spirjakin D, Akbari S, Somov A and Passerone R 2016 *Sens. Actuators, A* **238** 112
[3] Zahid Kausar A S M, Wasif Reza A, Saleh M U and Ramiah H 2014 *Renew. Sust. Energ. Rev.* **38** 973
[4] Akbari S *Proc. Fed. Conf. on Computer Science and Information Systems (Warsaw)* pp 987
[5] Somov A, Suchkov A, Karelin A, Mironov S, Baranov A and Karpova E 2015 *IEEE T. Ind. Inform.* **11** 1660
[6] Yang S H 2013 *Wireless Sensor Networks: Principles, Design and Applications* (Springer Science & Business Media) p 58
[7] Baranov A, Spirjakin D, Akbari S and Somov A 2015 *Sens. Actuators, A* **233** 279
[8] Somov A, Baranov A, Spirjakin D and Passerone R 2014 *IEEE Sens. J.* **14** 2056
[9] Tentzeris M M, Georgiadis A and Roselli L 2014 *Proc. IEEE* **102** 1644
[10] Basset P, Blokhina E and Galayko D 2016 *Electrostatic Kinetic Energy Harvesting* (Wiley-ISTE)
[11] Vankecke C, Assouere L, Wang A, Durand-Estebe P, Caignet F, Dilhac J M and Bafleur M 2015 *IEEE T. Power Electr.* **30** 3215
[12] Frezzetti A, Manfredi S and Pagano M 2013 *Proc. Int. Conf. on Clean Electrical Power (Alghero)*(IEEE) p 503
[13] Lopez-Lapena O, Teresa Penella M and Gasulla M 2011 IEEE *T. Ind. Electron.* **59** 1588
[14] Simjee F I, Chou P H 2008 *IEEE T. Power Electron.* **23** 1526
[15] Kim H, Min Y J, Jeong C H, Kim K Y, Kim C and Kim S W 2013 *IEEE T. Circuits-II* **60** 331
[16] Anwar S, Efstathiadis H and Qazi S 2012 *Handbook of Research on Solar Energy Systems and Technologies*, ed Anwar S (IGI Global) p 99
[17] Noguchi T, Togashi S and Nakamoto R 2002 *IEEE T. Ind. Electron.* **49** 217
[18] Lopez-Lapena O, Teresa Penella M and Gasulla M 2010 *IEEE T. Ind. Electron.* **57** 3129
[19] Li S, Wang W, Wang N, O'Mathuna C and Roy S 2015 *Micro Energy Harvesting* ed Briand D, Yeatman E and Roundy S (John Wiley & Sons) p 377
[20] Alippi C and Galperti C 2008 *IEEE T. Circuits-I* **55** 1742
[21] Brunelli D, Moser C, Thiele L and Benini L 2009 *IEEE T. Circuits-I* **56** 2519
[22] Yu H and Yue Q 2012 *Proc. Int. Conf. on Future Energy, Environment, and Materials (Hong Kong)*(Elsevier) p 1027
[23] Varley J, Martino M, Poshtkouhi S and Trescases O 2012 *Proc. 38th Annual Conf. on IEEE Industrial Electronics Society (Montreal, QC)*(IEEE) p 4806
[24] Wang W S, O'Donnell T, Ribetto L, O'Flynn B, Hayes M and O'Mathuna C 2009 *Proc.1st Int. Conf. on Wireless Communication, Vehicular Technology, Information Theory and Aerospace and Electronic Systems Technology (Aalborg)*(IEEE) p 36
[25] Weddell A S, Merrett G V and Al-Hashimi B M 2011, *IEEE T. Circuits-I* **59** 1196

IOP Conf. Series: Materials Science and Engineering **151** (2016) 012033 doi:10.1088/1757-899X/151/1/012033

Numerical modeling of perovskite solar cells with a planar structure

S P Malyukov[1], A V Sayenko[1] and A V Ivanova[2]

[1]Southern Federal University, Institute of Nanotechnologies, electronics and Equipment Engineering, Department of Electronic Apparatuses Design, Nekrasovskii per. 44, 347900 Taganrog, Russian Federation
[2]National Research Nuclear University MEPhI (Moscow Engineering Physics Institute), Kashirskoe shosse 31, 115409 Moscow, Russian Federation

Corresponding author e-mail address: spmalyukov@sfedu.ru

Abstract. The paper is devoted to the research and development of high-efficiency solar cells with a planar perovskite n-i-p structure. A numerical model of this solar cell in the drift-diffusion approximation based on Poisson equation and continuity equations provided to determine their photoelectric characteristics and design optimization. The author considers the spectral photogeneration, bulk and surface recombination, transport charge carriers in perovskite and their collection by the electron and hole transport layers. As a result of the simulation, it was obtained efficiency dependence on perovskite absorber material thickness and lifetime (diffusion length) of the charge carriers. It is found that in addition to absorption coefficient optimal perovskite thickness is determined largely by the charge carrier diffusion length, and it has the upper limit in thickness of 500-600 nm.

1. Introduction

Recently, the alternative materials and technologies are actively searching for the production of low-cost and highly efficient solar cells. Research of absorber materials new types has led to the use of perovskite semiconductor in the dye-sensitized solar cell (DSSC) design. Solar cells based on titanium dioxide (TiO_2) / perovskite semiconductor heterojunction showed high power conversion efficiency of 15-20% [1,2]. The organometal perovskite materials with crystal structure ABX_3 similar to the crystal lattice of calcium titanate ($CaTiO_3$) are useful for photovoltaic applications. The main advantages of perovskite such as the methylammonium lead iodide ($CH_3CN_3PbI_3$) are the lack of energy-intensive and complex coating processes on various substrates (crystallization temperature 100-150 °C), as well as their inherent band gap (1.5-2.3 eV) with direct transitions,high and balanced charge carrier mobility (0.1-10 $cm^2 \cdot V^{-1} \cdot s^{-1}$), relatively long diffusion length (0.1-1.5 μm) and a high absorption coefficient (more 5×10^4 cm^{-1}) [3].

Simulation is an important and necessary step in the development and manufacture of new semiconductor devices, such as solar cells. The simulation reduces the costs of the experimental researches and the parameters optimization. The operation principle of perovskite solar cells differs from substantially silicon solar cells and DSSC [1-3], and therefore requires the development of new physical and topological models for determination the photoelectric characteristics, research, and design optimization and increase solar cell efficiency. The thickness of perovskite absorber material is one of the most important parameters affecting the solar cell performance. In addition, the charge carrier diffusion length (perovskite quality) in the n-i-p solar cell structure must be large enough to photogenerated charge carriers effectively collection [4,5].

In this paper, the numerical model based on drift-diffusion system semiconductor equations was proposed. It allows taking into account the solar cell material parameters with photovoltaic behavior. The model can be used for optimization of perovskite absorber material thickness and research of its qualities (charge carrier lifetime and diffusion length) influence on photoelectric characteristics.

2. Development of a numerical model

Research and perovskite absorber material optimization were carried out with the use of the developed physical and topological one-dimensional model of perovskite solar cell with planar n-i-p structure. Figure 1 shows simulated solar cell design, where the absorber material is a relatively widely studied perovskite $CH_3NH_3PbI_3$ combination with an electron (TiO_2) and hole (Spiro-OMeTAD) transport layers [1-3]. As the substrate, we can use glass, sapphire or polyethylene naphthalate (PEN) coated by transparent conductive oxide (TCO). We assumed that the photogenerated charge carriers in perovskite are free electrons and holes, and perovskite semiconductor is intrinsic. Photogenerated electrons and holes diffuse and drift through the perovskite absorber and transport layers n- and p-type. In this case, the electron and hole transport layers are considered as perfect conductors for the major carriers.

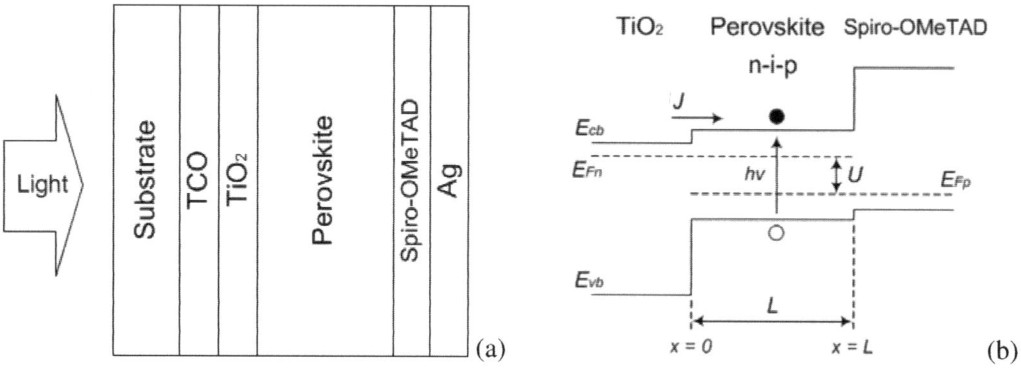

Figure 1. Design (a) and the energy diagram (b) of perovskite solar cell with planar n-i-p structure.

The model is based on drift-diffusion equations semiconductor system with stationary differential equations of continuity for the determination electrons and holes concentrations, and Poisson equation for the potential calculation in perovskite [4,6]:

$$D_n \frac{d^2n(x)}{dx^2} - \mu_n \left(\frac{dn(x)}{dx}\frac{d\varphi(x)}{dx} + n(x)\frac{d^2\varphi(x)}{dx^2}\right) + G(x) - R(x) = 0, \quad (1)$$

$$D_p \frac{d^2p(x)}{dx^2} + \mu_p \left(\frac{dp(x)}{dx}\frac{d\varphi(x)}{dx} + p(x)\frac{d^2\varphi(x)}{dx^2}\right) + G(x) - R(x) = 0, \quad (2)$$

$$\frac{d^2\varphi(x)}{dx^2} = \frac{q(n(x)-p(x))}{\varepsilon\varepsilon_0}, \quad (3)$$

where n and p is the electron and hole concentrations; D_n and D_p are the electron and hole diffusion coefficient; μ_n and μ_p are the electron and hole mobility; x is the coordinate on Figure 1; φ is the electrostatic potential within the absorber layer; q is the elementary charge; ε is the perovskite relative permittivity; ε_0 is the permittivity of vacuum; G is the electron and hole photogeneration rate; R is the electron and hole recombination rate.

The charge carrier photogeneration rate were determined in the spectral range of perovskite absorption on the base of Bouguer-Lambert law [5]:

$$G = \eta\alpha(1-t)\Phi e^{-\alpha x}, \quad (4)$$

where η is the photons generation of electron-hole pairs coefficient; α is the perovskite absorption coefficient; t is the reflectance from the front surface; Φ is the photon flux density in the spectral absorption range of perovskite.

Solar spectrum AM1.5 approximation was made by thermal radiation at temperature 5780 K (Planck's law) [5,7]:

$$F = \int_{280}^{800} K\left(\frac{R_S}{r_0}\right)^2 \left(\frac{2\pi c}{\lambda^4}\frac{f}{e^{hc/\lambda kT}-1}\right) d\lambda\,, \tag{5}$$

where λ is wavelength; h is Planck constant; c is the velocity of light in vacuum; f is the shading factor; R_S is the solar radius; r_0 is the average distance between the Sun and the Earth; T is the temperature; K is the Earth's atmospheric mass effect on the solar power density for AM 1.5 spectrum.

We take into account the direct interband recombination (non-radiative recombination) and recombination at impurity centers (bulk and surface recombination), i.e. deep levels located near the middle of the band gap (traps). In perovskite trap recombination (the crystal structure Schottky's and Frenkel's point defects [6]) is the dominant mechanism that limits the solar cell performance. Consequently, recombination rate in the perovskite was determined by using the following analytical models:

- the Shockley-Read-Hall model of carrier trap recombination [6]:

$$R_{SRH} = \frac{np-n_i^2}{\tau_n(n+n_i)+\tau_p(p+n_i)}\,, \tag{6}$$

where τ_n and τ_p are the electrons and holes lifetime; n_i is the intrinsic carrier concentration;

- the model of the direct interband non-radiative recombination [6]:

$$R_{dir} = C\left(np - n_i^2\right), \tag{7}$$

where C is the inter band recombination coefficient.

To determine the generated current density in the drift-diffusion models, it can be traditionally used the following form [4]:

$$J_n(x) = -qD_n\frac{dn(x)}{dx} - q\mu_n n(x)\frac{d\varphi(x)}{dx}\,, \tag{8}$$

$$J_p(x) = qD_p\frac{dp(x)}{dx} - q\mu_p n(x)\frac{d\varphi(x)}{dx}\,, \tag{9}$$

where J_n and J_p are the electron and hole components of the current density.

The output voltage of the perovskite solar cell defined as the potential difference at the ends of the perovskite, i.e. between the quasi-Fermi level for electrons at $x = 0$ and quasi-Fermi level for holes at $x = L$ (Figure 1):

$$U = E_{F_n} - E_{F_p} = E_g + kTln\left(\frac{n(0)}{N_c}\right) + kTln\left(\frac{p(L)}{N_v}\right), \tag{10}$$

where N_c and N_v are the effective density of states in the conduction and valence bands.

The solution of model equations system was carried out numerically (finite difference method) using an iterative method Hummel in short circuit, open circuit and operating conditions of the solar cell.

3. Results

The simulation photovoltaic characteristics of perovskite solar cells were carried out. The electrons and holes diffusion coefficients are 0.017 $cm^2 \cdot s^{-1}$ and 0.011 $cm^2 \cdot s^{-1}$ [8], respectively, and same lifetime in the range of 0.05-1.3 μs, which corresponds to the diffusion length of 0.29-1.48 μm (characteristic for organometal perovskites). The perovskite absorber should be thick enough to absorb a large quantity of photons, but too thick a perovskite layer creates more recombination centers and prevents of photogenerated charge carrier transport due to the decrease of the electric field in the perovskite. Thus, the optimum perovskite thickness should be in a balance of its absorption length and diffusion length. As a result of the simulation, the dependence of power conversion efficiency from the absorber perovskite thickness at a different charge carrier lifetime was obtained. Figure 2 (a) it is shown that for solar cells with a lifetime of 0.1 μs the optimum thickness is about 325 nm. If the lifetime increased to 0.3 μs, the optimum thickness is around 500 nm. A further lifetime increase causes a slight increase in the optimal thickness. In the case of a large diffusion length (lifetime is more than 0.3 μs) we can see a slight decrease of efficiency in a thick absorbent layer because it reduces the electric field in the perovskite. When the lifetime is less than 0.3 μs increase in the perovskite thickness leads to an increase of recombination losses and significant efficiency reduction.

Figure 2 (b) shows the optimum perovskite thickness of the lifetime and diffusion length. The optimal thickness of the perovskite is largely determined by the charge carrier lifetime and an upper limit of 500-600 nm with bulk and surface recombination.

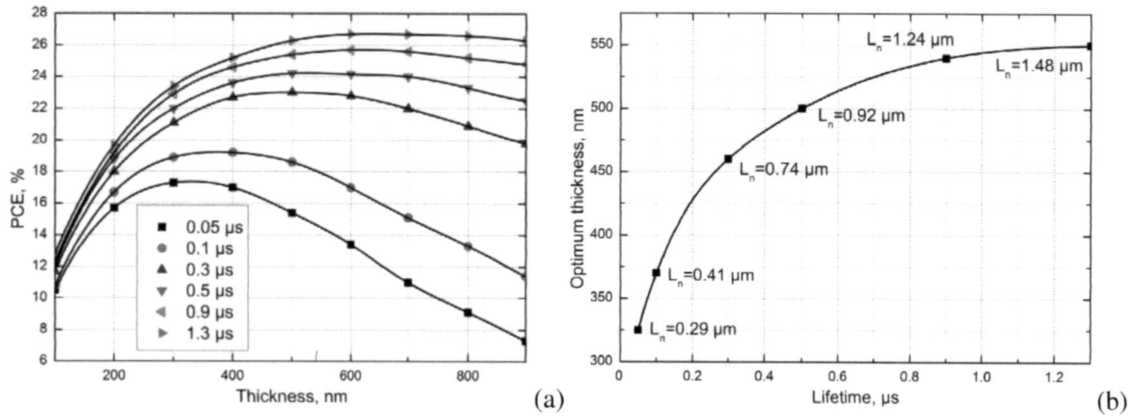

Figure 2. The efficiency dependence on thickness and lifetime (a).The optimum thickness dependence on lifetime and diffusion length (b).

Thus, in this paper, a numerical model of perovskite solar cell in drift-diffusion approximation was developed. Simulation results show that the absorber perovskite thickness and its quality (lifetime, diffusion length) significantly influence on the solar cell efficiency. It can be obtained efficiency 25-27 % with optimized design of perovskite solar cell (large diffusion length of 1-1.5 μm, the optimal thickness of 500-600 nm). The efficiency corresponds to the best silicon solar cells. The developed model allows us to understand of perovskite solar cells operation mechanism, and it contributes to the further improvement of their efficiency.

Acknowledgments
The study was financially supported by the Russian Foundation for Basic Research (project №16-38-00204mol_a).

References

[1] Wang P, Guo Y, Yuan S, Yan C, Lin J, Liu Z, Lu Y, Bai C, Lu Q, Dai S and Cai C 2016 *Res.Chem.Intermediat.* **42** 625
[2] Jung H S and Park N G 2015 *Small* **11** 10
[3] Gratzel M 2014 *Nature Materials* **13** 838
[4] Sun X, Asadpour R, Nie W, Mohite A D and Alam M A 2015 *IEEE J. Photovoltaics* **5** 1389
[5] Malyukov S P, Kulikova I V, Sayenko A V and Klunnikova Yu V 2014 *J. Phys. Conf.* **541** 12060
[6] Soucase B M, Pradas I G and Adhikari K R 2016 *In book: Perovskite Materials - Synthesis, Characterisation, Properties, and Applications* ed. by Pan L (In Tech) p 445-488
[7] Sha W E I, Ren X, Chen L and Choy C H 2015 *App. Phys. Lett.* **106** 221.
[8] Stranks S D, Eperon G E, Grancini G, Menelaou C, Alcocer M J P, Leijtens T, Herz L M, Petrozza A and Snaith H J 2013 *Science* **342** 341-344.

IOP Conf. Series: Materials Science and Engineering **151** (2016) 012034 doi:10.1088/1757-899X/151/1/012034

Point-of-load switched-capacitor DC-DC converter for distributed power systems

V E Shunkov[1], O N Kus[2], V Y Prokopyev[1], V A Butuzov[1,3], Y I Bocharov[3] and V E Shunkov[4]

[1]Megarad, LLC, Nikolaeva street 11, 630090 Novosibirsk, Russian Federation
[2]OKB Fifth Generation, LLC, Nikolaeva street 11, 630090 Novosibirsk, Russian Federation
[3]National Research Nuclear University MEPhI (Moscow Engineering Physics Institute), Kashirskoe shosse 31, 115409 Moscow, Russian Federation
[4]Scientific Research Institute of System Analysis, Russian Academy of Sciences, Nakhimovsky prospect 36-1, 117218 Moscow, Russian Federation

Corresponding author e-mail address: yibocharov@mephi.ru

Abstract. Integrated circuit (IC) of point-of-load DC-DC converter is presented. It is intended mainly for use in distributed power systems as a stand-along IC or as an on-chip block of the system-on-chip. The converter is based on multiphase switched-capacitor architecture and implemented in commercially available, 180 nm bulk-CMOS process. The paper discusses features of output voltage control in the multiphase system over a wide range of input and output voltages and load currents as well as output noise reducing techniques. The results of the test samples evaluation are presented.

1. Introduction

A rapid growth of electronic equipment, operating with limited power resources such as mobile battery powered devices and on-board equipment of spacecraft requires increasing the efficiency of power supply system. One of the main challenges in the design of such equipment is to convert an output of electrochemical sources to interconnect bus voltages and from them to the variety of supply voltages of the modern integrated circuits (ICs). The use of ICs manufactured in different technologies makes it necessary to have up to four or five or more supply voltages in the same unit. Point-of-load (POL) power supplies solve the challenge of high peak current demands and low noise margins, required by microcontrollers or application specific ICs, by placing individual linear regulators or DC-DC converters close to their point of use.

The requirement for energy efficiency leads to the widespread replacement of linear voltage regulators (LVR), traditionally used for the final DC-DC power conversion, onto the switched-mode power suppliers (SMPS), which allow achieving higher efficiency in a wide range of input and output voltages and load currents [1]. The most promising solution for on-chip SMPS is a switched capacitor DC-DC converter. In recent years, several papers have demonstrated the use of different types of integrated capacitors (MOS, stackable metal-insulator-metal – MIM and trench) to achieve a performance competitive with SMPS, using discrete inductances, at powers up to 1 W [2–4].

This paper presents a project of a fully integrated switched capacitor SMPS designed to convert supply rail voltages 3.3 V and 5 V as well as voltages of different types of lithium cells to a wide range of supply voltages of submicron ICs.

The rest of the paper is organized as follows. In Section 2 architectures of switched-capacitor converters are discussed. In Section 3 we consider circuit solutions, which were used in developed IC. In Section 4 the experimental results are provided. Finally, the paper is concluded in Section 5.

2. Switched-capacitor DC-DC converters

2.1. Architecture

In general, the architecture of switched-capacitor DC-DC converter is quite simple. It is a set of capacitors and switches for their switching [1]. Voltage conversion is performed due to a redistribution of a charge in the capacitor array. In the first phase, capacitors are charged from the input voltage source. In the second phase, they are discharged to the load. A voltage conversion ratio is determined by the topology of the capacitor circuitry in the first and second phases. For a commonly used parallel-series capacitor connection in down converter circuit with n capacitors the conversion ratio ranges from $1/(2n–1)$ to $(2n–2)/(2n–1)$. As a rule, switched-capacitor DC-DC converters use a reconfigurable architecture with a variety of topologies [2]. This ensures high efficiency in a wide range of input and output voltages.

2.2. Sources of losses

Design factors normally limit the available set of conversion ratios by two or three values. This leads to efficiency depending on the input and output voltage levels. The residual voltage is dissipated on the output resistance of the circuit, just like in a linear regulator. When parameters are far from optimal, the loss of efficiency can be tens of percent, and one of the tasks of the developer is the proper selection of a set of the conversion ratios. The output impedance of the switched-capacitor SMPS at low switching frequencies inversely proportional to the product of the effective switched capacitance on the switching frequency, which is about 1 MHz for circuits with external discrete capacitors and up to several tens of megahertz for fully integrated solutions. Higher output impedance limits the competitiveness of switched-capacitor SMPS against inductive SMPS. The third major source of losses is the parasitic capacitors associated with technological features of manufacturing of on-chip capacitors. For MOS and MIM capacitors parasitic capacitance from bottom plate to the substrate can reach several percents of the nominal capacity. This may reduce the efficiency of the converter by 10–15% in some ranges of operating voltages and currents [1]. The design of fully integrated SMPS must also take into account losses due to MOS-switches gate-capacitance recharging because they operate at much higher frequencies than in the inductive SMPS or capacitive SMPS with discrete external capacitors. A Proper account of all major sources of loss is the basis for optimization of the capacitive SMPS to achieve maximum efficiency in the desired range of operating voltages and currents.

2.3. The control techniques

There are two phases of regulation. The first phase is a selection of conversion topology for a course adjustment. In the second phase, the control of the SMPS output resistance is performed for a fine-tuning. Topology selection is implemented by a feedforward control circuit based on the data on input and output voltage levels. Fine adjustment of the output voltage is performed by a feedback circuit, which regulates the output resistance of the circuit and thus the voltage drop across it. Regulation of the output resistance can be carried out in several ways [1]. The techniques based on a frequency conversion control are most frequently used. A number of such methods have been developed, including a continuous-time regulation that uses voltage controlled oscillator in a feedback loop, and a digital control [5].

3. Designed DC-DC converter

A design goal was to develop a capacitive DC-DC converter suitable for stand-alone use as well as for incorporation into the system-on-chip. Its main parameters are as follows: input voltage of 2.8 V to 5.5 V, the output voltage of 1.2 V to 3.3 V, the output current up to 300 mA, design rules – 180 nm.

3.1. Architecture of the converter

Figure 1 shows the architecture of the designed switched-capacitor DC-DC converter.

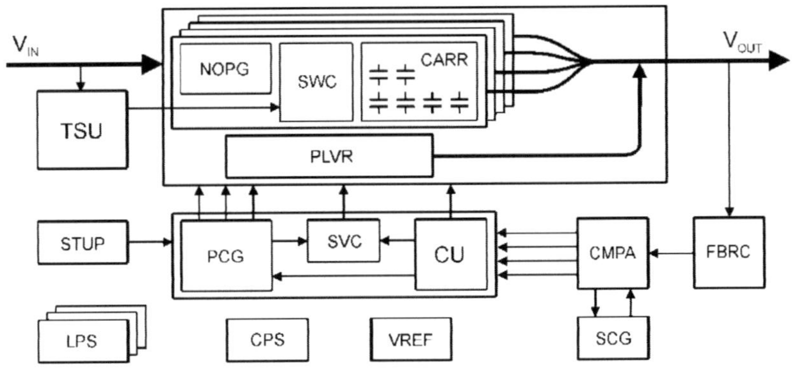

Figure 1. Converter architecture. Basic units indicated: TSU – topology selection unit; NOPG – non-overlapping phase generator; SWC – switches control; CARR – capacitor array; PLVR – power linear voltage regulator; STUP – startup; PCG – primary clock generator; SVC – selector of values of switched capacitors; CU – control unit; CMPA – comparator array; FBRC – feedback resistor circuitry; LPS – logic power supply linear regulators; VREF – voltage reference; CPS – comparators power supply linear regulator; SCG – secondary clock generator.

3.2. Basic conversion cell

Designed converter includes 512 identical basic conversion cells, divided into 8 groups, the corresponding conversion phases. Each group contains four subgroups including 16 basic cells in each of them. This solution is necessary to control the total value of switch capacitors and provide a multiphase operation of the converter, which reduces the output noise.

The basic cell provides voltage conversion ratios of 1/3, 1/2 and 2/3. On-chip linear regulator implements the ratio of 1/1. Set of topologies used allows, on the one hand, to ensure high efficiency in a wide range of input and output voltages, and on the other hand, it allows the use of simple basic cell consisting of two capacitors and nine switches (figure 2).

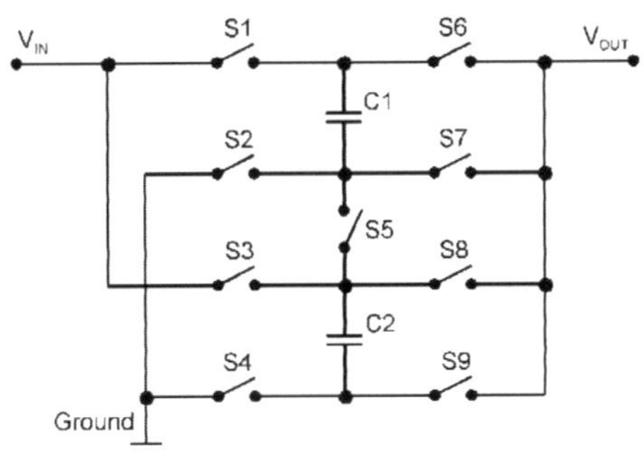

Figure 2. A basic conversion cell.

To achieve the conversion ratio of 2/3, both capacitors are connected in parallel between input and output in a charging phase and sequentially between the output and ground in a discharging phase. Ratios of 1/3 and 1/2 are similarly implemented using a series-parallel connection of capacitors. Increase the number of capacitors and switches in the cell in order to more accurately adjust the conversion ratios is limited to increasing losses in the switches and increasing of the cell output impedance. Figure 3 shows simulated diagrams of the efficiency against the input voltage for the capacitive SMPS, taking into account losses, and for the ideal linear regulator at the output voltage of 1.8 V. You can clearly see the points of SMPS topology changes.

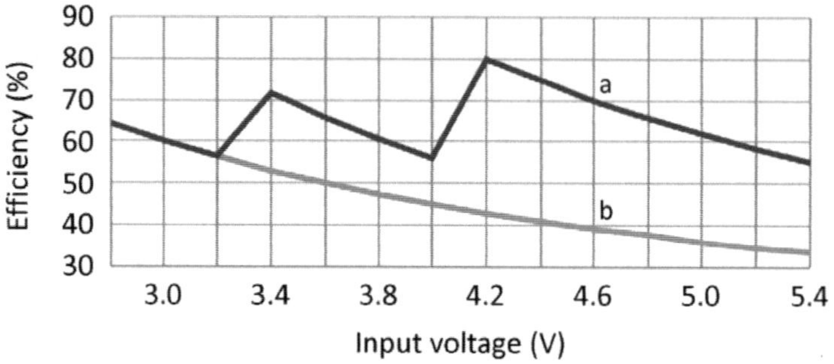

Figure 3. The efficiency of (a) capacitive SMPS and (b) ideal LVR depending on the input voltage.

If there is a small difference between input and output voltages, a built-in linear regulator is the primary power supplier. Otherwise, it is connected in parallel to the main capacitive converter, when the output current reaches a value of more than 60 % of the maximum level.

Figure 4 shows the efficiency of designed capacitive SMPS and ideal LVR, depending on the output current at the input voltage of 3.3 V and the output voltage of 1.8 V. One can see quite a high efficiency at medium and low currents. Reduced SMPS efficiency at high currents is explained by the increasing share of the current of LVR integrated on a chip. The decline in efficiency at the currents of less than 30 mA is caused by the increase of the relative share of losses on the recharging of MOSFET gate capacitances in switches.

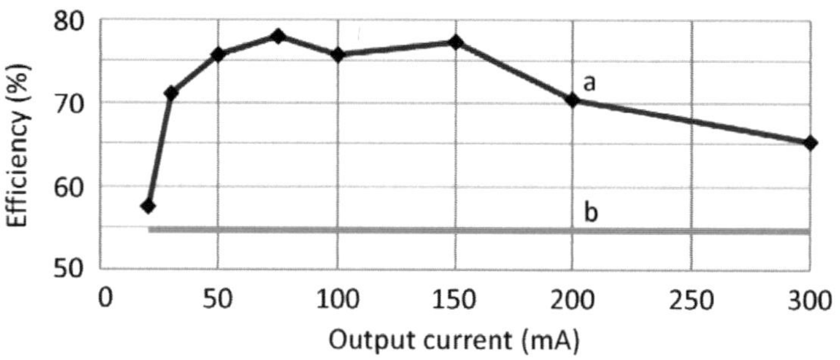

Figure 4. The efficiency of (a) capacitive SMPS and (b) ideal LVR depending on the output current.

3.3. The control unit

The control unit of the converter consists of three parts: topology selection unit, a conversion frequency control circuit and a selector of values of switched capacitors. The topology selection unit uses a feedforward regulation, while the other two – a feedback loop regulation. A topology selection unit contains a resistive circuit, which is supplied with the input signal. It forms dynamically threshold voltages for three comparators. Their output signals are used for topology selection.

The feedback block comprises a controlled resistive circuit that forms threshold voltages for four comparators that determine the output voltage position in one of five ranges [6]. If the output voltage is in a third range, the frequency of controlled oscillator is constant. If it is in first or second ranges, frequency rises. If the voltage is in fourth or fifth ranges, frequency decreases. If the output voltage does not differ by more than 5 % of the nominal value, comparators operate at a lower frequency for power saving. If the voltage is outside of these limits, the clock frequency is increased to ensure a more rapid response to transients.

The generator generates 32 clock frequencies in the range from 3 to 35 MHz. Since a conversion frequency has a lower limit, the operating current also had to be limited to some minimum value. The regulation would not be possible at lower currents. To solve this problem, the converter incorporates the block of control of the total value of the switched capacitance [7]. If the frequency is minimal, when receiving a control signal to the further decrease of the frequency, a half of currently used cells will be disconnected.

Similarly, a control signal to increase the frequency above the maximum value causes a doubling the number of cells used in conversion. The circuit has five levels of switchable capacitance, corresponding to the inclusion of 512, 256, 128, 64 and 32 basic cells. This solution can significantly improve the efficiency at low and medium output currents, as shown in Figure 4.

3.4. Power supplies of the converter

The supply voltage of control logic is 1.8 V. The supply voltage of comparators is 2.5 V. To power these circuits chip contains four linear regulators: one for the comparators, the other two for the digital clock generator and the conversion frequency generator, and a fourth – for the rest of the digital logic circuits.

The reference voltage is generated by a built-in band-gap voltage source. It is powered by a pre-regulator, which features low-noise and produces a voltage of 2.5 V. The core of band-gap reference has a conventional architecture, but instead of used diode-connected vertical p-n-p bipolar transistors, it uses a dynamic threshold voltage MOSFET (DTMOS) transistors [8].

4. Implementation and measuring results

For experimental studies aimed at the test of the effectiveness of proposed solutions, a test chip has been designed and manufactured. It comprises a converter of reduced power and a number of dedicated test structures of the main functional blocks of the converter. The test chip has been manufactured in commercially available, 180 nm bulk-CMOS process with five metal layers and stacked MIM-capacitors having a specific capacity of 3 fF/mm^2. Figure 5 shows the measured output voltage depending on the input voltage for several test samples. It can be seen that the converter has an output voltage error of less than ± 6 % over most of the operating range.

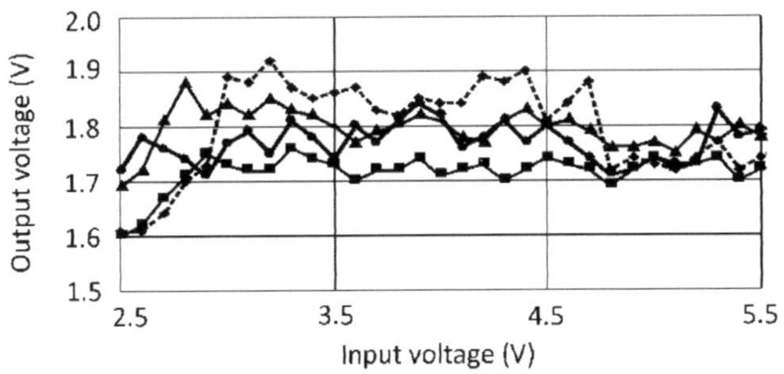

Figure 5. Measured output voltage versus the input voltage for four samples.

Figure 6 shows reference voltage output versus temperature for two test samples.

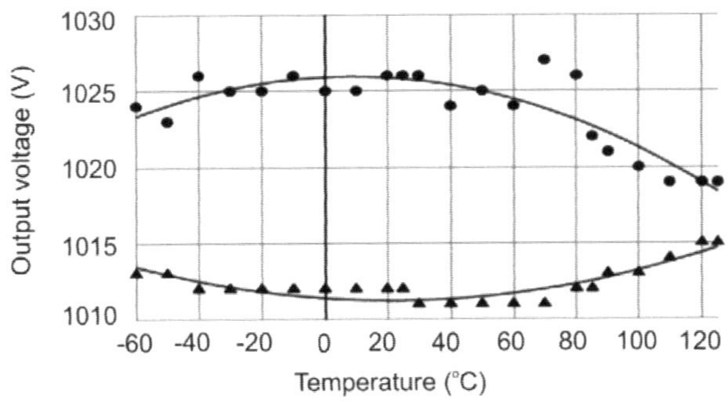

Figure 6. Temperature dependences of reference voltages of two samples.

Measurements showed that the voltage reference temperature drift is less than 0.2 mV/°C in a temperature range from -60 °C to +125 °C, while the spread of the reference voltage about the nominal value of 1.02 V from chip to chip is less than 20 mV.

Figure 7 shows a photo of the test version of designed IC.

Figure 7. Photo of the chip of switched-capacitor DC-DC converter.

5. Conclusion

A step-down (back) switched-capacitor DC-DC regulator for converting rail voltages of 5 V and 3.3 V as well as lithium electrochemical cell voltages to modern IC power supply voltages has been designed and evaluated. The designed IC is fully compatible with commercially available CMOS technology. As compared with the widely available capacitive SMPS it has a smaller noise level because of the multiphase architecture. It is suitable as POL power supply for stand-alone applications, and for the integration into systems on a chip, where its use will reduce power consumption compared to the built-in power supply, based on a linear regulator. The effectiveness of the proposed design solutions has been confirmed by the results of experimental studies of the test chip containing basic functional units of IC prototype.

Acknowledgments

The Space Technology and Telecommunications Cluster of "Skolkovo" Foundation support this work. Also, this work was supported by the Competitiveness Program of NRNU MEPhI.

References

[1] Van Breussegem T and Steyaert M 2013 *CMOS Integrated Capacitive DC–DC Converters* (New York: Springer Science)
[2] Andersen T M, Krismer F, Kolar J W, Toifl T, Menolfi C, Kull L, Morf T, Kossel M, Brandli M, Buchmann P and Francese P A 2014 *Proc. 29th Annual IEEE Applied Power Electronics Conference (Fort Worth, TX)* p 1448–1455
[3] Ng V W and Sanders S R 2013 *IEEE T Power Electr* **28** 4335
[4] Seeman M D, Ng V W, Le H P, John M, Alon E and Sanders S R 2010 *IEEE 12th Workshop on Control and Modeling for Power Electronics (*Boulder, CO) p 1–7
[5] Meyvaert H, Van Breussegem T and Steyaert M 2011 *IEEE Energy Conversion Congress and Exposition (Phoenix, AZ)* p 3234–3241
[6] Jeon H, Kim Y B and Kim K K 2012 *55th IEEE International Midwest Symposium on Circuits and Systems (Boise, ID)* p 1060–1063
[7] Kudva S S and Harjani R 2013 *IEEE J Solid-St Circ* **48** 1910-1920
[8] Gromov V, Annema A J, Kluit R, Visschers J L and Timmer P 2007 *IEEE T Nucl Sci* **54** 2727-2733

About the choice of initial voltages in the synchronous sense amplifier for CMOS memory

Y M Gerasimov[1], **N G Grigoryev and A V Kobylyatskiy**

National Research Nuclear University MEPhI (Moscow Engineering Physics Institute), Kashirskoe shosse 31, 115409 Moscow, Russian Federation

Corresponding author e-mail address: mgng@mail.ru

Abstract. It was shown that the choice of initial voltage sat the nodes of CMOS memory synchronous sense amplifiers has a significant effect on the amplifier data sensitivity. The research of the sense amplifier sensitivity with various initial conditions was performed. According to the results we gave some recommendations about the choice of initial voltages in synchronous sense amplifiers.

1. Introduction

A sense amplifier (SA) is the most important part of memory devices and is used to convert small data signals to digital signals during the read cycle [1-3]. SA are divided into synchronous (triggered) and a synchronous. The first ones are based on clocked latch circuitry, the last ones – on using asynchronous circuitry without synchronization.

In this work, we analyze the differential synchronous SA without data regeneration in memory. They differ from asynchronous in lower power consumption, the absence of signal output glitch and better embed ability in synchronous memory blocks. In literature, it is paid in sufficient attention to the influence of the SA initial voltages on its most important parameter –the read data sensitivity. In this paper, the sensitivity of synchronous SA with various initial voltages is analyzed.

2. Basic circuits

Many different schematics of synchronous SA without data regeneration are based on basic triggered cell (figure 1a) and it dual version (figure 1b).

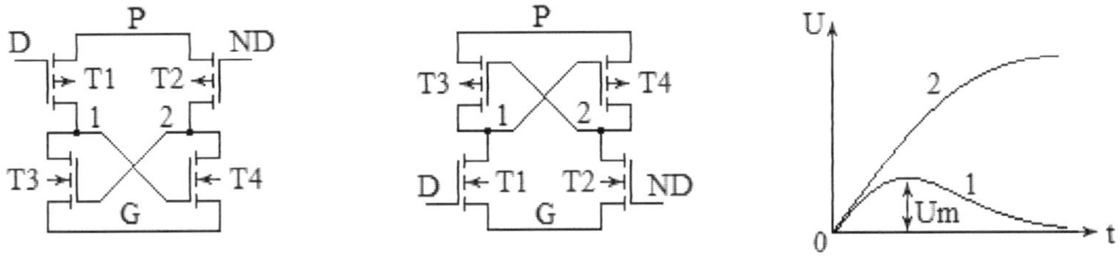

Figure 1. Basic circuits of SA (a, b) and switching (c)

The read data comes to the differential inputs D and ND, according to that data SA starts to switch (Figure 1c). Initially, the potentials of internal nodes 1 and 2 are set the same, and the electrical conditions of symmetrical sides of trigger cell become the same. During the switching, there is a regenerative process, which results in the voltage rise (transient) with amplitude U_m. The amplitude of this transient depends on transistor sizing, loading circuitry and the range of differential input signal. It should be rather low because the loading circuits can be switched and produce undesirable pulses.

The most important parameter of the sense amplifier is the sensitivity U_S. The sensitivity is a minimum input voltage (differential signal at the inputs D and ND), which makes sense amplifier to direct switch.

The performed analysis showed that the sensitivity of SA is determined not only by the transistor parameter variation in the symmetrical sides of SA trigger but also by the difference of loading capacitances in the nodes 1 and 2. So during the design of memory reading circuitry the maximum symmetry of the SA nodes 1 and 2 and the first stages of load, connected to these nodes, should be provided.

The sensitivity is also affected by the initial voltages in the internal nodes 1 and 2 of the sense amplifier (figure 1).

For the subsequent analysis, we chose a basic cell with n-channel input transistors T1-T2 in figure 1b. Practically this variant is the most prevalent in readout data circuitry, as it is consistent with the standard CMOS memory cells [3].

The effect of the initial voltages dependence is caused by the impact of the p-channel transistors specific transconductance k_p variation, the difference of the initial voltages in nodes 1 and 2, the difference of p-channel transistors threshold voltages and the capacitances in nodes 1 and 2 on the SA sensitivity. The experimental analysis showed that the component of sensitivity, related with the specified parameters variation, is:

$$U_S = U_C\left[\frac{1}{3}\left(1+\overline{U}+\overline{U}^2\right)\delta k_p + K\left(1+\overline{U}\right)(\Delta U_{thp}+\Delta U_0) + \left(1-\frac{1}{3}\left(1+\overline{U}+\overline{U}^2\right)\right)\delta C\right], \quad (1)$$

where U_C is the factor of the common-mode voltage level at the SA inputs, $\overline{U}$ is the parameter of the initial voltages in nodes 1 and 2, K is a factor of specific transconductance relation.

As can be seen from (1), the impact of the nodal capacitances variation is minimized when $\overline{U}=1$. This occurs when the initial voltage is equal to the voltage when the nodes 1 and 2 are shorted in a static mode. However, this case is not used because of the significant currents and output noise.

The impact of the initial voltages differences ΔU_0 in nodes 1 and 2 on the sensitivity can be excepted by the complete potential equalization in these nodes until the sense amplifier activation. It can be provided by using the equalizing transistors TS1-TS2 in combination with transistor TS (figure2).

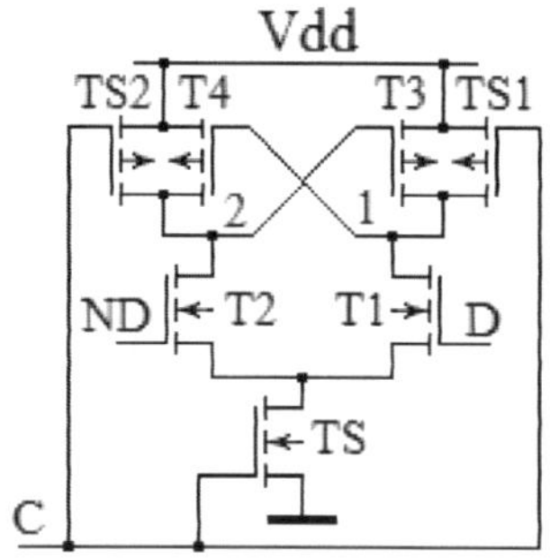

Figure 2. The equalization of the initial voltages by using the extra transistors TS1-TS2 with TS.

The parameter $\overline{U}$ in (1) depends from the initial voltage $U(0)$ and the supply voltage U_{dd}:

$$\overline{U} = \alpha\left[U_{dd} - U_{thp} - U(0)\right], \quad (2)$$

where α is the parameter of specific transconductance relation of p- and n- channel transistors and the common-mode voltage at the inputs.

As follows from (2), the parameter $\overline{U}$ becomes maximum when $U(0) = 0$. According to (1), this means that the specific transconductance variation and the p-channel transistors threshold voltages variation with zero initial voltage cause the strongest influence on the sensitivity.

3. Experimental results

The experiment to determine the influence of initial voltages in the SA internal nodes on the sensitivity was provided using two variants of sense amplifier: with maximum (figure 3a) and minimum (figure 3b) initial conditions.

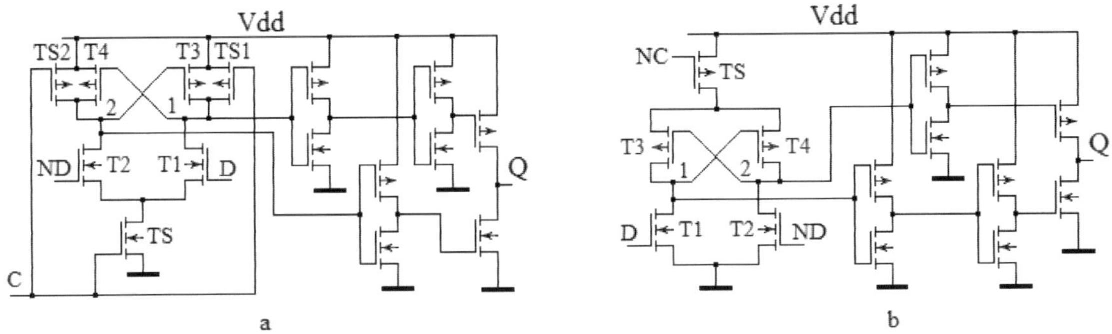

Figure 3. Circuits of SA with maximum (a)and minimum (b) initial voltages

Both variants are based on the basic trigger cell on the transistors T1-T4. In the first one the initial voltages of the nodes 1 and 2 are formed at the level of supply voltage U_{dd} thought the transistors TS, TS1, TS2, while in the second one – at the level of the ground bus by using the input transistors T1-T2 and TS. In the first case, SA is activated by applying the positive voltage to the clock input C, and in the second – by the negative voltage at the input NC.

The results of the experimental sensitivity determination in the sample of 326 amplifiers of the first type and the sample of 74 amplifiers of the second type are presented in figure 4.

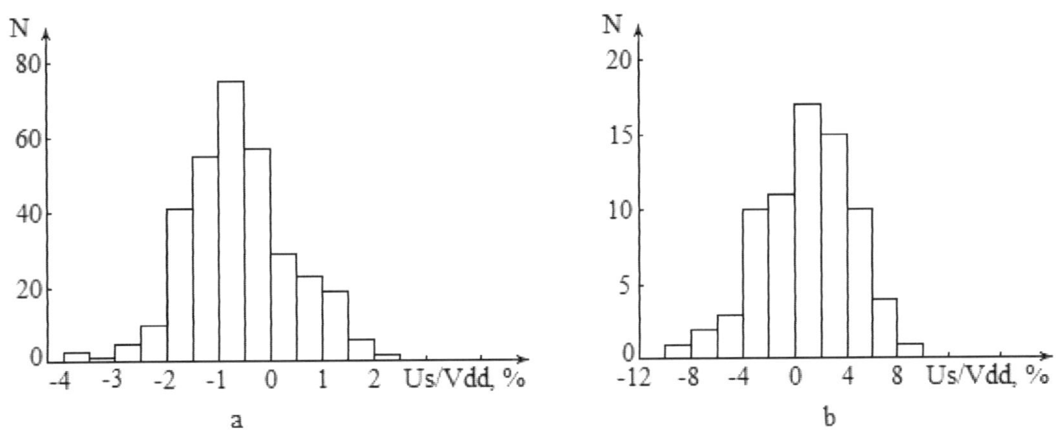

Figure 4. Normalized sensitivity distribution histogram: a – SA type 1, b – SA type 2

As can be seen from figure 4, both histograms have some sensitivity offset related to 0, not exceeding 1% from the supply voltage U_{dd}. This offset results from the topological asymmetry of the trigger cell nodes because it is impossible to provide the absolute symmetry of cross-connected circuit layout.

The standard deviation of the normalized sensitivity related to the average value -0.6% for SA type 1 is 1.08%. For the second type of SA, the standard deviation is 3.4 times worse and reaches the value 3.72% from the average value 0.96%.

Therefore, during the readout circuits design preference should be given to the amplifiers based on the circuit in figure 3a. An important factor influencing the yield ratio is to determine the delay of SA activation relatively to the moment of memory cell choice and data reading, ensuring the correct amplifiers switching for all of the parameters variations. The experimental values of the SA sensitivity allow reasonably choose the input signal level at the moment of the amplifier activation.

4. Conclusion

Thus, the initial voltage in the internal nodes of the synchronous SA significantly impacts on its most important parameter – sensitivity. According to our experimental data, the SA sensitivity may vary in 3-4 times depending on the initial voltages. Therefore, the way to maximize the sensitivity is to choose correctly the initial voltage in its internal nodes during the design of synchronous SA schematics.

References

[1] Gerasimov Y M, Grigoryev N G and Karmazinkiy A N 1982 *Microelectronics* **11** 5
[2] Wicht B, Paul S and Schmitt-Landsiedel D 2001 *IEEE J Solid-St. Circ.* **11 285**
[3] Pavlov A and Sachdev M 2008 *CMOS SRAM Circuit Design and Parametric Test in Nano-Scaled Technologies* (Springer Science and Business Media)

IOP Conf. Series: Materials Science and Engineering **151** (2016) 012036 doi:10.1088/1757-899X/151/1/012036

Low-power low-jitter PLL clock synthesizer for microprocessors with clock range 200-768 MHz

V Baykov and A Garmash

National Research Nuclear University MEPhI (Moscow Engineering Physics Institute), Kashirskoe shosse 31, 115409 Moscow, Russian Federation

Corresponding author e-mail address: aagarmash@mephi.ru

Abstract: In this article the results of the clock synthesizer development are shown. Different variants of voltage repeater for the system of automatic frequency tuning were analyzed. It was shown that for the purpose of energy consumption and jitter reducing the repeater on peripheral transistor can be used. The synthesizerwas created on the technology with design rules 180 nm. Scaling for the technology with design rules 90 nm is also possible.

1. Introduction

Frequency synthesizers (FS) with phase locked loop (PLL) are widely used in modern high-performance digital, switching and wireless systems. Usage of complex functional (IP) blocksof SC PLL is the effective way of clock signals distribution chain construction and reducing energy consumption in it. For this chain, SC can be both sources and distributor (knot of chain) of clock signals.

Using PLL as the sources of synchronous signal demands high stability of frequency. Its jitter should be as low as possible. Phase deviation of a clock pulse from the reference signal can cause a malfunction in system times of date setup and date holding. This situation is most dangerous in communication systems and systems with the import of date from external sources. In both cases very precision synchronization must be provided.

PLL jitter reduction questions are discussed in many publications and some methods of solving these problems have been proposed [1-2]. In this work jitter decreasing was achieved not only by means of known methods but also due to simplification of analog part of PLL. Peripheral n-channel MOF transistor was used as voltage repeater, what allowed 5-times are reducing of phase jitter.

2. Interface

For perspective Micron technologies with design rules, 90 nm was constructed a functional block of IPPLL. The purpose of the design was solving synchronization problems in high-performance microprocessors which clock frequency is in the range of 5-800 MHz. Characteristics of IP-block were optimized for the range of power source voltage 1.0-1,2 V and the range of working temperature 60-125 $^{\circ}$C.

FS-block designed to be integrated into VLSI system on chip and is the source of the clock signal for distribution tree of clock signal propagation. The symbol of IP-block PLL is shownin Figure 1, the appointment of external connections is described in Table 1.

IOP Conf. Series: Materials Science and Engineering **151** (2016) 012036 doi:10.1088/1757-899X/151/1/012036

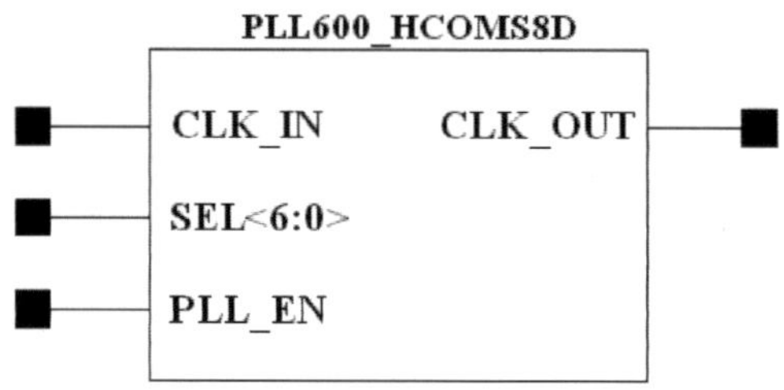

Figure 1. Symbol of IP-block PLL.

Table 1. Appointment of external connections of IP-block.

Pin name	Pin type I – in, O –out	Pin function
CLK_IN	I	Input frequency
PLL_EN	I	PLL on/of
SEL[6:0]	I	PLL programming
CLK_OUT	O	Output frequency

3. PLL realization feature

Common structural scheme of IP block FSPLL [3] is shownin Figure 2. It consists of digital and analog components. Digital part includes frequency divider with programmable coefficients K and L, which value are determined by the control signal. Analog part includes: stabilized block of bias voltage (is not shown); faze-frequency detector (PFD); charge pump; 2-d order capacitive filter of low frequency (LPF); integrated capacity (C); damping circuit C_d, R_d; voltage repeater (1); voltage control frequency generator (VCO).

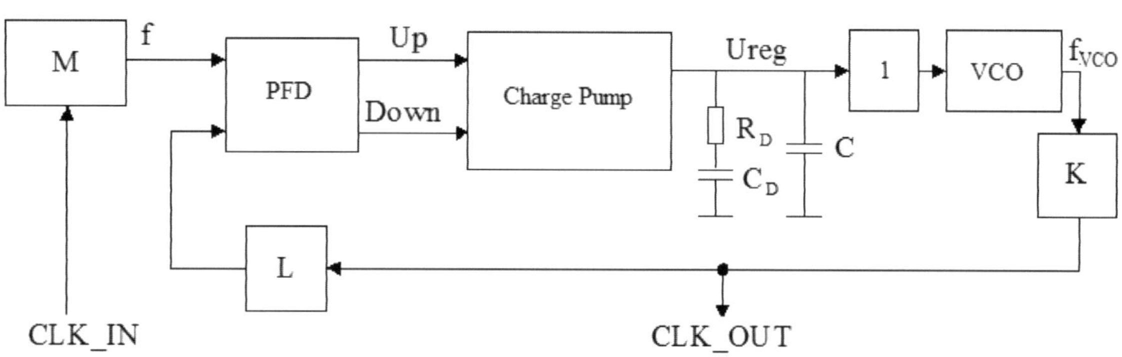

Figure 2. Common structural scheme of IP block SC PLL.

Earlier authors substantiated [4-5] that good decision for microprocessor's PLL is inverter generator and voltage repeater shown in Figure 3. For the technology with design rule, 180 nm the jitter of this scheme is in acceptable limits [6].

However, computer simulation of this circuit solution has shown that for 90-nm technology the value of phase jitter can reach critical limit 100 psec. It is unacceptable for microprocessors with a clock frequency in the range 200-780 MHz and high. As it turned out the main source of jitter in this design is voltage repeater. Transistor's electrical noise in it causes voltage oscillations U_{ref} (Fig.2). Whereas U_{ref} is the control voltage for frequency generator, the frequency of thelast also changes proportionally to the voltage of transistor's noise.

IOP Conf. Series: Materials Science and Engineering **151** (2016) 012036 doi:10.1088/1757-899X/151/1/012036

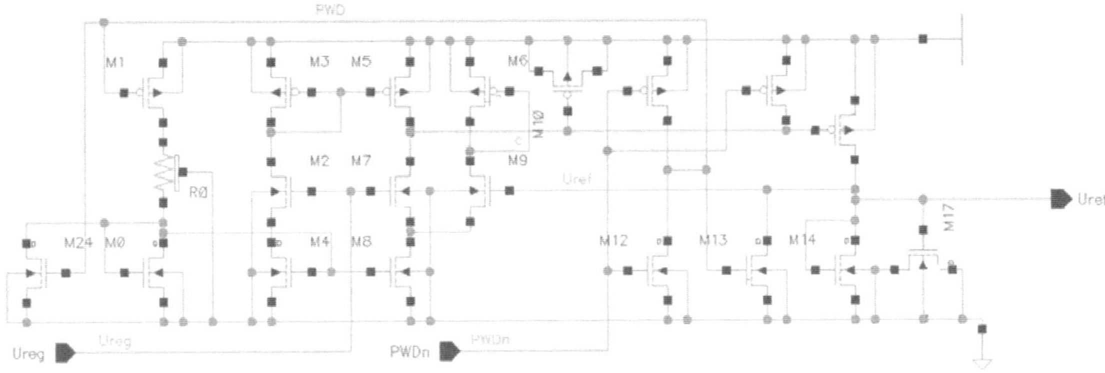

Figure 3. Voltage repeater.

One of the methods of noise reduction is the increase of transistor's size. Correction of transistors area (increase in 1.5 times) allowed reducing of phase jitter in 4.5 times. For normal condition, its value did not exceed anacceptable level of 18.5 psec. However, this correction inevitable reduces the coefficient of noise suppression for power bus.

To solve the problem of jitter reduction without disturbing the noise suppression, theperipheral transistor was used as voltage repeater. For this purpose charge pump block was modernized (Fig 4) and adapted for peripheral voltage supply 2.5 V (PVDD).

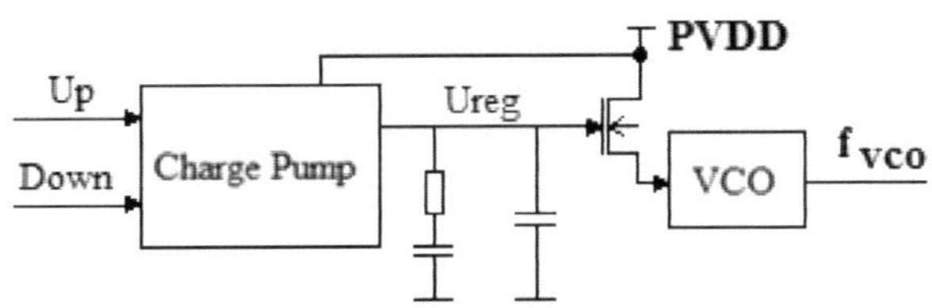

Figure 4. Modernized voltage repeater.

4. Measurement

On Fig. 5 the plots of therelation between noise reduction coefficient and frequency for 3 variants of voltage repeater are shown. The best characteristic in working frequency range has the repeater on theperipheral transistor.

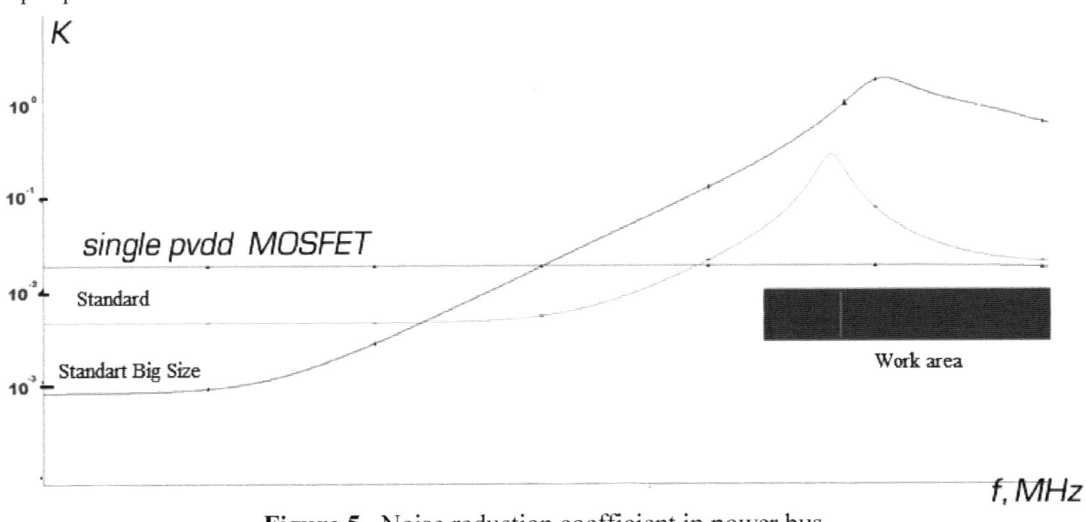

Figure 5. Noise reduction coefficient in power bus.

In Table 2 the results of measurement of phase noise and phase jitter of FSPLL for 3 variants of voltage repeater. The best characteristics also belong to variant with aperipheral transistor.

Table 2. Results of measurement.

Repeater variant	Phase noise (for 1 MHz)	Phase jitter (sigma)
Standard	**- 72 dB**	**82.8 ps**
Standard with increased area	**- 85 dB**	**18.6 ps**
Single peripheral transistor	**- 87,7 dB**	**15.4 ps**

5. Conclusion

Results of computer simulations and tests shown that the usage of voltage repeater on theperipheral transistor in PLL can significantly reduce phase jitter. It is very important for technologies with high transistor's noise level. The disadvantage of this technical decision is the necessity of additional lines of peripheral voltage and ground supply.

Developed FSPLL possesses following characteristics:

- Frequency range 4-762 MHz;
- Reference frequency 4-6 MHz:
- Phase jitter – not more 16 ps;
- Setting time of frequency - not more than 50 μs;
- Mean current consumption for PLL EN = 0 – not more than 65 μA;
- Mean current consumption for PLL EN = 1 – not more than 1 mA.

References

[1] Mansuri M 2003 *Low-Power Low-Jitter On-Chip Clock Generation* (Dissertation presented to obtain the Ph.D degree in Electrical Engineering / University of California, Los Angeles) http://www.ece.tamu.edu/~spalermo/ecen689/pll_thesis_mansuri_ucla_2003.pdf
[2] Moore R 2001 *Agilent Technologies*
[3] Garmash A A and Biykov V D, 2015 *Electronic Engineering, series 3: Microelectronics*, **1** 157
[4] Garmash A A, Biykov V D, Samonov A A and Sevryukov A N, 2005 *Problems of development of advanced microelectronic systems* IPPM RAS p 366-372
[5] Biykov V D, Garmash A A and Dubinskiy A V, 2015 *Problems of Advanced Micro- and Nanoelectronics Systems Davelopment* IPPM RAS, p 12
[6] Hajimiri A, Limotyrakis S and Lee T 1999 *IEEE J. Solid-St. Circ.* **34** 790-804

Pseudomorphic HEMT quantum well AlGaAs/InGaAs/GaAs with AlAs:δ-Si donor layer

A N Vinichenko and I S Vasil'evskii

National Research Nuclear University MEPhI (Moscow Engineering Physics Institute), Kashirskoe shosse 31, 115409 Moscow, Russian Federation

Corresponding author e-mail address: ivasilevskii@mail.ru

Abstract. A comparison between the electron transport and optical properties of pseudomorphic high electron mobility quantum well with conventional AlGaAs donor layer and with AlAs delta-Si doped donor layer is presented. The structure with AlAs donor layer exhibits a mobility rise combined with the electron concentration decrease. We address this effect to the suppression of electron scattering on remote donor impurities and the decreased doping efficiency of Si atoms embedded in pure AlAs.

1. Introduction

Pseudomorphic quantum well heterostructure with high electron mobility is the most widely used material for microwave compound semiconductor electronics. The advantages of PHEMT are high electron velocity and density, high enough breakdown electric field and excellent manufacturability. The best performance of the monolithic microwave integrated circuits devices based on PHEMT technology are shown for X, Ku and Ka bands [1-3]. PHEMT technology benefit is low-noise amplifiers and other related MMICs. Despite layout variety of PHEMTs the common features for this type of structure are AlAs content in the barrier and Schottky layers in the range of 0.15÷0.25 and InAs content in the quantum well (channel) in the range of 0.15÷0.25. Typical electron transport properties of PHEMTs are electron concentration in the range $(1\div3)\cdot10^{12}$ cm^{-2} and electron mobility in the range 5000÷6500 cm^2/(V·s) at room temperature [4,5]. The choice of InAs content in the $In_yGa_{1-y}As$ ternary is ruled by the compromise between the quantum well depth and electron effective mass on the one hand and the elastic lattice deformation limit for the strained $In_yGa_{1-y}As$ layer on the other. The reason of AlAs content in $Al_xGa_{1-x}As$ is different and related to the compromise between the barrier height and electron trapping in $Al_xGa_{1-x}As$:Si layer due to DX-centers for the x > 0,25 [6,7]. DX centers also can affect the transistor performance by the interaction of hot carriers [8]. AlAs also rarely used because of its low stability to atmosphere ambient. It used even as etch stop layer in some cases to achieve certain recess depth [9]. Some study revealed the enhanced electron mobility in AlGaAs/GaAs HEMTs with AlAs spacer and barrier [10]. High electron mobility and low scattering are essential for low noise application. So the attempts to enhance of electron mobility at room temperatures for PHEMTs are useful. AlAs/GaAs heterojunction has significant conduction band discontinuity for Γ-band electrons and AlAs has X-band minima itself. DX-center energy as ~30 meV above X-minima was reported [11] so it corresponds to ~ 12 meV below the Γ-band and thus no trapping effect expected for the heterostructure with AlAs donor layer.

IOP Conf. Series: Materials Science and Engineering **151** (2016) 012037 doi:10.1088/1757-899X/151/1/012037

We have investigated the PHEMT quantum well structure electron properties with nonuniform QW barrier, containing AlAs:δ-Si doped donor layer in comparison with the conventional PHEMT with uniform AlGaAs barrier.

2. Electron band structure

The conduction band profile and the subband structure were calculated for PHEMTs with conventional AlGaAs donor layer (left) and with AlAs delta-Si doped donor layer (right) by solving the Schrödinger and Poisson equations self consistently for the electrons in Γ-band. The δ-doped layer had a nominal width of 3 nm. The conduction band profiles shown in Fig.1, zero energy corresponds to Fermi level.

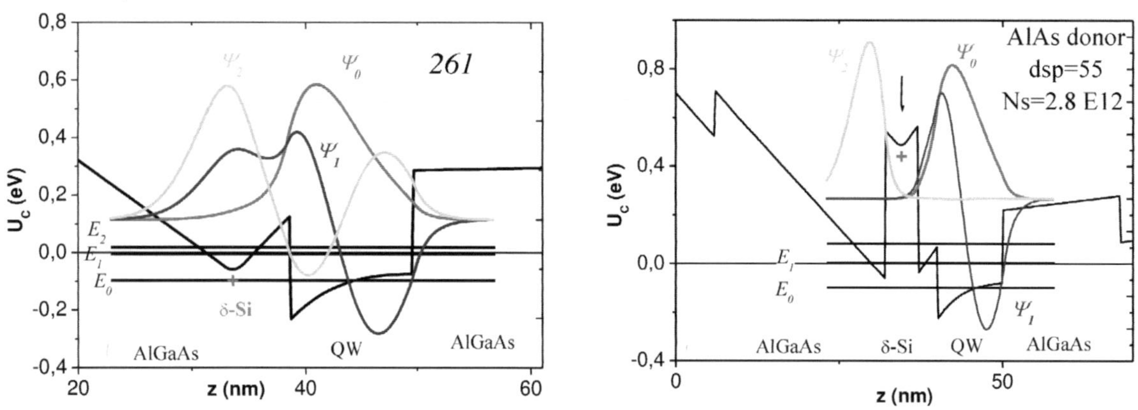

Figure 1. Conduction band U_c and electron wave functions profiles, subband energy levels for PHEMTs with conventional AlGaAs donor layer (left) and with AlAs delta-Si doped donor layer (right).

Conduction band profile is modified by AlAs: δ-Si barrier and U_c increased considerably around the Si donor area. Thus Si donor level in AlAs must be substantially above Fermi level, whereas for the conventional uniform AlGaAs barrier it can be near or even below E_F. In [12] Si activation energy in bulk Si doped AlAs is 57 meV, but should decrease in delta-doped layer due to very high donor density.

It can be seen from Fig.1 that the introduction of AlAs: δ-Si as donor layer changes the electron subband energies and wavefunction spatial profiles. For the ground state subband the confinement in the channel area increases without a remarkable shift in subband energy E_0. As for the excited states, in the uniform barrier-donor AlGaAs the Ψ_1 and Ψ_2 wavefunctions are hybridized towards the extended area of V-shaped potential well around Si donors and InGaAs channel. AlAs donor serves as high potential barrier thus suppressing a tunnel coupling between InGaAs channel and the low-energy triangle well to the left side of AlAs. Thereby the second subband E_1 wavefunction becomes completely confined in InGaAs QW. The third subband shift is most valuable and its energy increases by ~75 meV.

3. Sample growth and preparation

Pseudomorphic $In_{0.2}Ga_{0.8}As$ quantum well structures with uniform $Al_{0.24}Ga_{0.76}As$ donor layer (sample 261) and with AlAs delta-Si doped donor layer (sample 24B) were grown by molecular-beam epitaxy on semi-insulating (001) GaAs substrates. Quantum well width L_{QW} was 10 nm. The samples were delta doped by Si from the upper side through 5.5 nm spacer with concentrations of $3.2 \cdot 10^{12}$ cm^{-2} for the sample 261 and 1.33 times higher for the sample 24B. The last has 6 nm AlAs with delta-Si doping at the center keeping the same delta-layer position related to QW as in the sample 261.

For the electron transport measurements samples were processed by photolithography and mesa etching followed by the AuGe/Ni contact metallization and rapid thermal annealing at 390 °C. Hall bar lateral probe distance were 3.5 times of channel width.

4. Results and discussion

4.1. Electron transport properties

Electron transport properties were measured at the different temperatures in the range 4.2 K to 295 K. Both samples demonstrate the metallic behavior of the sheet resistance. Electron mobility rises while the temperature is decreased, but the mobility is higher in the sample 24B. The parameters are listed in Table 1.

Table 1. Electron transport properties of the samples at the different temperatures.

Sample #	Donor layer	295 K		77 K		4.2 K	
		nH, 10^{12} cm^{-2}	μH, cm^2/Vs	nH, 10^{12} cm^{-2}	μH, cm^2/Vs	nH, 10^{12} cm^{-2}	μH, cm^2/Vs
261	Standard	1.55	7 160	1.23	28 500		
24B	**6 nm AlAs: δ Si**	0.94	**7 320**	0.83	**30 470**	0.83	**33 150**

Despite the increased doping level in the sample 24B it demonstrates the decreased by ~ 2.2 times Hall electron concentration compared to the standard sample 261. Nevertheless the concentration variation with the temperature decrease shown to be lower in for the sample with doped AlAs. The reduced electron concentration in QW with doped AlAs can be addressed by low efficiency of Si atoms incorporation in Al sites of AlAs during the doping. Despite the decreased electron density in QW their mobility is sufficiently higher and reaches the values that are typical for the InP HEMT quantum well structures. We attribute this effect to the suppression of electron scattering on remote ionized donor impurities through the tunneling depression of the QW ground state wavefunction through the AlGaAs barrier. It is worth notice that the energies of the two lowest quantum state in the structures mainly refer to the quantum well and are weakly dependent on the conduction band energy at the donor area. Thus the observed change of mobility isn't connected to any intersubband scattering effects.

4.2. Photoluminescence spectroscopy

The photoluminescence (PL) spectra of the samples have been measured at T = 77 K. The results are reported in Fig. 2.

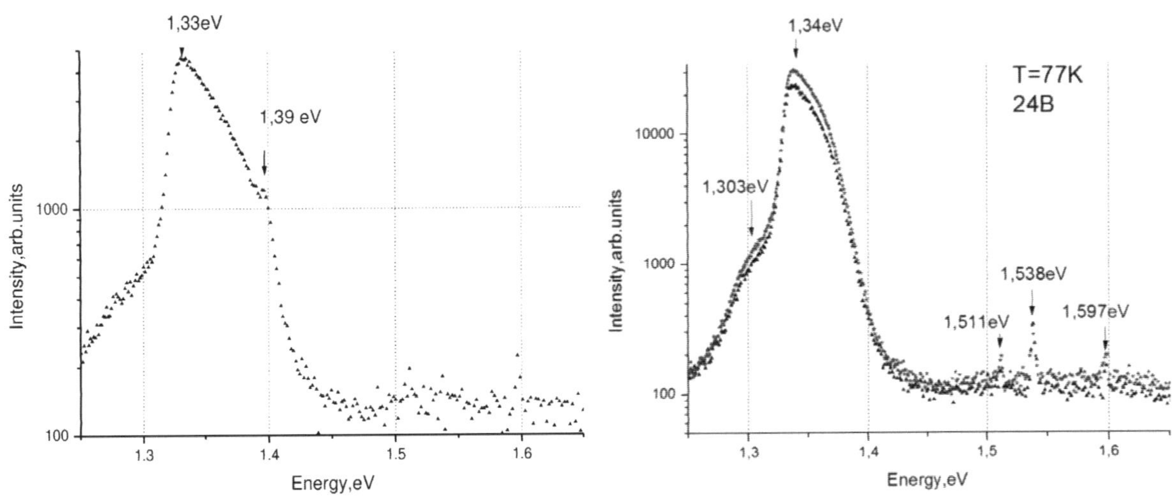

Figure 2. Photoluminescence spectra of PHEMTs with conventional AlGaAs donor layer (left) and with AlAs delta-Si doped donor layer (right) at T = 77 K.

All PL spectra of the $In_{0.12}Ga_{0.88}As$ quantum wells exhibit a pronounced maximum in the energy range 1.30÷1.40 eV, which is somewhat below the transition in bulk GaAs at 1.508 eV. For the PHEMTs with conventional AlGaAs donor layer (left) and with AlAs delta-Si doped donor layer (right) the peaks are relatively broad and the PL intensity rise differs for the different sample, indicating the presence of several transition energies. For a sample with PHEMTs with AlAs donor layer the PL peak is less broad, which shows that the upper electron levels are unoccupied. Furthermore, one can see a significant enhancement of PL intensity for the sample 24B. It can originate from the enhanced confinement of photoinduced carriers in the QW area.

5. Conclusion

A comparison between the electron transport and optical properties of pseudomorphic high electron mobility quantum well with conventional AlGaAs donor layer (left) and with AlAs delta-Si doped donor layer is presented. The structure with AlAs donor layer exhibits a mobility rise combined with the electron concentration decrease. We address this effect to the suppression of electron scattering on remote donor impurities and the reduced doping efficiency of Si atoms embedded in pure AlAs.
Photoluminescence spectroscopy shows the decrease of main peak width in the sample with Si-doped AlAs donor layer and the significant enhancement of PL intensity.

Acknowledgements

This work was supported by the Competitiveness Program of NRNU MEPhI.

References

[1] Cho S-J, Wang C and Kim N-Y 2014 *J. Microelectronic Engineering* **113** 11
[2] Chao P, Hu W, Liu J, Hoffmann W and Swanson A 1999 *Proc. GaAs ManTech Conf. (Vancouver)*(GaAs MANTECH, Inc.)
[3] Chang C, Chang A, Chen H, Chen C and Chao P 2002 *Proc. GaAs ManTech Conf. (San Diego)* (GaAs MANTECH, Inc.)
[4] Remashan K and Radhakrishnan K 2004 *Microelectronic Engineering* **75** 127-136
[5] Galiev G B, Vasil'evskiĭ I S, Klimov E A, Mokerov V G and Cherechukin A A 2006 *Semiconductors* **40** 1445-1449
[6] Shubert E and Ploog K 1984 *Phys. Rev.* B **30** 7021
[7] Mooney P M 1990 *J. Apl. Phys.* **67** R1
[8] Dieci D, Canali C, Menozzi R, Pavesi M and Cetronio A 1999 *J. Apl. Phys. Lett.* **74** 1147
[9] Mahajan A, Fay P, Afara M and Adesida I 1998 *IEEE T. Electron. Dev.* **45** 338-340
[10] Holland M C, Skuras E, Davies J H, Larkin I A, Long A R and Stanley C R 1995 *J. Cryst. Growth* **150** 1215-1219
[11] Kasu M, Fujita S and Sasaki A 1989 *J. Appl. Phys.* **66** p 3042
[12] Chand N, Henderson T, Klem J, Masselink W T, Fischer R, Chang Y C and Morkoc H 1984 *Phys. Rev.* B **30** 4481

IOP Conf. Series: Materials Science and Engineering **151** (2016) 012038 doi:10.1088/1757-899X/151/1/012038

Schematic techniques for improving precision and radiation hardness of current feedback operational amplifiers

A A Lebedev, A E Nazarenko, M S Vakhnenko, I I Liakhchilina, N M Klokov and A A Komlev

National Research Nuclear University MEPhI (Moscow Engineering Physics Institute), Kashirskoe Shosse 31, 115409 Moscow, Russian Federation

Corresponding author e-mail address: Lebedev741@mail.ru

Abstract. The circuit design method of increasing the accuracy of the current feedback amplifier (CFA) by incorporating dynamic shunt into the Wilson current mirror is considered.

1. Introduction
Typical integrated circuits (ICs) of CFAs have a high slew rate and settling time and low bandwidth so they are widely used in real-time mode [1,2]. The disadvantage of such amplifiers is in the order of large error in the reproduction of voltage gain with a closed loop. Their advantages are the relative independence of the bandwidth of the voltage gain with feedback (K_u must be less than 10).

This paper discusses the method of improving the accuracy of CFA and reducing the response time to the impact of a single heavy charged particles by incorporating a dynamic shunt into the Wilson current mirror.

2. Description of a typical circuit and transimpedance gain in op amp CF with open loop
The block diagram of CFA is shown in Fig. 1. It consists of input and output buffers. The current mirrors (CMs) are used as the input buffer load. The CMs' output is supplied with a reflector on the output buffer, which provides work in a low-resistance load of 100 Ohms.

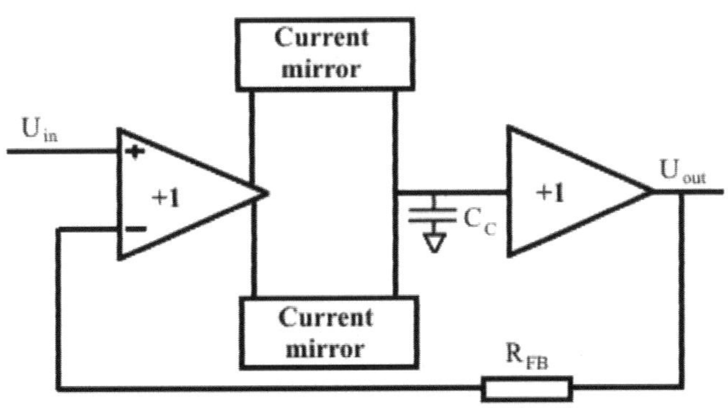

Figure 1. The block diagram of CFA

Fig. 1 shows that the output of CFA is constantly connected through a resistor R_F to the non-inverting input. This connection enables work of the op amp with the feedback in the current repeater non-inverting mode. For example, applying the positive 1V signal to the input with a small delay we get the equal 1 V signal at the output: at the same time through resistor R_F flows currently $I_F = U_{out} / R_F = 1\ mA$.

Despite the fact that in the classical sense there is a concept of negative feedback voltage, in the scientific literature, it is referred to as the feedback current, and therefore the gain of CFA with open loop will be determined by the so-called trans-impedance [3]: $Z_i = R_i.C_i$, where R_i is the output resistance of CMs, C_i is the stray capacitance between CMs' output and the input of output buffer (Z_i - trans-impedance gain - "resistance conversion factor" measured in V/A or Ohm). Therefore Bode diagram for the CFA is the dependence of Z_i on frequency. Moreover, since R_F – enabled constantly, for the experimental determination of Z_i it is necessary to divide a constant and a variable component of the input signal by connecting to the output capacitor large quantities. Bode diagram for this case is shown in Fig. 2.

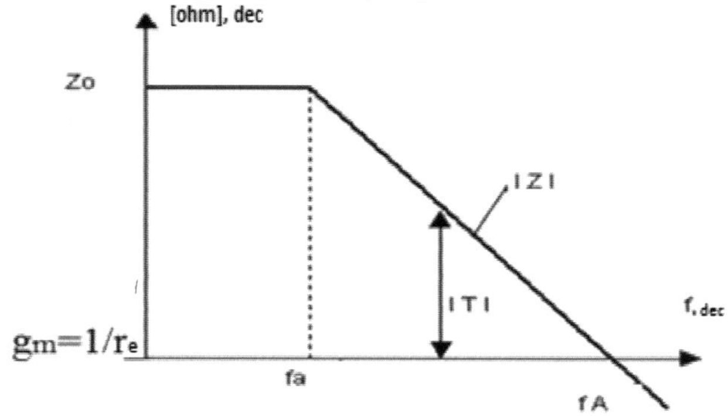

Figure 2. The Bode diagram of CFA with close loop ($f_a=1/2\pi R_i C_i$ – the dominating pole frequency, $f_A=Z_i f_a/R_{FB}$ – unity gain frequency)

3. Methods of increasing trans-impedance gain

The load of the input repeater (Fig. 1) is a Widlar current mirror, which allows increasing the output impedance on the order in comparison with CM, collected on two transistors. The maximum gain of this scheme is about 80 dB. Further increase in the gain can be achieved by using a static or dynamic CM, shown in Fig. 3.

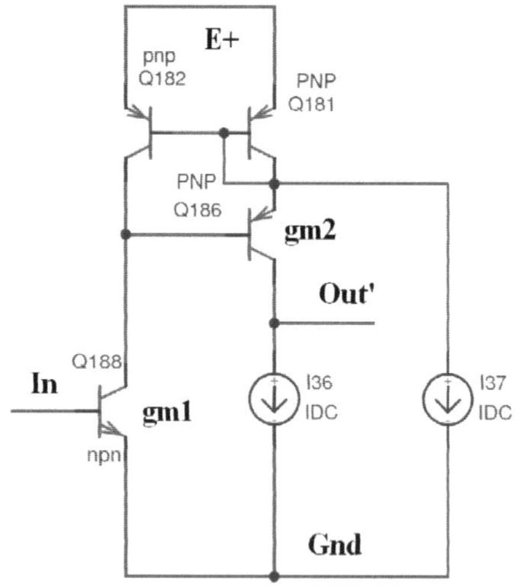

Figure 3. The upper half of the Wilson current mirror with I37 shunt

From Fig. 3 it follows that Q186 transistor is arranged on a common base with a dynamic shunt, which decreases its static current n times, the output impedance (Q186 collector resistance), respectively, is increased n times compared with Widlar CM.
The gain can be described by the following expression:

$$K_U = K_{UQ186} \cdot K_{U2}, \tag{1}$$

where:

$$K_{U2} = \frac{g_{m1}}{g_{m2}} \approx 5, \tag{2}$$

$$K_{UQ186} = (r_{Q186} \parallel R_{in.buf.}) \cdot g_{m1}. \tag{3}$$

Considering that $\beta_n = \beta_p = 100, (r_{Q186} \parallel R_{in.buf.}) \approx 1000\ kOhm,\ I^0_{K188} = 1mA$, we get: $K_U = 5 \cdot \frac{1000000}{25} = 200000.$

Thus, the main problem of obtaining gain 100dB at one stage with the help of static (dynamic) shunt is solved.

4. CFA key features

The concept of the developed CFA is shown in Fig. 4, and its frequency and transient response - in Fig. 5 and 6, respectively. The Fig. 5 shows that the unity gain frequency is 2.4 GHz with a phase margin of 40°.

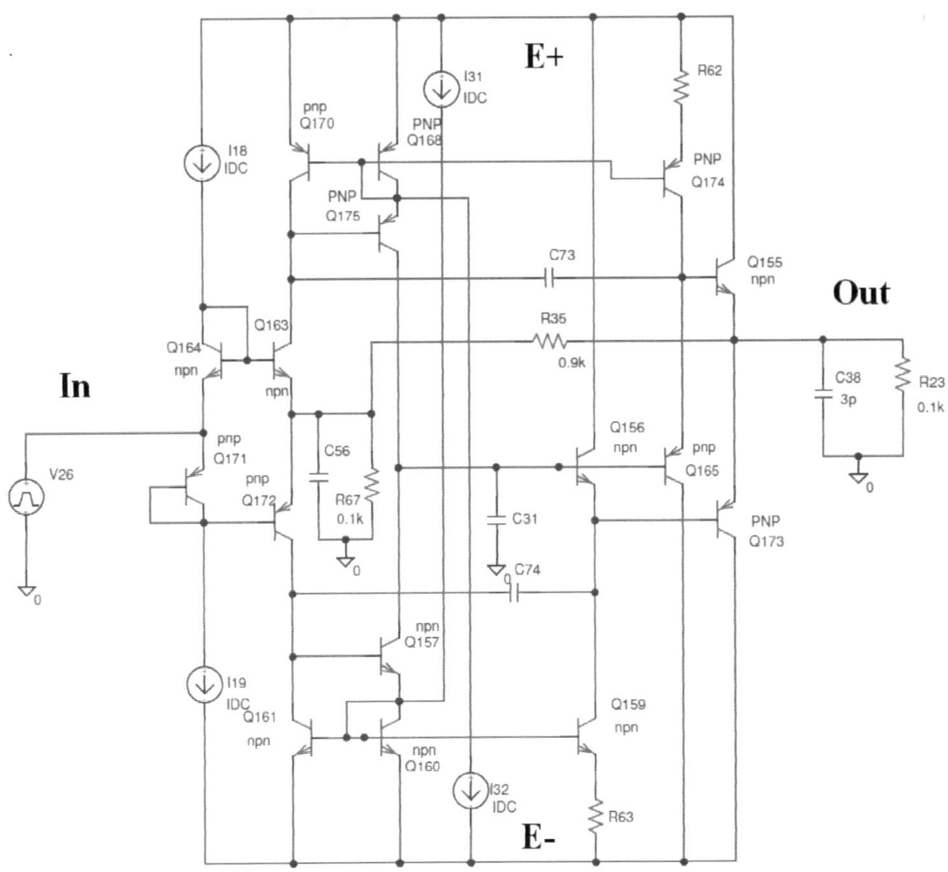

Figure 4. The concept of the developed CFA

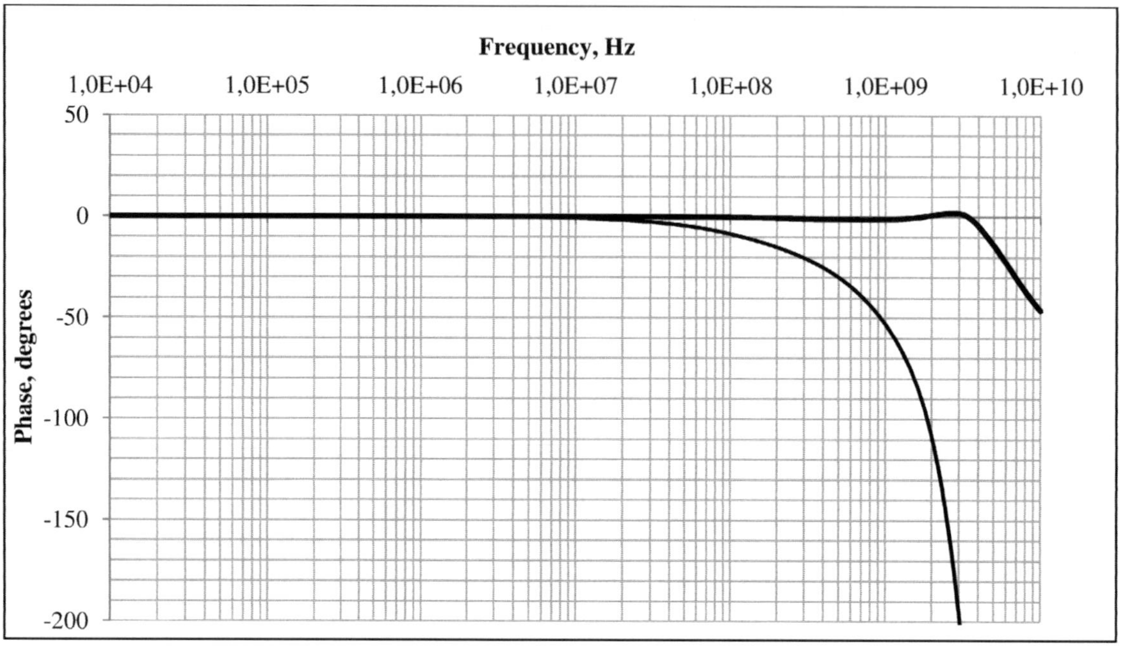

Figure 5. The Bode diagrams in the repeater mode

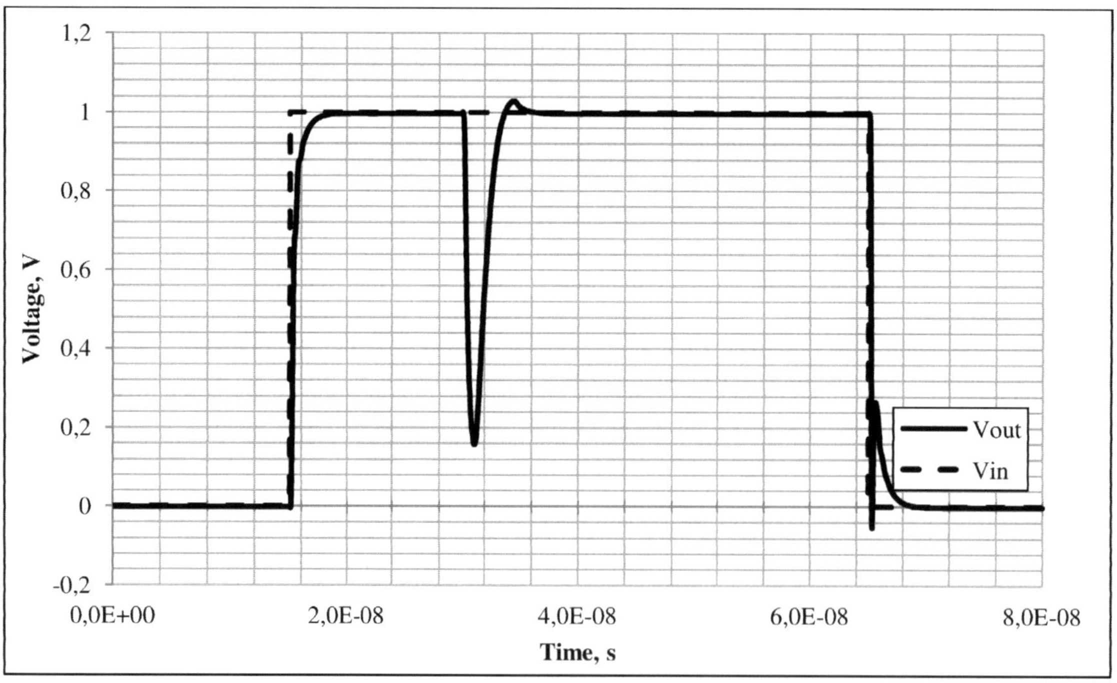

Figure 6. Transient response under the impact of HCP

5. Response to the impact of HCP

Increased bandwidth has reduced single event time (SET) from 5 ns for a typical CFA to 3 ns. However, it should be noted that the further reduction of SET after hitting hard charged particle (HCP) a single critical node (Q159) is connected with the design of the CM automatically generating additional current while the current amplification factor β is degradation [4].

IOP Conf. Series: Materials Science and Engineering **151** (2016) 012038 doi:10.1088/1757-899X/151/1/012038

6. Conclusion

The developed CFA circuit based on Wilson CM with static (dynamic) shunt allows to improve accuracy characteristics (trans-impedance increased by 3-5 times compared with the same of Wilson CM), to expand the bandwidth to 2.5 GHz with a phase margin of 40 °, and also to reduce SET from 5 to 3 ns.

Acknowledgements

This work was supported by the Competitiveness Program of NRNU MEPhI.

References

[1] Bales J 1997 *IEEE J. State circuits* **32** 1490
[2] Intersil Inc 1995 *Current Feedback Amplifier Theory and Application* (Application Note 9420.01)
[3] Franco S 1993 *IEEE Int. Symp. Circuits Syst.* **2** 1050
[4] Lebedev A A, Felitsyn V A, Komleva V A and Komlev A A 2016 *IFMBE Proceedings* **55** 537

Using ionization response maps for SET characterisation in UHF mixers

D V Savchenkov[1] and A G Kuznetsov[2]

[1] National Research Nuclear University MEPhI (Moscow Engineering Physics Institute), Kashirskoe shosse 31, 115409 Moscow, Russian Federation
[2]Specialized Electronic Systems, Kashirskoe shosse 31, 115409 Moscow, Russian Federation

Corresponding author e-mail address: dvsav@spels.ru

Abstract. SEE sensitivity of integrated circuits is characterized by their SEE cross-section vs. linear energy transfer dependences– σ(LET) which are obtained through ion tests. Often those σ(LET) dependencies have such few data points that it's much trouble to approximate them correctly. In those cases σ(LET) can be complemented withSEE cross-section vs. laser energy σ(J) obtained from laser tests. At the same time, σ(J) rarely follow exactly the shape of σ(LET) due to the metallization non-uniformly over chip area. Nevertheless, σ(J) can be corrected against this non-uniformity by examining the ionization response maps ΔU(x, y). In this work we demonstrate that such corrected σ(J) correlates better with σ(LET) by the example of single event transients (SET) in two types ofUHF mixers.

1. Introduction

The goal of almost every single event effect (SEE) test is to estimate the frequency of certain single event effect in the device under test (DUT) in space. This frequencyν can be obtained from the SEE cross-section vs. ion's linear energy transfer σ(LET) and the ion's LET differential spectrum in space φ(LET):

$$\nu = \int_0^\infty \varphi(LET) \cdot \sigma(LET) \cdot d(LET), \quad (1)$$

where σ(LET) is usually obtained from ion accelerator tests. However, the available set of LET values is always limited. Sometimes this leads to very few data point in the final σ(LET) dependence. In such cases, laser test can help in the reconstruction of the whole σ(LET) more precisely.

2. Experimental technique

Let's consider an example. Only three LET values were available in the single event transient (SET) test for the GaAs microwave mixer which was not sufficient for σ(LET) approximation. To supplement more data to these three data points, a series of laser scanning has been performed (SET locations at different laser energies are shown in Fig. 2) using FEMTO-T laser facility with 870 nm wavelength and 1.0 ps pulse duration [1]. Thus a SET cross-section (σ) vs. laser energy (J) dependence was obtained. Finally, these to dependencies were overlaid one with another (Fig. 1). This means that σ(J) curve was scaled both horizontally and vertically (or "shifted" in logarithmic scale) with certain scale factors K_{LET}[MeV·cm^2/(mg·nJ)] and $K_\sigma = 0.1$ so that finally we have $K_\sigma \cdot \sigma(K_{LET} \cdot J)$ instead of σ(J).

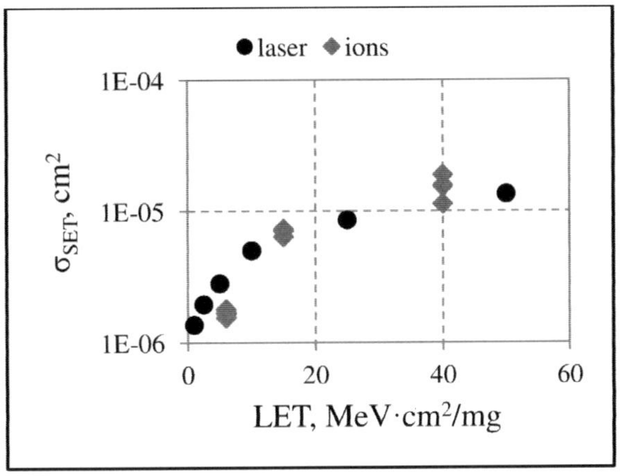

Figure 1. σ(LET) and $K_\sigma \cdot \sigma(K_{LET} \cdot J)$ for GaAs UHF mixer. K_{LET} = 5 MeV·cm^2/(mg·nJ), K_σ = 0.1.

However, one can note the certain difference between laser and ions data at small LETs. Furthermore, factor K_σ is too small which tells us that we have laser SEE cross-section rather overstated in comparison to the ions' one. The hypothesis is that the σ(J) curve obtained with a laser is not quite correct. The fact is that metallization layers covering the chip are opaque for the laser beam. The amount of laser light passing through those layers depends on the metallization density [3,4]. The latter varies over the chip's area, so the amount of laser energy passing through it at different locations varies as well. This means that if we are scanning the laser beam across the chip with certain pulse energy, the energy deposited in the chip's SEE-sensitive area is different depending on the location of the laser spot. So we cannot assign any definite energy to the SEE cross-section we obtain in such a scanning.

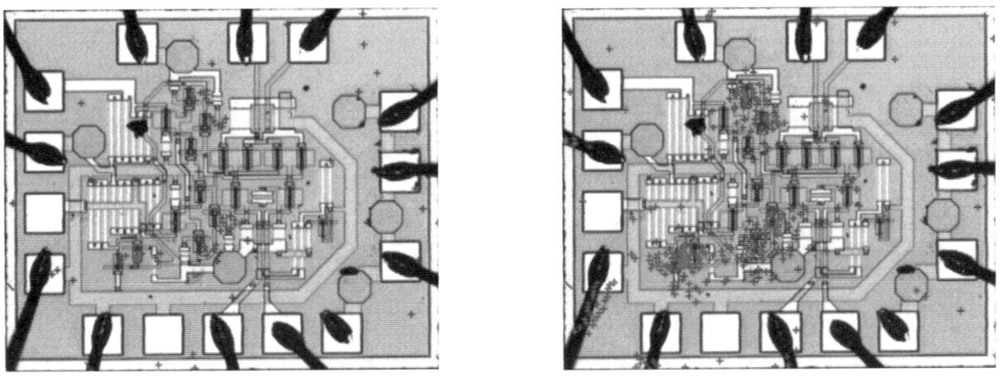

J = 0.2 nJ J = 5.0 nJ

Figure 2. SET locations at different laser energies for GaAs UHF mixer's chip.

[2] demonstrates the method of the SEE threshold LET evaluation based on the determination of the SEE threshold laser energyJ_{th} and the examination of the so-called ionization response voltage. Ionization response is the voltage pulse which one can observe if he (or she) puts the device to test the so-called photodiode mode and expose it to the laser pulse. Photodiode mode is realized when one connects all input pins of the DUT to the power pin keeping the outputs floating. Power pin and the ground pin of this device are the electrodes of the photodiode, and the ionization response is the voltage pulse which this photodiode generates when being exposed to laser light.

The key point is that the amplitude of the ionization response pulseΔU_{IR} is inversely proportional to the laser energy losses which are due to the metallization. Thus the product $J_{th}\cdot\Delta U_{IR}$ is directly proportional to the SEE threshold LET:

$$LET_{th} = K_{LET} \cdot (J_{th} \cdot \Delta U_{IR}), \quad (2)$$

whereK_{LET} is the proportionality coefficient which is the same for every location on the chip.

If we knew the SEE threshold LET in every point of the chip, we could reconstruct the σ(LET) curve with as many data points as we want. Indeed for determining SEE cross-section for a particular LET, we need just to sum all the sensitive areas with threshold LETs smaller than that LET.

So we have to determine J_{th} and ΔU_{IR} for every point of the chip – $J_{th}(x, y)$ and $\Delta U_{IR}(x, y)$. First, we define the grid of points on the chip. To determine J_{th} in each point of the grid, we can use a dichotomy process whose flowchart is shown in Fig. 3. The error of the finalJ_{th} value in this process is $\Delta J_{th} = J_{start}/2^{N_{max}}$, where J_{start} is the starting laser energy, N_{max} is the maximum number of iterations in the dichotomy process.

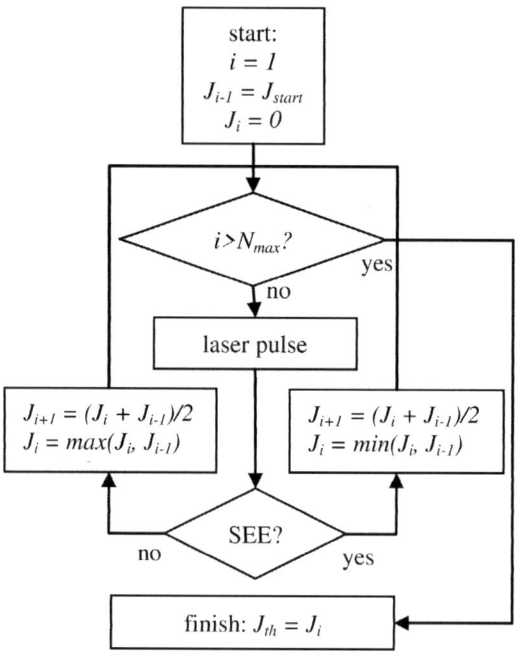

Figure 3. The process of SEE threshold energy determination through dichotomy. J_{start} is a certain starting energy which is set up each time the dichotomy process is started; N_{max} is the maximum number of iterations in the process.

The $J_{th}(x, y)$ map for GaAs UHF mixer is shown in Fig. 4. From such map, one can obtain σ(J) curve with the far larger amount of data points ($2^{N_{max}}$) than that from a series of scanning (Fig. 5). Nevertheless, that curve still doesn't take into account optical losses of laser energy due to the metallization and thus it still doesn't follow exactly the shape of σ(LET) curve.

The next step is to determine $\Delta U_{IR}(x, y)$. Ionization response map for GaAs UHF mixer is shown in Fig. 7. Multiplying $J_{th}(x, y)$ and $\Delta U_{IR}(x, y)$, we obtain a value which is directly proportional to the SEE threshold LET (see (2)). The $J_{th}(x, y)\cdot\Delta U_{IR}(x, y)$ map is shown in Fig. 8. Again from this map we can extract the dependence $\sigma(J\cdot\Delta U_{IR})$ though for this time we know that this dependence must follow the shape of the σ(LET) curve. Moreover, this is actually the case as shown in Fig. 9 where we see σ(LET) overlaid with $\sigma(J\cdot\Delta U_{IR})$ scaled along the horizontal and vertical axis by coefficients K_{LET}, K_σ respectively. Note that K_σ appeared to be exactly 1.

IOP Conf. Series: Materials Science and Engineering **151** (2016) 012039 doi:10.1088/1757-899X/151/1/012039

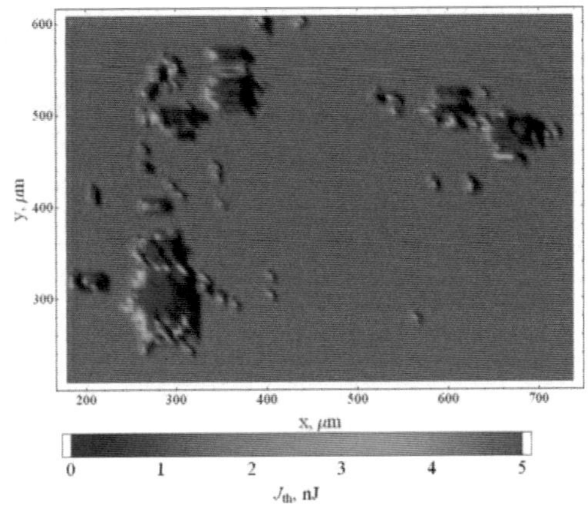

Figure 4. SET threshold energy vs. location (x, y) on the chip for GaAs UHF mixer. d = 2.5 μm, step = 5 μm.

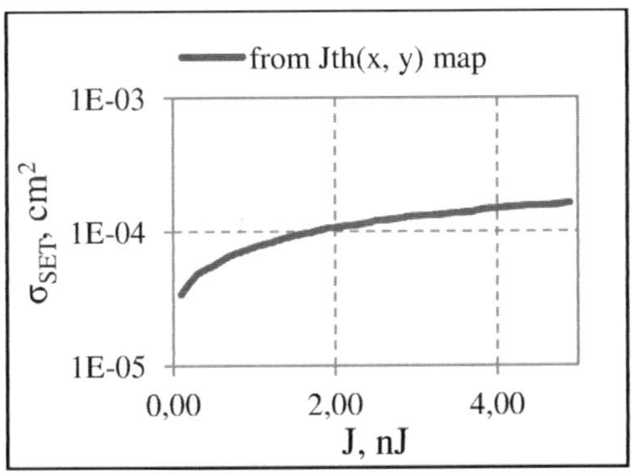

Figure 5. SET cross-section vs. laser energy obtained from J_{th}(x, y) map.

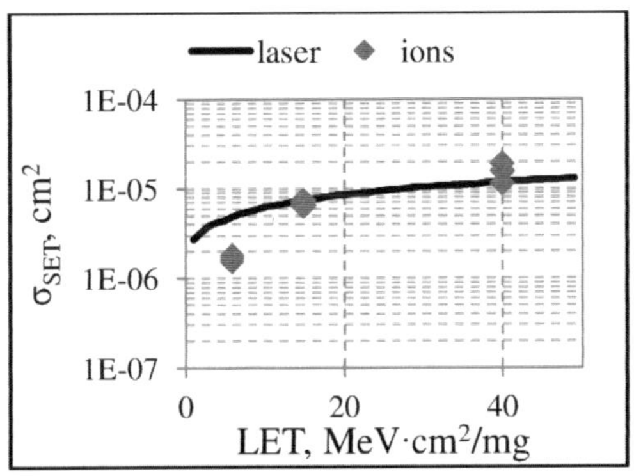

Figure 6. σ(LET) and $K_\sigma \cdot \sigma(K_{LET} \cdot J)$ for GaAs UHF mixer. K_{LET} = 10 MeV·cm²/(mg·nJ), K_σ = 0.1.

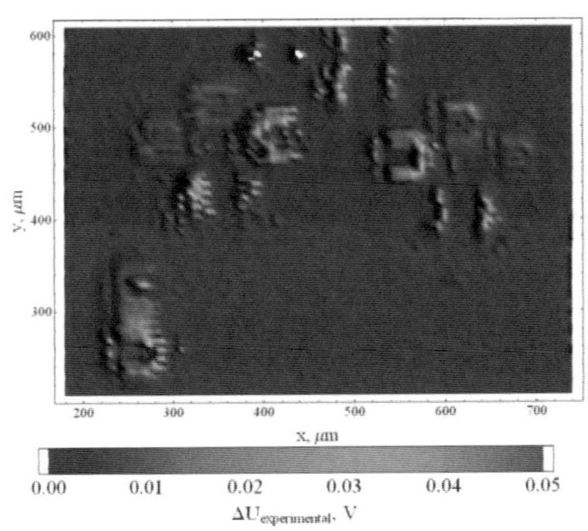

Figure 7. $\Delta U_{experimental}$ vs. location (x, y) on the chip for GaAs UHF mixer. J=5 nJ, R=50 Ohm, d = 2.5 μm, step = 5 μm.

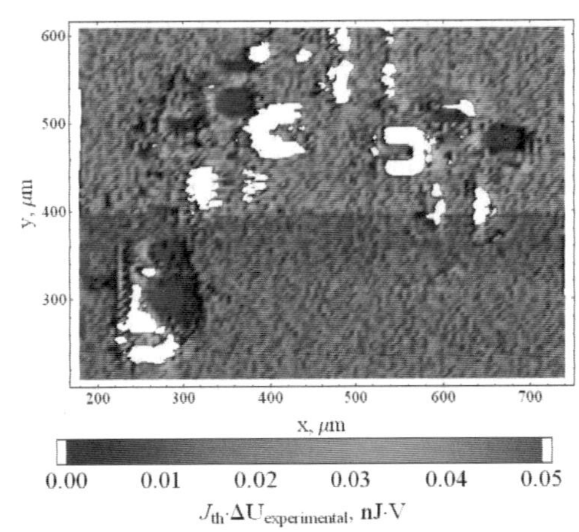

Figure 8. $J_{th} \cdot \Delta U_{experimental}$ vs. location (x, y) on the chip for GaAs UHF mixer.

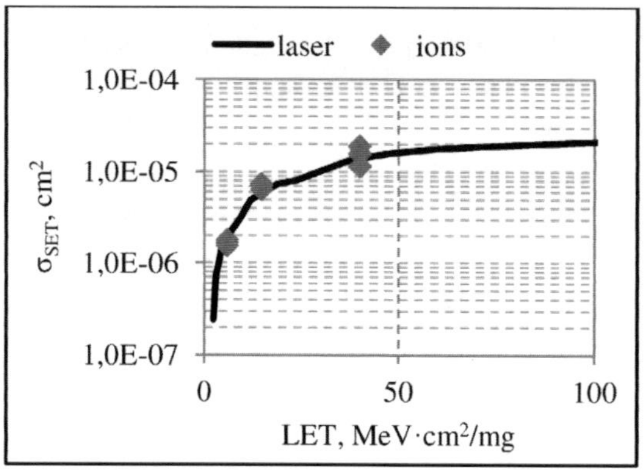

Figure 9. σ(LET) and $K_\sigma \cdot \sigma(K_{LET} \cdot J \cdot \Delta U_{experimental})$ for GaAs UHF mixer. $K_{LET} = 2.4 \cdot 10^5$MeV·cm^2/(mg·nJ·V), $K_\sigma = 1.0$.

The technique described above has been applied to a SiGeUHF mixer. In this case, only a part of the whole chip area was tested which was found to be SET sensitive. $J_{th}(x, y)$, $\Delta U_{IR}(x, y)$ and $J_{th}(x, y) \cdot \Delta U_{IR}(x, y)$ maps are shown in Figures 10, 11 and 12 respectively. $\sigma(J)$ and $\sigma(J \cdot \Delta U_{IR})$ overlapped with σ(LET) are shown in Figures 13 and 14.

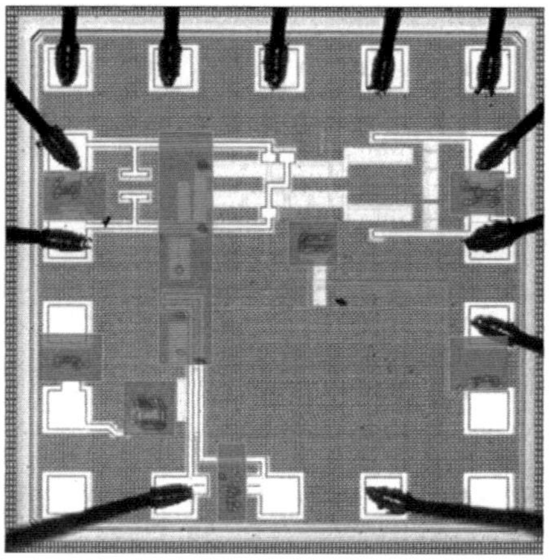

Figure 10. SET threshold energy vs. location (x, y) on the chip for SiGe UHF mixer.

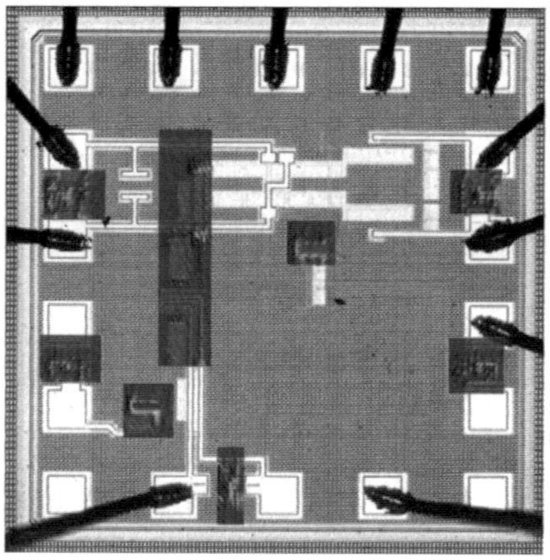

Figure 11. $\Delta U_{experimental}$ vs. location (x, y) on the chip for SiGe UHF mixer. J = 5 nJ, R = 50 Ohm, J = 5 nJ, d = 2.5 um, step = 5 um.

3. Conclusion

Laser SEE tests usually complement ion tests because they (laser tests) provide continuous energy range in contrast to the ions' LETs. Thus laser test provides far more data points in SEE cross-section vs. energy dependence than the accelerator test can provide for SEE cross-section vs. LET. Nevertheless, the laser curve is distorted by the effect of chip metallization nonuniformity. The technique presented in this paper allows not only to correct the SEE cross-section curve against this nonuniformity but also provides almost any number of data points in this curve.

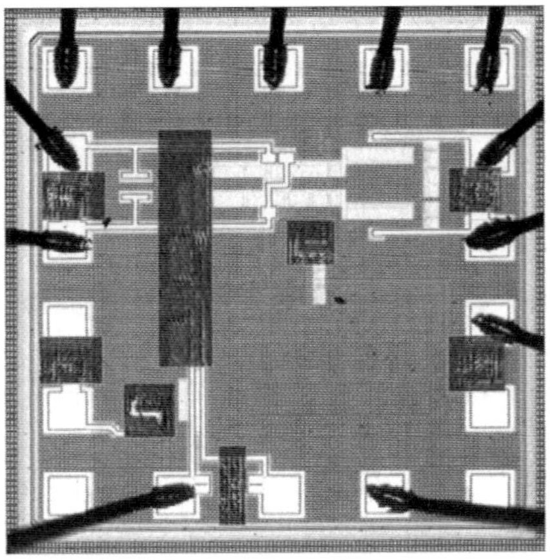

Figure 12. $J_{th}\cdot\Delta U_{experimental}$ vs. location (x, y) on the chip for SiGe UHF mixer.

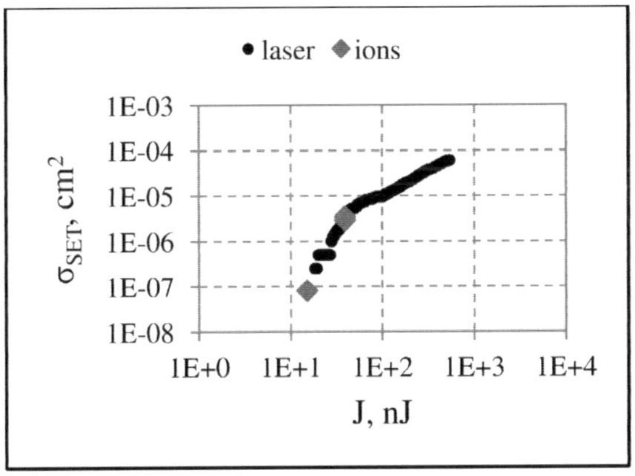

Figure 13. σ(LET) and $K_{\sigma}\cdot\sigma(K_{LET}\cdot J)$ for SiGe UHF mixer. K_{LET} = 110 MeV·cm^2/(mg·nJ), K_{σ} = 1.0.

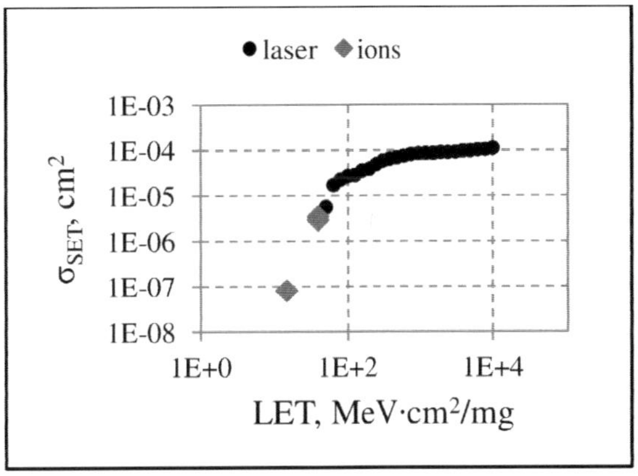

Figure 14. σ(LET) and $K_{\sigma}\cdot\sigma(K_{LET}\cdot J\cdot\Delta U_{experimental})$ for SiGe UHF mixer. $K_{LET} = 1.0\cdot10^4$ MeV·cm^2/(mg·nJ·V), K_{σ} = 1.0.

Acknowledgements
This work was supported by the Competitiveness Program of NRNU MEPhI.

References

[1] Egorov A N, Chumakov A I, Mavritskiy O B, Pechenkin A A, Telets V A and Yanenko A V 2015 *Radiation Effects Data Workshop (Boston)* (IEEE) art. no. 7004570 pp 247-250

[2] Chumakov A I, Pechenkin A A, Savchenkov D V, Tararaksin A S, Vasil'ev A L and Yanenko A V 2011 *Proc. 12th Eur. Conf. on Radiation and Its Effects on Components and Systems (Sevilla)* (IEEE) pp 449–453.

[3] Miller F 2011 *Applicability of laser methodologies to analog, digital and power electronics* (RADLAS, Suresnes, France).

[4] Savchenkov D V, Chumakov A I, Petrov A G, Pechenkin A A, Egorov A N, Mavritskiy O B and Yanenko A V 2013 *Proc. of 14th Eur. Conf. on Radiation and Its Effects on Components and Systems (Oxford)*(IEEE).

Impact of recombination on heavy ion induced single event upset cross-section

K S Zemtsov, A M Galimov, M E Gorchichko and I V Elushov

National Research Nuclear University MEPhI (Moscow Engineering Physics Institute), Kashirskoe shosse 31, 115409 Moscow, Russian Federation

Corresponding author e-mail address: cirilll@list.ru

Abstract. The experimental dependence of the single event upset (SEU) cross-section versus linear energy transfer (LET) function in the nanoscale (with feature size less than 100 nm) memories is explained with the proposed recombination-limited charge yield, which is significantly dependent on LET.

1. Introduction

Modern high-scaled integrated circuits have a very high sensitivity to single event effects (SEE) due to their low critical charges and small sizes of microelectronic components [1]. The feature size of cells on digital memory circuits becomes smaller than heavy ion tracks, which leads to multiple cell upsets (MCUs), when a single particle can cause more than one simultaneous errors [2]. Low critical charges of memory cells are a consequence of supply voltage lowering in high-scaled memories. As a result, the noise margin of integrated circuits reduces and circuits become more susceptible to ionizing radiation.

If deposited charge of the heavy ion is greater than a critical charge of the memory cell, it may lead to SEU. It is supposed that deposit charge is proportional to deposit energy from heavy ion, but it is right only for a moment of interaction. In fact, effective charge yield reduces because of recombination. For many cases of ionizing radiation, reducing of charge yield can be neglect because of the low density of electron-hole pairs, but heavy ion's hit make the high density of electron-hole pairs in slim ion's track. In this case reducing of charge yield because of recombination is significant.

We in this paper derive an experimental form of the SEU cross-section versus LET function of nanoscale memory. The experimental form is explained by the recombination-limited charge yield dependence on LET.

2. Cross-section interpolation and electron–hole recombination at the high LET

2.1. Logarithmic approximation

The SEE sensitivity characterization methods include the interpolation of average cross-section dependence by means of the Weibull function [3]:

$$\sigma(\Lambda) = \sigma_{SAT}\left(1-\exp\left[-\left(\frac{\Lambda-\Lambda_C}{W}\right)^s\right]\right). \tag{1}$$

The Weibull function has four parameters, where two of them (the saturation cross-section σ_{SAT} and the threshold (critical) value of LET (Λ_C)) are main parameters and other two (determine the shape of the cross-section curve in a subthreshold region W and the shape of a transition region between the subthreshold and the above-threshold parts of the curves s) are adjusting.

Recent studies show the absence of saturation in the cross-sections up to 120 MeV·cm²/mg (see figure 1, taken from [4]). Therefore, a use of Weibull interpolation leads to problems related to an uncertainty of main parameter σ_{SAT}, which requires a distinct saturation of the curve, and extraction procedures.

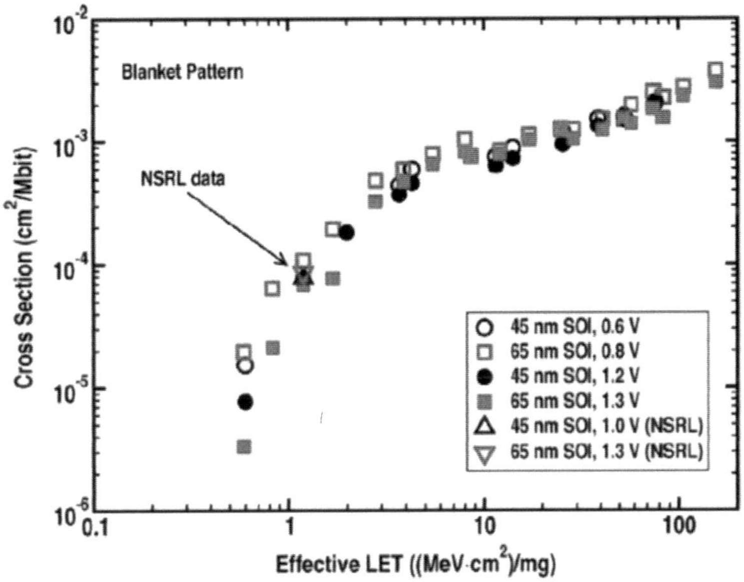

Figure 1. Experimental data points of a cross-section in a wide range by LET [4].

A lack of saturation means, properly, that the Weibull parameter σ_{SAT} is dependent on a range of LET, especially on the maximum experimental LET value. This dependence produces ambiguity of parameters extraction procedure.

The more precise way to describe average cross-section dependence is using cross-section versus LET curve logarithmic approximation, which proposed in [5]:

$$\sigma(\Lambda) = K_d W \ln\left[1+\exp\left(\frac{\Lambda-\Lambda_C}{W}\right)\right], \quad (2)$$

where K_d is the differential slope of quasi-linear part of the above-threshold cross-section versus LET curve, W is an adjusting parameter for the subthreshold region.

2.2. *Effective charge yield dependence at high LETs as recombination limited process*

Equation (2) implies a quasi-linear shape of $\sigma(\Lambda)$ in the above-threshold region. This assumption appropriate for relatively narrow LET ranges used in measurements [6]. For wide LET ranges (up to 120 MeV·cm²/mg) the dependence in the above-threshold region has sub-linear shape. This conclusion is made from detailed analysis of experimental data [4]. Such a sub-linear dependence may be explained by reducing the electron–hole charge yield for the high-LET ions due to increased recombination at high injection levels:

$$\sigma(\Lambda) = \eta_{eff}(\Lambda) K_d W \ln\left[1+\exp\left(\frac{\Lambda-\Lambda_C}{W}\right)\right], \quad (3)$$

where $\eta_{eff}(\Lambda)$ is an effective charge yield, which assumed to be a decreasing function of the injection level, dose rate and LET, because this parameter is limited by the recombination process of the charge carriers excess.

The initial bulk electron–hole concentration can be estimated as follows:

$$\Delta p_0 = \Delta n_0 = \frac{\Lambda \rho}{\varepsilon_{e-h} S_{track}} \equiv K_\Lambda \Lambda, \tag{4}$$

where ρ is the silicon mass density, S_{track} is an effective transversal area of the ion track, ε_{e-h} is an energy required for generation of an electron–hole pair (3.6 eV in silicon).

The effective ion-induced electron–hole concentration is the difference between the initial value Δn_0 and the value recombined during the charge collection time. An effective charge yield dependence on LET $\eta_{eff}(\Lambda)$ can be obtained from this equation:

$$\eta_{eff} \Delta n_0 = \Delta n_0 - \tau_C R, \tag{5}$$

where τ_C is the charge collection time, R is recombination rate, which contains bimolecular recombination R_B and Auger recombination R_A:

$$R = R_A + R_B. \tag{6}$$

The Auger recombination rate can be estimated as follows:

$$R_A = c_{nA} n^2 p + c_{pA} n p^2, \tag{7}$$

where $c_{nA} \cong 2.28\times10^{-31}$ cm$^{-6}\cdot$s^{-1} and $c_{pA} \cong 9.9\times10^{-32}$ cm$^{-6}\cdot$s^{-1} are Auger recombination constants [7]. We use it here in a simplified form:

$$R_A = 2C_A \Delta n_{eff}^3 , \tag{8}$$

where $C_A = 1.6\times10^{-31}$ cm$^{-6}\cdot$s^{-1}, $n = p = \Delta n_{eff}(\Lambda) = \eta_{eff}\,\Delta n_0(\Lambda)$. The bimolecular recombination rate can be estimated as follows:

$$R_B = C_B np = C_B \Delta n_{eff}^2, \tag{9}$$

where $C_B \approx 10^{-12}$ cm$^{-4}\cdot$s^{-1} is coefficient of bimolecular recombination. From (5) - (9) we can get self-consistent non-linear equation for effective charge yield:

$$\eta_{eff} = 1 - f_A \eta_{eff}^3 - f_B \eta_{eff}^2, \tag{10}$$

where functions f_A and f_B are defined as follows:

$$f_A(\Lambda) = 2\tau_C C_A \Delta n_0^2 = 2\tau_C C_A K_\Lambda^2 \Lambda^2 \equiv \left(\frac{\Lambda}{\Lambda_A}\right)^2, \tag{11}$$

$$f_B(\Lambda) = \tau_C C_B \Delta n_0 = \tau_C C_B K_\Lambda \Lambda \equiv \frac{\Lambda}{\Lambda_B}. \tag{12}$$

Equation (10) has the exact solution:

$$\eta_{eff} = \frac{1}{6f_A}\left(2^{2/3}\left(27f_A^2 + 9f_Af_B - 2f_B^3 + 3\sqrt{3}\left(f_A^2\left(27f_A^2 - f_B^2(1+4f_B)+2f_A(2+9f_B)\right)\right)^{1/2}\right)^{1/3} - \right.$$
$$\left. -2f_B - \frac{2^{4/3}\left(3f_A - f_B^2\right)}{\left(27f_A^2 + 9f_Af_B - 2f_B^3 + 3\sqrt{3}\left(f_A^2\left(27f_A^2 - f_B^2(1+4f_B)+2f_A(2+9f_B)\right)\right)^{1/2}\right)^{1/3}} \right) \tag{13}$$

Results of modeling equation (3) with calculated charge yield from (13) are shown in figure 2.

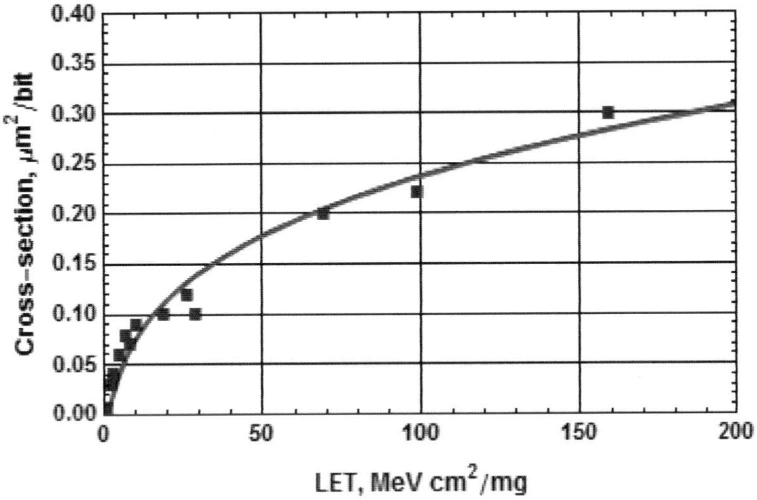

Figure 2. Cross-section vs LET for IC with SOI technology node 45-65 nm [4] interpolated with equation (3) at K_d = 1.4×10^{-2} μm^2/(MeV·cm^2/mg), W=0.35 MeV·cm^2/mg, Λ_C =1.2 MeV·cm^2/mg, τ_C =1 ns, S_{track} = 8.5×10^{-11} cm^2.

3. Conclusion

We have presented a new approach for charge carriers recombination's influence on cross-section versus LET curve. This approach improves traditional SEE sensitivity characterization methods by accurately describing effective charge yield dependence at high LETs as limited recombination process. The sub-linear SEU cross-section dependence as a function of the ion LET is explained by reducing the electron–hole charge yield for the high-LET ions due to increased recombination. Increasing recombination at high injection levels is considered as Auger and bimolecular recombination processes in this approach. Using this new proposed way, modeling results show distinct accordance with experimental data, which provide evidence of improved precision.

Acknowledgements

This work was supported by the Competitiveness Program of NRNU MEPhI.

References

[1] Kurinec S K and Iniewski K 2014 *Nanoscale Semiconductor Memories: Technology and Application*, (CRC Press, Taylor & Francis Group)
[2] Jedec standart № 89A 2001 *Measurment and reporting of alpha particles and terrestrial cosmoc ray-induced soft errors in semiconductor devices* (Arlington: Jedec Solid Ststa TA)
[3] Petersen E L, Pickel J C, Adams J H and Smith E C 1992 *IEEE T. Nucl. Sci.* **39** 1577-1599
[4] Heidel D F, Marshall P W, Pellish J A, Rodbell K P, LaBel K A, Schwank J R, Rauch S E, Hakey M C, Berg M D, Castaneda C M, Dodd P E, Friendlich M R, Phan A D, Seidleck C M, Shaneyfelt M R and Xapsos M A 2009 *IEEE T. Nucl. Sci* **56** 3499-3504
[5] Zebrev G I and Zemtsov K S 2016 *Nucl. Instrum. Meth.* A **827** 1-7
[6] Zebrev G I, Gorbunov M S, Useinov R G, Emeliyanov V V, Ozerov A I, Anashin V S, Kozyukovand A E and Zemtsov K S 2015 *Nucl. Instrum. Meth.* A **775** 41-45
[7] Sze S M 2007 *Physics of Semiconductor Devices Third Edition* ed. by Ng K K (New York: Wiley–Interscience)

IOP Conf. Series: Materials Science and Engineering **151** (2016) 012041 doi:10.1088/1757-899X/151/1/012041

Inner filled constructive material effect on shielding effectiveness of screened electronics in resonance mode

V I Butin[1,2] **and P Ya Kundyshev**[1]

[1]All-Russia Research Institute of Automatics (VNIIA), Sushchevskaya st. 22, 127055 Moscow, Russian Federation
[2]National Research Nuclear University MEPhI (Moscow Engineering Physics Institute), Kashirskoe shosse 31, 115409 Moscow, Russian Federation

Corresponding author e-mail address: butin.valentin@gmail.com

Abstract. In this paper, the results of an investigation of the resonance regime of shielding of cylindrical conducting enclosure as unmanned aerial vehicle hull model with an inner filling of common constructive components are presented. Calculations of the hull shielding effectiveness so as the induced currents in the inner placed transmission line were done in FEKO on the basis of the confirmed accuracy of the simulation. It was found that the increasing the level of the hull inner filling would result in a shift of the shielding effectiveness resonances to lower frequencies. The difference between the shielding effectiveness of the hollow hull and 90%-filled can be of 50-100 dB.

1. Introduction

The resonance process of UAV conducting hull can cause both in the uprising of electromagnetic (EM) field penetration through UAV hull and an increase of the induced currents in the inner signal circuits [1,2]. It is therefore of interest to predict the reaction of UAV to EM impact in the resonance regime. The actual UAV consists of the variety of electronics including onboard flight management systems, radio receivers and transmitters, functional systems for monitoring etc. that can be susceptible to microwave EM exposure. The inner filled constructive materials (device housing, supporting frame) will result in a redistribution of the penetrated EM field within the hull. EM fields inside the filled and the hollow hull can differ significantly. Thus, it is of concern to investigate how the level of hull filling will effect on its shielding properties in the resonant regime.

The problem of the resonance determination in the shielding enclosure (e.g., box) with metal filling like boards, plates in EM field irradiation (microwaves) has widely been considered in literature (e.g. [3,4]). The purpose of this study is to determine the effect of UAV hull filling consisting of cylindrical metal blocks on its shielding effectiveness (SE) and the induced currents in two-wire transmission line placed inside the hull. On the basis of the high accuracy of FEKO simulation confirmed by the results of the pattern experiments the hull SE and the induced current will be investigated.

The paper is organized as follows. In section II, the results of the pattern experiments so as the brief functionality of FEKO simulation tool are outlined. In section III, the description of UAV hull model with filling for calculating SE and the induced currents in the two-wire transmission line is presented. In section IV, the results of FEKO simulation so as the analysis and discussion are given. The conclusions are in section V.

2. Results of pattern experiments

It was demonstrated in [1] that the measured SE of UAV hull model within the frequency range 100-1000 MHz agrees with FEKO simulation accurately. In addition, FEKO is capable of simulating the currents induced in the ohmic loads of the electric circuits [5]. In order to verify this simulation process the pattern experiment of the induced currents in two-wire transmission line under microwave irradiation was conducted. The measurement setup is shown in figure 1.

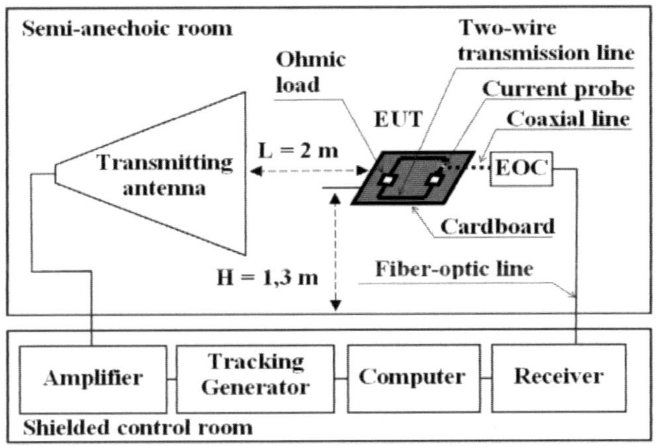

Figure 1. The measurement setup.

The equipment under test (EUT) was two-wire transmission line made of two unshielded one-core wires 0.35 mm^2 connected with two ohmic loads (r = 1 Ohm, R - variable) placed on the dielectric basement (cardboard). The wires were stretched out at full length (200 mm) and separated in parallel at 35 mm. The equipment typical for the measurement of the induced currents in microwave irradiation including the transmitting antenna, a current probe placed in the vicinity of the load, coaxial line, electrical-optical converter (EOC), fiber-optic line, receiver, computer, tracking generator and amplifier are also shown in figure 1. The results of the experiments and simulation are presented in figure 2. The analysis indicates that the measured data are consistent with the simulation within the frequency range 100-1000 MHz with high accuracy.

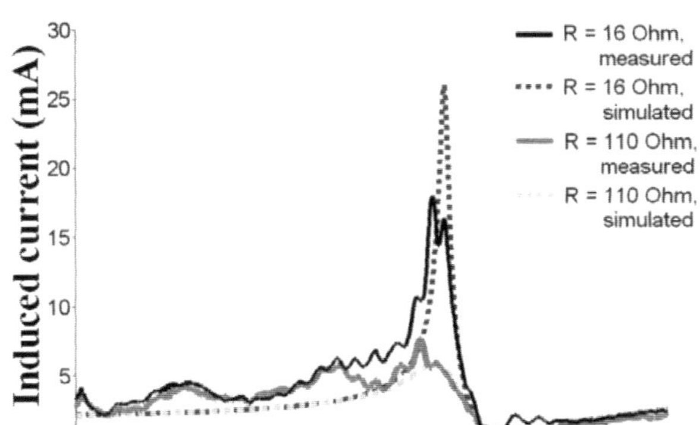

Figure 2. The measured and simulated induced currents in the two-wire transmission line for R = 16 Ohm and 110 Ohm.

Based on high reliability of FEKO simulation of the pattern UAV hull SE and the currents induced in two-wire transmission line it is adequate to simulate these parameters for new untested objects with similar constructive parameters in considered frequency range. In addition, it is sufficient to study in detail UAV hull inner filled constructive materials effect on SE and the induced currents in the two-wire transmission line using FEKO.

3. Simulation of hull with filling

UAV hull model was a hollow metallic cylinder with a height d = 632 mm, radius ρ = 171 mm. The through hole of radius ρhole = 10 mm was located on the underside end face (see figure 3). The thickness of the cylinder walls was 1 mm. The volume inside the cylinder was modeled as air medium.

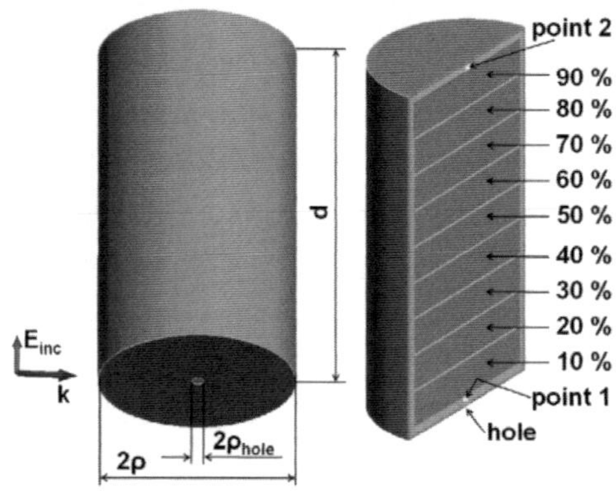

Figure 3. UAV hull model (left) and the view of the cut with filling (right).

The filling was a solid cylindrical block made of metal with a radius 160 mm and a variable height. It was situated coaxial with the hull main cylindrical axis (see figure 3 right). The height of the block was chosen so that the certain space inside the hull was occupied. For example, when modeling a 10% filling of the entire internal volume of the hull the block height was approximately 67 mm, 20% - 2×67 mm, 30% - 3×67 mm etc. The plane electromagnetic wave of 100-1000 MHz frequency range was incident on the hull with the electric intensity vector E_{inc} parallel to the hull axis (see figure 3 left). The shielding effectiveness in the form SE = 20 lg (E_{inc}/E_{point}) was calculated at two points on the axes of the hull. Point 1 and point 2 were put at a distance of 15 mm from the inner surface of the ends on the bottom and the top of the hull, respectively. Point 1 was located, thus, closer to the hole, and point 2 was further from it.

The investigation of the induced currents was provided by adding two-wire transmission line modeled as the rectangular line of a length l = 230 mm and a separation 20 mm to the previous geometry (see figure 4).

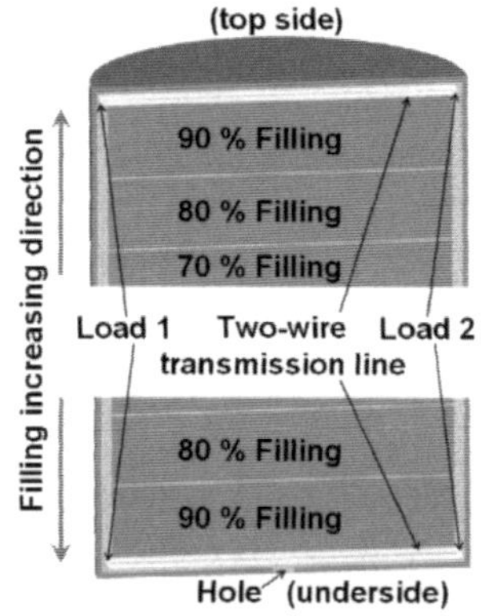

Figure 4. The model of two-wire transmission line inside the hull with different level of filling.

The line was placed in series in two planes with point 1 and point 2 parallel to the end faces of the cylindrical hull. The ports with ohmic loads R_{Load1} = 1 Ohm and R_{Load2} = 100 Ohm were placed in the middle of the narrow sides of the rectangular line. The induced currents in the ports ILoad1 and ILoad2 were calculated and presented for convenience as reduced to the incident electric intensity in the form $I_{reduced} = 20 \lg [(I_{Load} \cdot Z)/(E_{inc} \cdot l)]$, where Z = 377 Ohm is the wave impedance. The increasing of the filling level was modeled from the top down for the line on the underside (in the point 1 plane). On the contrary, the filling increasing direction for the line on the top side (in the point 2 plane) was modeled from the bottom up (see figure 4).

4. Results

The results of the simulation of the hull with different level of filling are shown in figures 5-7.

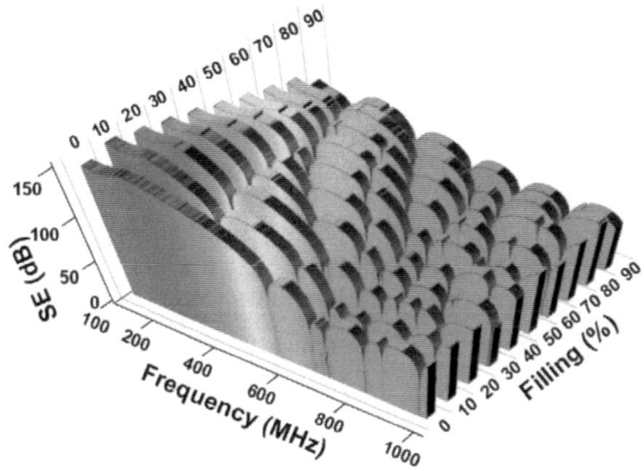

Figure 5. The 3-D bar plot of the simulated SE of the hull with different level of filling in point 2.

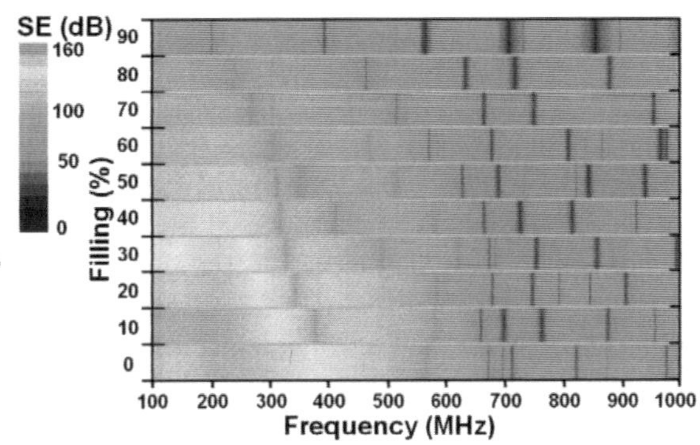

Figure 6. The contour plot of the simulated SE of the hull with various level of filling in point 2.

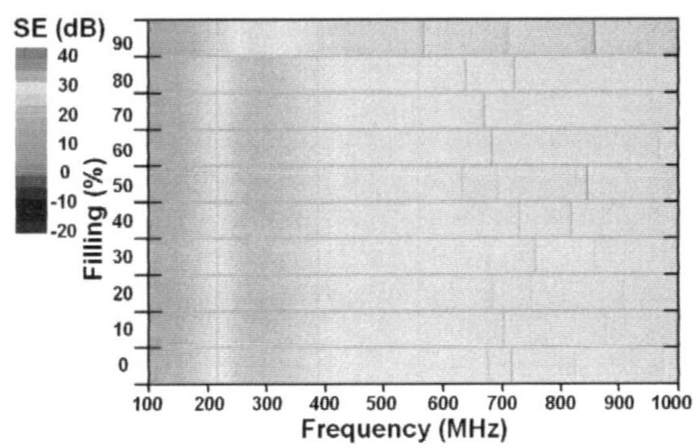

Figure 7. The contour plot of the simulated SE of the hull with different level of filling in point 1.

It follows from the data in figures 5 and 6 that the sequential increasing of the filling level from 0% to 90% will result in shifting the first SE resonance minimum to a lower frequency range for point 2. When the space inside the hull is decreasing EM field will be forced out to free air medium so the higher EM intensity in left space will be produced. Increasing the level of the filling will result in SE decreasing from 50-100 dB to 0 dB at one certain frequency. SE in point 1 will not change significantly if the filling level is increasing (see figure 7). It can be explained by the proximity of the hole that is constant reradiating antenna inside the hull so the resonance modes cannot be adequately excited in its vicinity.

The results of the simulation of the induced currents in two-wire transmission line placed on the top side and underside of the hull with different level of filling are shown in figure 8.

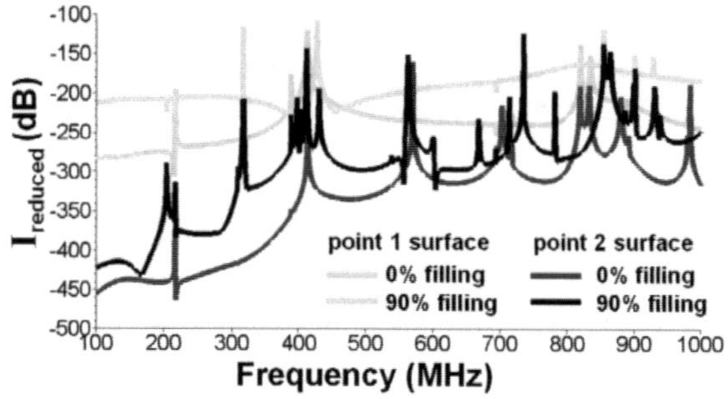

Figure 8. The simulated reduced current in 1 Ohm load of two-wire transmission line placed inside the hull with different filling level.

It is evident from figure 8 that the filling increasing will result in the rising of the currents induced in loads of the wire. By analogy with SE resonance minimum the currents maximum will be shifted to the lower frequency range. The currents maximum at 90% hull filling can be 50-100 dB greater than ones for the hollow hull (0% filling). Furthermore, when comparing two lines it is significant that the proximity to the hole produces the currents in the underside line to be on average higher than ones in the top side line. However, it can be seen that above 500 MHz the currents induced in the top side line placed far from the hole can exceed ones in the underside line at resonance peaks.

5. Conclusions

In this paper, the results of an investigation of the shielding effectiveness of cylindrical conducting enclosure as UAV hull model with an inner filling of typical constructive components like devices housing, supporting frame at the beginning of the resonance regime have been presented. On the basis of the high accuracy of FEKO simulation the hull SE and the induced currents in transmission line placed inside the hull were studied. It was found that SE minimum so as the current maximum induced in a two-wire transmission line would be shifted to the lower frequency range if increasing the level of the hull filling. SE difference between the hollow hull (0% filling) and 90% hull metal filling can be of 50-100 dB. The difference of about 100 dB will be for the currents induced in the inner placed line. The induced currents in the line and EM intensity within the vicinity of the hole inside the hull can on average be higher than ones in the far from the hole where resonance modes will effectively be excited. However, at resonance peaks the currents induced in the line placed far from the hole can exceed ones in the line close to the hole.

References

[1] Butin V and Kundyshev P 2015 *Proc. Asia-Pacific Int. Symp. on Electromagnetic Compatibility* p 33-36
[2] Butin V 2013 *Proc. Asia-Pacific Int. Symp. on Electromagnetic Compatibility* p 1-4
[3] Butin V and Kundyshev P 2012 *Technologies of Electromagn. Compat.* **4**(43) 7-17
[4] Thomas D W P, Denton A C, Konefal T, Benson T, Christopoulos C, Dawson J E, Marvin A, Porter S J and Sewell P 2001 *IEEE T. on Electromagn. Compat.* **43** 161-169
[5] Butin V 2014 *Technologies of Electromagn. Compat.* **1/2**(48) 16-20

IOP Conf. Series: Materials Science and Engineering **151** (2016) 012042 doi:10.1088/1757-899X/151/1/012042

Colloidal quantum dots for fluorescent labels of proteins

P Gladyshev[1], V Kouznetsov[2], C Martinez Bonilla[2], S Dezhurov[3], D Krilsky[3], A Vasiliev[1,4], O Morenkov[5], V Vrublevskaya[5], P Tsygankov[6], S Ibragimova[1] and A Rybakova[3]

[1]State University "Dubna", Universitetskaya st. 19, 141982 Dubna, Moscow region, Russian Federation

[2]Universidad Industrial de Santander, Carrera 27, Calle 9, 68001000 Bucaramanga, Santander, Colombia

[3]Research Institute of Applied Acoustics, 9 Maya st. 7A, 141980 Dubna, Moscow region, Russian Federation

[4]National Research Center "Kurchatov Institute", Akademika Kurchatova pl. 1, 123182 Moscow, Russian Federation

[5]Institute of Cell Biophysics, Institutskaya st. 3, 142290 Pushchino, Moscow region, Russian Federation

[6]Bauman Moscow State Technical University, 2-ya Baumanskaya st. 5/1, 105005 Moscow, Russian Federation

Corresponding author e-mail address: pglad@yandex.ru

Abstract.
The work is devoted to the synthesis of colloidal quantum dots (QDs) and their bioconjugates with proteins. Various QDs were obtained as well with synthesis method in an organic solvent followed by hydrophilization and functionalization or synthesis in aqueous phase provides obtaining hydrophilic QDs directly. Particular attention is paid to the synthesis of QDs as fluorescent tags in the near infrared where minimum absorption occurs and the fluorescence of biological tissue and synthetic materials used in analytical systems. A method for the QDs synthesis of type fluorescent core/shell CdTeSe/CdS/CdZnS-PolyT with mixed telluride, selenide cadmium core with a high quantum yield and high resistance to photoaging. It is shown that these quantum dots may be effectively used in the immunoassay.

1. Introduction

Quantum dots (QDs) - semiconductor nanocrystals, due to their small size, exhibit extraordinary optical and electronic properties, which are due to quantum effects are physically limited in space of the electron cloud. It is important that these properties are determined not only by the type of crystalline material of but their size. This makes it possible to synthesize the quantum dots with desired optical and other physical properties. Rapid progress has been associated with the use of QDs nanotechnology as fluorescent labeling of proteins and other biomolecules fluorescent labels instead of organic bioconjugates selectively binding to biomarkers led to their widespread use in a variety of biological and biomedical research [1]. Various aspects of the synthesis and use of QDs are discussed in the review [2]. However, the compounds that allow providing long-term stability and a simultaneous detection of several biomarkers. The unique optical properties of quantum dots (QDs), the possibility of obtaining water-soluble QDs and their potential use of QDs in medical diagnostics has not been fully disclosed and must be a lot to do to use it to the fullest. In this regard, it is important to continue the work on the synthesis and surface modification of QDs and development on their basis of new bioconjugates for immunochemical analysis. The problem of potential toxicity and high stability of the QDs can be solved by the use of new non-toxic materials [3,4] and the formation on QDs the additional encapsulating layers [5]. QDs have a wide spectral range of emitted light that can be tuned to NIR where low background occurs. The development of increasingly sophisticated methods to create NIR-emitting QDs is one of the directions of the colloidal synthesis of materials with unique optical properties [6,7]. New developed NIR-emitting QDs should provide a high brightness and photostability for sustainable high-sensitivity detection of analytes in complex biological systems. The use of quantum dots (QDs) fluorescent labels in the NIR spectral region is promising in areas such as molecular biology, diagnostics, and visualization. In this spectral region, there is a minimal absorption and fluorescence of biological tissue and synthetic materials used in analytical systems and thus occurs the maximum ratio of the analytical signal/noise ratio. One of the key limitations of most NIR phosphors is their limited stability, in particular, instability to photooxidation. The method of synthesis of fluorescent core/shell based on a mixed cadmium telluride selenide core with high quantum yield and high resistance to photodegradation was developed. It is shown that these QDs can be effectively used in immunoassay [6,7].Despite the achievements, there is still much to be done in a reliable and reproducible functionalizing the QDs surface and the development of effective QDs bioconjugates. This work is devoted to the synthesis colloidal QD and their bioconjugates with proteins.

QDs have a wide spectral range of emitted light that can be tuned to NIR where low background occurs. The development of increasingly sophisticated methods to create NIR-emitting QDs is one of the directions of the colloidal synthesis of materials with unique optical properties [1,2]. New developed NIR-emitting QDs should provide a high brightness and photostability for sustainable high-sensitivity detection of analytes in complex biological systems.

2. The synthesis and properties of colloidal quantum dots for proteins labeling

In most cases, the synthesis of colloidal QDs performed in organic solvents in the presence of surface-active compounds and leads to the formation of water-insoluble hydrophobic QDs [2]. QDs for labeling proteins and protein conjugates must be hydrophilic and soluble in water. Therefore were developed methods for the synthesis CdTe QDs in water, which leads directly to the water-soluble QDs [8]. It is believed that the aqueous synthesis in comparison with the organic phase synthesis is more reproducible, cheaper and QDs are more soluble in water and biocompatible.

For strong bond QDs with proteins necessary to form covalent bonds between them. This suggests the presence of reactive groups capable of binding with the carboxyl or amino groups on the surface of the protein globules. As such groups on the surface of QDs can act primarily respectively amino and carboxyl groups. In both cases, the peptide bonds are formed between the protein and QDs. Various methods of hydrophilization, functionalization QDs and an introduction to their reactive groups were described [9,10].

The composition and structure of synthesized in this work of colloidal quantum dots are presented in Table 1. Colloidal QDs QD1 - QD6 were synthesized in the aqueous phase by the procedure similar to that described in [8], and QDs QD7 - QD12 were synthesized according to previously described methods [7,11,12] with subsequent functionalization.

Synthesized QDs properties are shown in Table 2. The method [13] was used to determine the average QDs size. The quantum yield QDs was measured by method [14]. As can be seen from Table. 2 fluorescence of Ds cover a wide spectral range, and these QDs can be used for labeling many objects, including biomolecules. For some QDs (QD1 - QD5), having the same structure, the difference between the spectral characteristics changes achieved synthesis time 1 to 10 hours, respectively QDs size. All other QDs have one or two additional shells, that increases the stability of QDs and provide a high quantum yield.

3. Conjugates of NIR-emitting QDs with proteins for Lateral Flow Assay

In connection with the development of new analytical detection platform especially dangerous infections IDORI [15] of particular interest to the author presents obtain conjugates of antibodies with NIR-emitting QDs. IDORI is based on the immunochromatographic analysis (IChA) – Lateral Flow Assay (LFA) and is using QDs as fluorescent labels. For this purpose were synthesized very bright and photostability NIR-radiating CdTeSe / CdS / CdZnS QDs on which were obtained by QDs QD10 - QD12 with COOH-functional groups.

Table 1. Composition and structure of the synthesized colloidal quantum dots.

Code	Composition	Structure	Code	Composition	Structure
QD1	CdTe-TGA	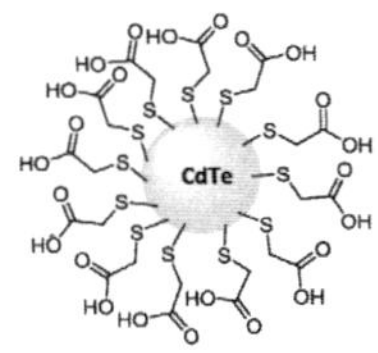	**QD7**	CdSe/CdS/ ZnS-PolyT	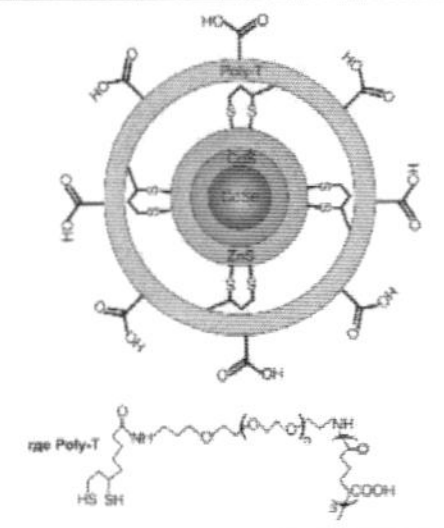
QD2	CdTe-TGA	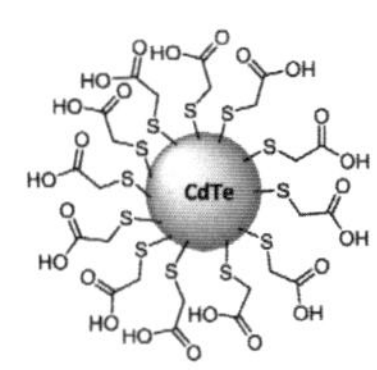	**QD8**	CdSe/CdS/ ZnS-PolyT-APS	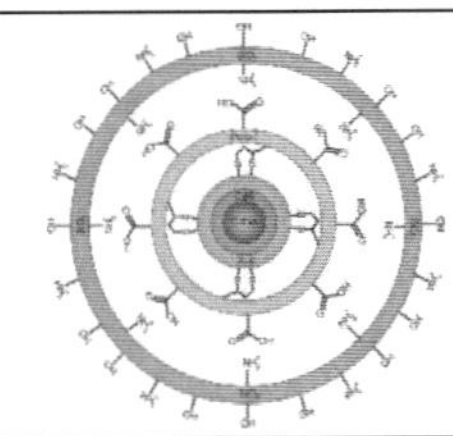
QD3	CdTe-TGA	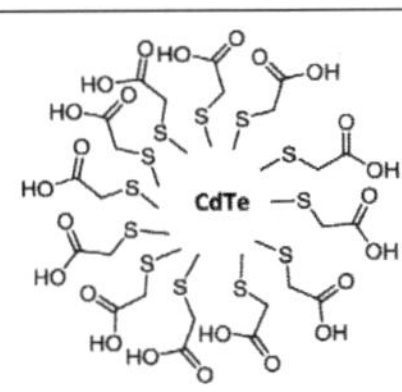	**QD9**	CdSe/CdS/ ZnS-PTVP	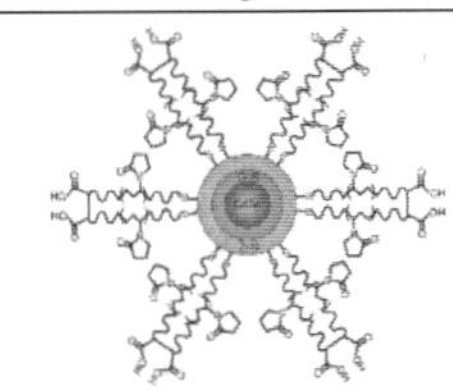
QD4	CdTe-TGA	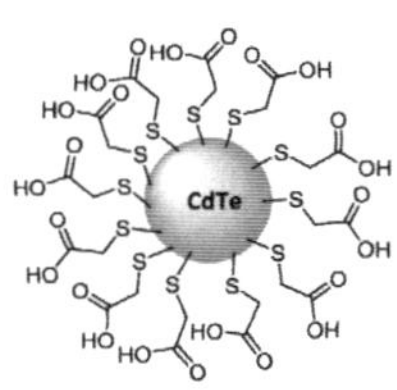	**QD10**	CdTeSe/CdS /CdZnS-PolyT	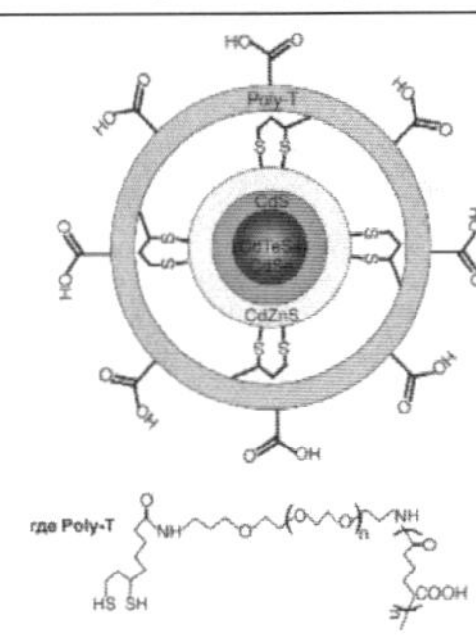
QD5	CdTe-TGA	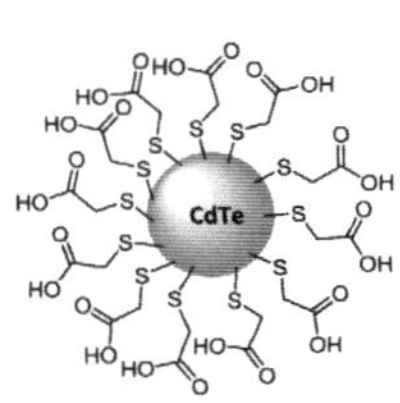	**QD11**	CdTeSe/CdS /CdZnS-PolyT-APS	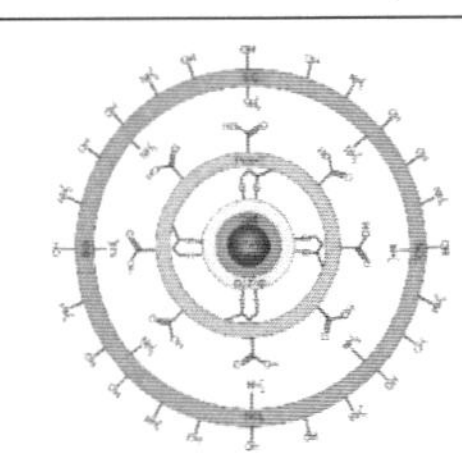
QD6	CdTe/ZnS-TGA	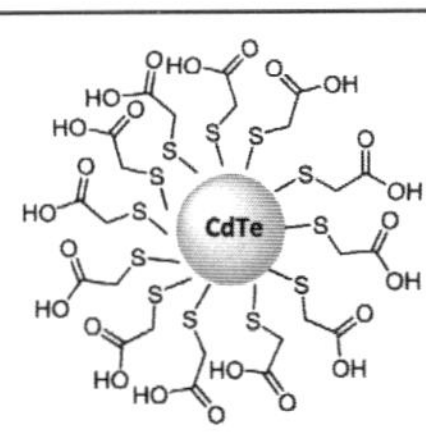	**QD12**	CdTeSe/CdS /CdZnS-PTVP	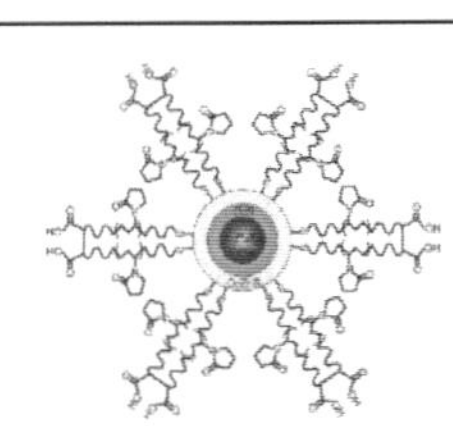

Table 2. Characteristics of the synthesized colloidal quantum dots.

Code	Composition	Particle size, nm	Excitation, nm	Max. fluorescence, nm	Quantum yield, %
QD1	CdTe-TGA	2-3	490-650	530	24
QD2	CdTe-TGA	3-4	490-650	570	37
QD3	CdTe-TGA	4-6	490-650	610	43
QD4	CdTe-TGA	6-7	490-650	630	40
QD5	CdTe-TGA	6-7	490-650	680	38
QD6	CdTe/ZnS-TGA	5-6	320-650	534	68
QD7	CdSe/CdS/ZnS-PolyT	10-20	300-590	600	40-50
QD8	CdSe/CdS/ZnS-PolyT-APS	15-40	300-600	610	30-50
QD9	CdSe/CdS/ZnS-PTVP	15-35	300-630	640	70-80
QD10	CdTeSe/CdS/CdZnS- PolyT	10-20	300-690	700	40-50
QD11	CdTeSe/CdS/CdZnS-PolyT-APS	15-40	300-710	720	30-50
QD12	CdTeSe/CdS/CdZnS-PTVP	15-35	300-730	740	70-80

The basis of the developed technique was taken from the paper [16]. At the same time, the thickness of the shell of wide bandgap semiconductors (CdS, ZnS) was increased to 7 or more monolayers. Building CdS shell not only significantly increases quantum yield, but also passivates considerably QDs. It was also proposed to use Te additives (3-7 mol. % with respect to Se) in the synthesis of nuclei. This increased the bathochromic shift of the QDs cores up to 70 nm relative to the results of [16]. The additional approach was to build the finishing shell of ZnS to passivate QDs in water. Additionally, ZnS is a more bio-compatible material compared to cadmium chalcogenides. Along with additional bathochromic shift (10-30 nm), the application of additional semiconductor shells leads to more symmetric emission peaks while retaining the quantum yield. The synthesis of CdTeSe / CdS / CdZnS / ZnS QDs involves several steps. First colloidal synthesis of CdTeSe was implemented. Then, these nuclei were coated sequentially with CdS, CdZnS and ZnS shells. Interjacent CdTeSe, CdTeSe / CdS, CdTeSe / CdS / CdZnS and end CdTeSe / CdS / CdZnS / ZnS QDs were purified by reprecipitation from toluene with methanol-butanol. The size of nanocrystals after the formation of shells increased from 4-5 to 10-11 nm. These QDs were functionalized with the COOH-group using thiol-modified PVP. Further QDs were covalently conjugated to a monoclonal antibody directed to glycoprotein B, pseudorabies virus (PRV-gB) using carbodiimide chemistry methods. The best system PRV-gB LFA was based on QDs with an emission maximum at 700 nm (QDs QD10), which showed a detection limit of 0.5-1 ng / ml PRV-gB in buffers and tissue lysates with a very high specificity. This COOH functionalized CdTeSe / CdS / CdZnS core-shell QDs can be successfully used to prepare conjugates with various proteins.

4. Conclusions

It was synthesized colloidal quantum dots, which can be used as labels for various proteins. It was demonstrated that fluorescent labels based on QDs are of great interest for diagnosis of diseases and immunoassay. Maximum fluorescence QDs can be realized in a wide spectral range. This makes it possible to realize the visible labels and hidden from the eyes NIR-labels. Highly bright and photostable NIR-emitting CdTe / QdS / CdZnS core / shell colloidal QDs with fluorescence maximum at about 700 nm were synthesized. QDs has a half-width of the fluorescence peaks of 60-70 nm, the quantum yield of 80% and a high photostability. QDs were functionalized with the COOH-group. It is shown that these functionalized QDs can be successfully used for the preparation of active conjugates with proteins for LFA applications.

Acknowledgements
The authors acknowledge the support from the Russian Scientific Foundation (project no. 15-19-00229).

References

[1] Medintz I L, Uyeda H T, Goldman E R and Mattoussi H 2005 *Nature Materials* **4** 435-446
[2] Kim J Y, Voznyy O, Zhitomirsky D and Sargent E H 2013 *Adv. Mater.* **25** 4986-5010
[3] Yang H, Jauregui L A, Zhang G, Chen Y P and Wu Y 2012 *Nano Lett.* **12** 540
[4] Seefeld S, Limpinsel M, Liu Y, Farhi N, Weber A., Zhang Y, Berry N, Kwon Y J, Perkins C L, Hemminge J C, Wu R and Law M 2013 *J. Am. Chem. Soc.* **135** 4412
[5] Liu Y, Gibbs M, Perkins C L, Tolentino J, Zarghami M H, Bustamante J and Law M 2011 *Nano Lett.* **11** 5349
[6] Petryayeva E, Algar W R and Medintz I L 2013 *Appl Spectrosc.* **67** 215-252
[7] Dezhurov S V, Trifonov A Yu, Lovygin M V, Rybakova A V and Krylsky D V 2016 *Nanotechnologies in Russia* **11** 337-343
[8] Wang Y and Liu S 2012 *J. Chil. Chem. Soc.* **57** 1109
[9] Sperling R A and Parak P W 2010 *J. Trans. R. Soc. A* **368** 1333-1383
[10] Da Cunha A S 2012 *Miguel Quantum Dots: Synthesis, Functionalization and Bioconjugation for Biological Applications* (Dissertation presented to obtain the Ph.D degree in Engineering Sciences and Technology Instituto de Tecnologia Química e Biológica | Universidade Nova de Lisboa Oeiras)
[11] Loginova Y F , Dezhurov S V , Zherdeva V V, Kazachkina N I, Wakstein M S and Savitsky A P 2012 *BBRC* **419** 54-59
[12] Yildiz I, McCaughan B, Cruickshank S F, Callan J F and Raymo F M 2009 *Mater. Chem. B.* **25** 7090-7096
[13] Schuetze B, Mayer C, Loza K, Gocyla M, Heggenc M and Epple M J. 2016 *Mater. Chem. B* **4** 2179-2189
[14] Laverdant J, Daney de Marcillac W, Barthou C, Chinh V D, Schwob C, Coolen L and Benalloul P 2011 *Pham Thu Nga, Agnès Maître Materials* **4** 1182-1193
[15] Gladyshev P P, Vasilyev A A, Morenkov O S and et al. 2016 *ANS "SibAK"* **45** pp 22-48
[16] Chen O, Zhao J, Chauhan V P, Cui J, Wong C, Harris D K, Wei H, Han H-S, Fukumura D, Jain R K and Bawendi M G 2013 *Nature Materials* **12** 445-451

IOP Conf. Series: Materials Science and Engineering **151** (2016) 012043 doi:10.1088/1757-899X/151/1/012043

Finishing technologies for processing of optical microelectronic items

K R Muratov[1], T R Ablyaz[1], E A Gashev[1], A M Khanov[1] and N V Varlamov[2]

[1]Perm National Research Polytechnic University, Komsomolsky prospekt 29, 614990 Perm, Perm krai, Russian Federation
[2]National Research Nuclear University MEPhI (Moscow Engineering Physics Institute), Kashirskoe shosse 31, 115409 Moscow, Russian Federation

Corresponding author's e-mail address: [1]karimur_80@mail.ru

Abstract. This article discusses comparative studies of the influence of specific pressure and speed of relative motion of tool and item on polishing capacity and quality of processed surface of lithium niobate single crystal. The equation for calculation of total depth of layer destroyed upon finishing and time of polishing required for its removal. Analysis of polishing tool has accounted for strength properties of the polishing material, removal capacity, and quality of processed surface of lithium niobate single crystal. It is experimentally established that upon basic pressure both for woven and unwoven polishers with the highest capacity the lower limit of grain coarseness of polishing suspension is 1/0.5 μm. A reserve of increase in shape accuracy and polishing quality is the use of polishers with equal thickness and structure of working surface.

1. Introduction

Recently the list of crystalline materials used in commercial microelectronics has been sufficiently expanded. Lithium niobate single crystals (LNC) are the most widely applied in integral optics due to high values of non-linear optical and electric optical coefficients, as well as possibility of commercial crystal growth and fabrication of plates. Integral optical phase modulators on the basis of LNC are at present widely applied in fiber optics gyroscopes, electric current sensors, and other devices. Upon mechanical processing of working surfaces of LNC plates finishing procedures are the most labor-consuming. The main technological difficulties of these procedures are stipulated by physicomechanical properties of material, as well as by high requirements to the quality of processed surface. Hence, mechanization and automation of finishing procedures together with steady quality and processing cap is an urgent task.

2. Formulation of the problem

Upon polishing, as upon abrasive finishing or fine grinding, the main process variables are the speed of relative motion of tool and processed surface, the contact pressure between them and processing time, depending on numerous factors, mainly on the aforementioned engineering and kinematic processing variables. The influence of polishing modes of brittle optical and ceramic materials on capacity and quality of processed surface was studied elsewhere [1, 2, 3, 4, 5], from where it follows that the dependences upon finishing and polishing are similar with quantitative differences.

In the course of abrasive finishing or fine grinding microcracks are generated in the surface layer of the processed item as a consequence of the non-spherical shape of abrasive grains, which roll over and collide. The microcracks intersect each other; indents are generated upon repeated impact of abrasive grains. The processed surface acquires peculiar dull relief, compose of bosses and cavities, microcracks remain under the surface in the bulk of the material. Therefore, ground glass surface or any other brittle material consists of visible relief layer and invisible subsurface fissured layer. The combination of these layers is known as destroyed or broken layer of the polished surface.

The researchers of finishing of brittle materials [2, 6, 7, 8, 9] established that the ratio of total depth F of destroyed and visible relief layer h for various grades of optical glass and minerals with close values of brittleness is qualitatively constant and in the first approximation equals to $F / h = K \approx 4$, Figure 1.

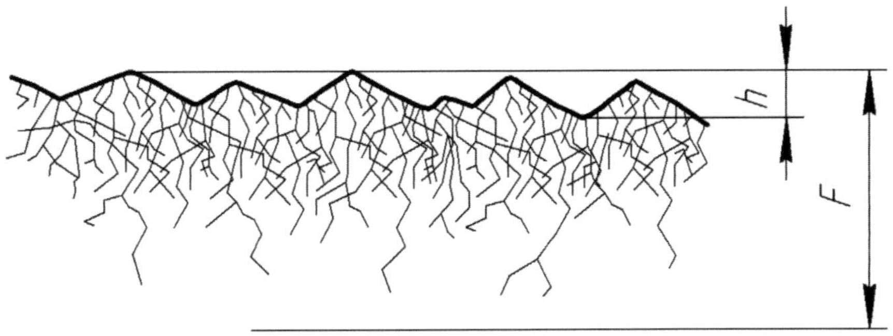

Figure 1. The structure of destroyed layer upon abrasive finishing or fine grinding.

It is experimentally established that the highest influence on the depth of relief layer is exerted by the size of abrasive grain and its hardness, and, to a lower extent, by the hardness of abrading tool and no influence by the pressure between tool/item and processing speed. Upon finishing of crystalline materials, for instance, single crystals of lithium fluoride and fluorite, the propagation direction of cracks and occurring stresses sharply differs from their direction in glass and depends on their structure. Thus, the microrelief of ground crystal surface also related with their structure and crystallographic direction of polishing [10]. The depth of destroyed layer at cube edge is slightly higher than at rhombic dodecahedron edge and by 1.5 times greater than in glass. The latter fact is obviously related to the value of angle at which cracks propagate in glass and crystals.

Perm University analyzed abrasive polishing of the surface of lithium niobate single crystal; the aim was to select the material of polishing tool as well as polishing technological conditions. Lithium niobate plates with the sizes of 12×36×1 mm and edge inclination of 10° to symmetry axis were used as samples. The main requirements to the quality of processed surface: surface coarseness R_a – not higher than 0.010 µm, no fractures, indents and notches at the plate edges.

3. Experimental

Tables 1 and 2 summarize the experimental results of the influence of specific pressure and speed of relative motion on polishing capacity and quality of processed surface of lithium niobate single crystal, obtained in Perm University. The experiments were performed with VelTex polisher (Buehler), ASM 1/0.5 basic suspension. Constant process variables: V_{avg} = 1 m/s, (60 m/min); pressure P = 65 kPa. Time of one polishing cycle T_{pc} = 6 min.

The experimental results demonstrated that polishing capacity (Q, µm/min) increases linearly actually in proportion to specific pressure. The coarseness of polished surface of the single crystal in the pressure range of 44 kPa – 104 kPa does not depend on pressure. It is established that in the region of minor specific pressures from 31 kPa and lower there is certain deterioration of coarseness as a consequence of insufficient polishing. Probably, at moderate specific pressures, firstly, elastic properties of polisher are more pronounced, and secondly, only coarse grains participate in the process. The pressure is restricted by polisher strength and danger of dry friction.

Table 1. Influence of specific pressure on polishing capacity and quality of processed surface of lithium niobate single crystal (V_{avg} = 1 m/s).

P, kPa (kg/cm^2)	Coarseness parameters, μm					Q, μm/min
	R_a	R_z	R_q	R_{max}	R_t	
31.8 (0.32)	0.002	0.013	0.002	0.015	0.015	1.5
44.5 (0.45)	-	-	-	-	-	2.0
65.4 (0.65)	0.001	0.009	0.002	0.009	0.009	3.0
75.6 (0.76)	-	-	-	-	-	3.5
83.3 (0.83)	-	-	-	-	-	4.0
104 (1.04)	0.001	0.010	0.002	0.010	0.011	5.0

Table 2. Influence of cutting speed on polishing capacity and quality of processed surface of lithium niobate single crystal (P = 6 kPa).

V_{cp}, m/s (m/min)	n_{pol}, rpm	Coarseness parameters, μm					Q, μm/min
		R_a	R_z	R_q	R_{max}	R_t	
0.5 (30)	57	0.001	0.009	0.002	0.009	0.009	1.1
1.0 (60)	140	0.001	0.009	0.002	0.009	0.009	3.0
1.5 (90)	227	-	-	-	-	-	4.2
1.94 (116)	300	0.001	0.010	0.002	0.011	0.012	5.7

The polishing capacity of the single crystal is directly proportional to the average speed of relative motion (Table 2). The increase the rate by about four times does not influence on the coarseness of polished surface of the single crystal. The speed is restricted by the same reasons as upon finishing (fine grinding): the danger of dry attrition and polisher strength.

The time of polishing cycle (T_{pc}) is determined mainly by removal capacity (Q) or speed of polishing-off and depth of layer destroyed by grinding. The obtained experimental results (Table 3) demonstrated that with increase in polishing time in the range from 3 min to 12 min the removal of material Δ_{Aavg}, μm increases proportionally and polishing capacity Q, μm/min remains unchanged, which completely agrees with the experimental results of polishing of optical glass in [4].

Table 3. Influence of polishing time on removal capacity

Polishing cycle time, T_{pc}, min	Polisher material			
	Woven: Bologna fabric P = 61 kPa, V_{avg} = 1 m/s		Unwoven: Microlap Remet and VelTex Buehler P = 39 kPa, V_{avg} = 1 m/s	
	Δ_{Aavg}, μm	Q, μm/min	Δ_{Aavg}, μm	Q, μm/min
3	-	-	5.7	1.9
6	21.9	3.6	11.7	1.9
9	32.4	3.6	16.3	1.8
12	41.9	3.5	22.0	1.8

The following polishing modes of lithium niobate single crystal can be recommended for the proposed polishers, *Microlap* (*Remet*) and *VelTex* (*Buehler*): polisher rotation speed 150-250 rpm and pressure 50-100 kPa, which provides the required quality and sufficiently high polishing capacity. When polishing cycle time varies from 3 min to 12 min at constant suspension supply, the polishing capacity remains unchanged.

With known removal capacity, maximum depth h of relief layer, which is determined by R_{max}, and ratio [10] between the depth of destroyed F and relief h layers for crystalline materials $F/h \approx 6$, it is possible to calculate approximately minimum polishing time. Herewith, it should be taken into account that the depth h (R_{max}) of relief layer measured by stylus method on profilograph is by 2-2.5 times underestimated in comparison with the actual depth determined by polishing-off [2, 6, 11].

Therefore, upon finishing abrasive processing of crystalline materials, in particular, plate edges of lithium niobate single crystal, in the first approximation, in order to calculate total depth of layer destroyed by finishing and minimum polishing time required for its removal the following expressions can be applied: $F \approx 12 \cdot R_{max}$ μm; $T_{un} \approx F / Q$ min.

A further aim of the study of polishing of lithium niobate single crystal was a selection of the material of polishing tool, as well as technological conditions of polishing. Abrasive was made of synthetic diamond in the form ASM 1/0.5 suspension and deionized water. The main requirements to the quality of processed surface: surface coarseness R_a – not higher than 0.010 μm, no fractures, indents and notches at the plate edges.

Figure 2 illustrates the scheme of motions upon mechanical polishing of single crystal on the polishing machine.

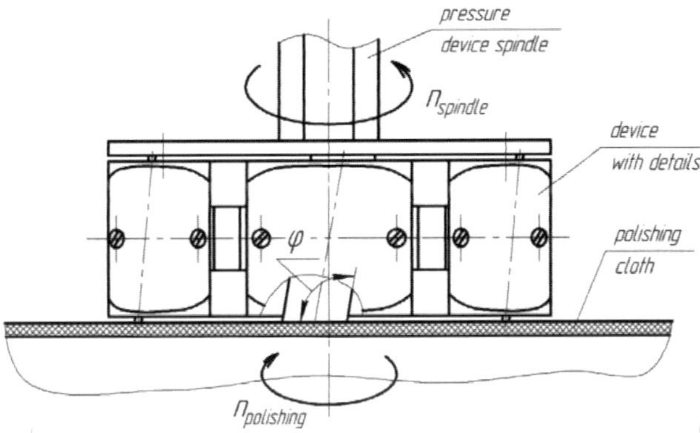

Figure 2. Schematic view of motions upon mechanical polishing of single crystal: n_{sp} - spindle RPM of device with details; $n_{pol.}$- polishing tool RPM.

A pack of processed plates is located in the grooves of the device at an angle of 10° to the vertical axis (see Figure 2). As a consequence, the angle between the processed edge and right lateral plane are obtuse, that is, higher than 90°, and between the edge and left plane it is lower than 90°, that is, acute angle. Taking into account that spindle rotation of loading device (automatic module) and device with items is not reversed and always directed counter clockwise, the groove inclination in the body is made so that the vector of cutting speed is directed towards obtuse angle φ. Thus, to provide similar direction of the vector of cumulative cutting speed it is necessary to provide counter-rotation of the polisher, that is, clockwise.

Such polishing kinematics makes it possible to increase significantly the polisher strength and provides sufficiently high cutting speed at moderate rotation frequencies of the polisher. It should be mentioned that upon polishing it would be unreasonable to apply the mode of the free rotation since in this mode it is impossible to provide the required direction of the vector of cutting speed.

The comparative experiments involved the following woven and unwoven materials as polishers, their structure is illustrated in Figure 3.

In addition, composite polishers on the basis of cellulose were tested [12], their use is justified by availability and moderate cost. However, some restrictions exist in operation with cellulose polishers. No resistance against water does not permit to use them in combination with water-based suspensions. Thus, processing by cellulose polishers was performed in combination with fat-based diamond pastes, washed by organic solvents.

Polishing was performed by the suspension on the basis of deionized water and micro powder of synthetic diamonds ASM with the grain size of 1/0.5 μm. The suspension concentration was 1/10. Experiments were carried out with basic polishing modes: average pressure P = 61 kPa, average cutting speed V_{avg} = 1 m/s, (60 m/min), polishing cycle time T_{pc} = 6 min. The results of comparative experiments on polishing capacity (Q) and quality of processed surface are summarized in Table 4. The resistance of polishers, Table 5, was estimated by cumulative machine time Σt_{mach} of operation until the occurrence of local points of physical wear on working surface or until a significant decrease in polishing capacity and occurrence of noticeable notches on the polished surface.

Figure 3. Polisher material (magnification ×45): a) Bologna fabric; b) 0.09 mm Capron mesh, mesh size ~ 0.1 mm; c) 0.14 mm polyethylene film; d) laminated fabric (80% polyamide, 20% polyurethane); e) synthetic tent cloth with hydrophobic coating; f) VelTex polisher (Buehler); g) Microlap polisher (Remet); h) SK-8 artificial leather.

Table 4. Influence of polisher material on polishing capacity of lithium niobate single crystal.

	Polisher material	Coarseness parameters, μm					Q,
		R_a	R_z	R_q	R_{max}	R_t	μm/min
Woven	Bologna fabric	0.004	0.014	0.005	0.017	0.018	3.23
Woven	0.09 mm Capron mesh, mesh size ~ 0.10 mm 0.12 mm;	0.001	0.010	0.002	0.012	0.013	0.61
Woven	laminated fabric (80% polyamide, 20% polyurethane)	0.002	0.011	0.002	0.013	0.014	2.97
Woven	synthetic tent cloth with hydrophobic coating	0.002	0.013	0.002	0.014	0.016	3.17
Unwoven	composite polisher on cellulose basis	0.001	0.007	0.001	0.008	0.008	0.29
Unwoven	0.14 mm polyethylene film	0.001	0.009	0.001	0.011	0.012	1.78
Unwoven	SK-8 artificial leather	0.001	0.005	0.001	0.007	0.007	1.12
Unwoven	Microlap polisher, Remet and VelTex polisher, Buehler	0.001	0.006	0.001	0.006	0.007	2.91

It can be seen in Table 4 that the highest removal capacity was demonstrated by synthetic woven hydrophobic materials, such as laminated fabric, as well as by proprietary unwoven polishers: Microlap and VelTex. However, regarding coarseness, the woven polishers are slightly behind the unwoven ones. This can be attributed to higher elasticity and homogeneity of the structure of unwoven materials, providing more uniform grain protrusion in abrasive interlayer, as well as by good adaptiveness of polishers to processed surface of lithium niobate single crystal [4, 8].

Comparison of cumulative machine operation time Σt_{mach} summarized in Table 5, shows that upon processing of edge surfaces of a single crystal with low contact area the polisher resistance is determined not only by material strength but its structure as well.

Table 5. Polisher resistance.

	Polisher material	Resistance Σt_{mach}, min
Woven	Bologna fabric laminated fabric (80% polyamide, 20% polyurethane) synthetic tent cloth with a hydrophobic coating	~ 40-50
Woven	0.09 mm Capron mesh, mesh size ~ 0.1 mm	~ 6-10
Unwoven	0.14 mm polyethylene film	~ 6-10
Unwoven	composite polisher on cellulose basis	~ 12-18
Unwoven	SK-8 artificial leather Microlap polisher, Remet and VelTex polisher, Buehler	More than 120

As a consequence of the action of sharp edges of processed plates, the fibers of woven polishers are broken, local fractures and cuts appear on the polisher surface. Destruction of polyethylene film and composite polishers on the basis of cellulose occurs in a similar. The highest resistance was shown by unwoven polishers with pile structure: SK-8 artificial leather, Microlap polishers (Remet) and VelTex polishers (Buehler). Even after two hours of operation steady quality of polished surface was maintained, the polisher material was suitable for further processing.

Numerous studies in polishing of optic glass and some crystalline materials established [4, 13, 14] that the speed of polishing off or polishing capacity is proportional to average grain size of polishing powders. Herewith, depending on physicomechanical properties of polisher and abrasive, there are upper and lower limits of grain size of the polishing powder. Thus, upon polishing of optical glass on pitch polisher by crocus grains finer than 0.4-0.5 μm no polishing takes place as a consequence of uncompressible water layer between polisher and processed surface [4]. The upper limit of grain size is restricted mainly by the quality of obtained surface.

In this regard polishing of lithium niobate by diamond suspension with the coarseness of 0.5/0 and concentration of 1/10 was tested in basic modes. The experiments were performed with polishers of different structure, thickness, and hardness of material: 0.14 mm polyethylene film, Bologna fabric, Microlap (Remet) and VelTex (Buehler). The obtained results are summarized in Table 6 and in Figs. 4-5. With the aim of comparison, the table shows data on polishing of single crystal by diamond suspension with the coarseness of 1/0.5 and the same concentration.

Table 6. Influence of suspension coarseness on polishing properties of lithium niobate single crystal.

Suspension	Polisher material	Coarseness parameters, μm			Q, μm/min
		R_a	R_z	R_{max}	
ASM 0.5/0	0.14 mm polyethylene film	0.001	0.009	0.010	2.44
	Bologna fabric	0.006	0.019	0.028	1.66
	Microlap polisher, Remet and VelTex polisher, Buehler	0.002	0.013	0.014	0.72
ASM 1/0.5	0.14 mm polyethylene film	0.001	0.009	0.011	1.78
	Bologna fabric	0.004	0.014	0.017	3.23
	Microlap polisher, Remet and VelTex polisher, Buehler	0.001	0.006	0.006	2.91

As can be seen in the table, upon polishing on thicker and softer (pile) polishers, such as laminated fabric, Microlap or VelTex, a decrease in suspension coarseness to 0.5/0 reduces significantly the speed of polishing off with the simultaneous deterioration of surface quality. Figures 4 and 5 shows photographs of polished surfaces of a single crystal with sufficiently large flat sections separated by sinuous lines looking like overlaps (Figure 3) or spotted slope cavities (Figure 5). In [3] such relief peculiarities of surface polished by fine abrasive are attributed to plastic flow of surface layers of crystals as a consequence of high local temperatures in the contact areas. Taking into account sufficiently high thermal stability of single crystals and relatively soft polishing modes, we suppose that such explanation of this fact is not consistent and requires additional experimental studies.

Our conclusions are confirmed by the results of polishing of single crystal by the same suspension but on 0.14 mm polyethylene film with more elastic and smooth working surface, summarized in Table 6 and in Figure 6.

Contrary to the previous case, upon polishing of single crystal on polyethylene film by ASM suspension with the coarseness of 0.5/0 the polishing capacity increased by about 30% and the coarseness parameters remained nearly unchanged. Herewith, the polished surface (Figure 6) has uniform smooth relief without noticeable defects. Moderate increase in polishing capacity at lower coarseness but the same weight concentration of the suspension can be attributed to increasing in the amount of cutting grains in the contact tool/processed surface.

Aiming at final substantiation of polisher selection we performed a comparative analysis of deviation from linearity of the surface of the single crystal. Deviation of shape upon mechanical polishing on SK-8 leather is by about three times lower than upon manual polishing. This is quite natural and can be explained by two reasons. Firstly, the proposed engineering equipment for fixation of plates of single crystals provides good self-adjustability of processed surfaces by at least three points. Secondly, processing of plate pack significantly increases contact surface area, provides more uniform removal over the surface and eliminates the generation of tumbles at the plate edges as it occurs upon manual polishing. Upon polishing on VelTex polisher (Buehler), the errors of geometry are even lower, which can be attributed to more uniform structure and thickness of working layer in comparison with SK-8 leather.

IOP Conf. Series: Materials Science and Engineering **151** (2016) 012043 doi:10.1088/1757-899X/151/1/012043

Meas. Instrum.:	Perthometer S2	Lt:	0,56 mm [N=5]
Drive Unit:	GD 25	Ls:	2,5 µm
Pick-up:	MFW-250 [-4,0 %]	VB:	±25 µm
		Vt:	0,50 mm/s
		Points:	5600

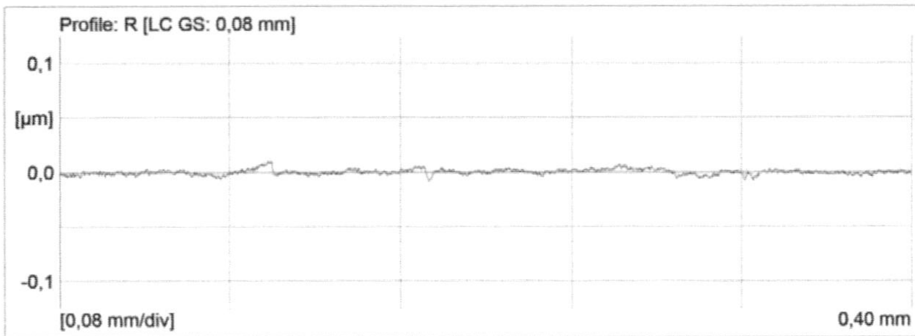

Profile: R [LC GS: 0,08 mm]	
Ra	0,002 µm
Rq	0,002 µm
Rz	0,013 µm
Rmax	0,014 µm
Rt	0,018 µm

Figure 4. Coarseness and surface pattern of single crystal after polishing by ASM 0.5/0 suspension on VelTex polisher (magnification ×50).

Figure 5. Surface pattern of single crystal after polishing by ASM 0.5/0 suspension on polisher made of Bologna fabric (magnification ×50).

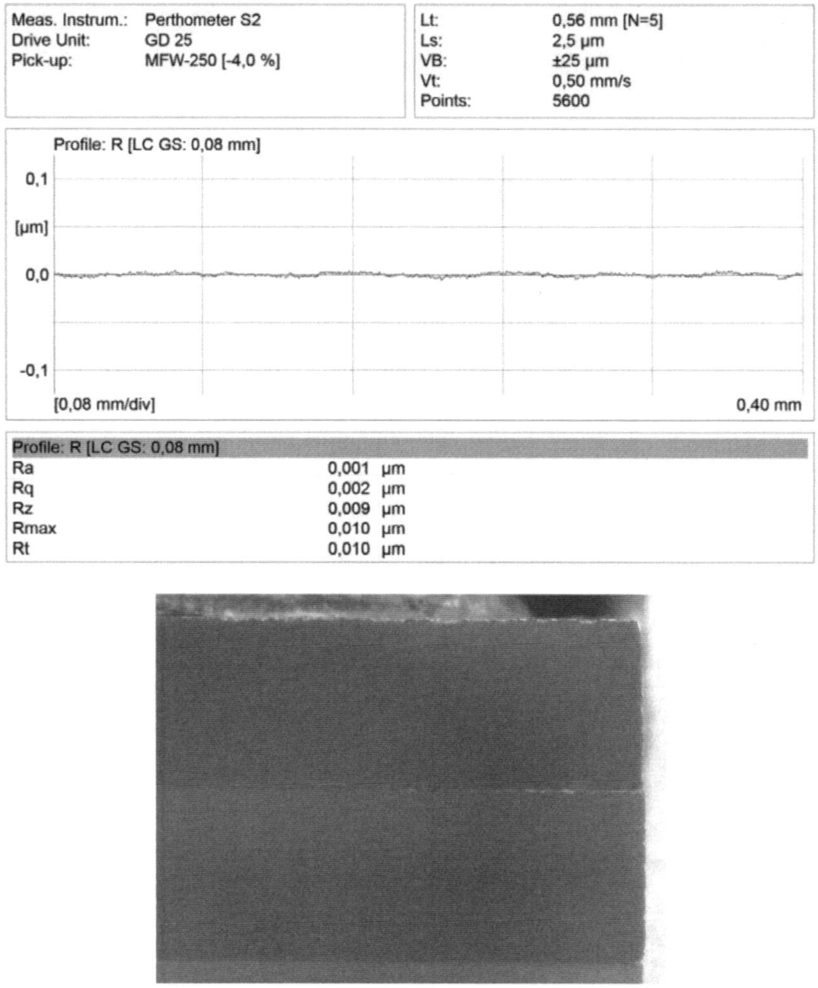

Figure 6. Coarseness and surface pattern of single crystal after polishing by ASM 0.5/0 suspension on polyethylene film (magnification ×50).

4. Conclusions

The following polishing modes of lithium niobate single crystal can be recommended for the proposed polishers, *Microlap* (*Remet*) and *VelTex* (*Buehler*): polisher rotation speed 150-250 rpm and pressure 50-100 kPa, which provide the required quality and sufficient polishing capacity. When polishing cycle time varies from 3 min to 12 min at constant suspension supply, the polishing capacity remains unchanged. Upon finishing abrasive processing of crystalline materials, in particular, plate edges of lithium niobate single crystal, in the first approximation, in order to calculate total depth of layer destroyed by finishing and minimum polishing time required for its removal the following expressions can be applied: $F \approx 12 \cdot R_{max}$ μm; $T_{un} \approx F / Q_{min}$.

Thus, it is experimentally established that upon basic pressure both for woven and unwoven polishers with the highest capacity the lower limit of grain coarseness of polishing suspension is 1/0.5 μm. Herewith, taking into account the best properties regarding resistance of polisher working surface and quality of processed surface of the single crystal, the following materials of overall considered list can be recommended for mechanized polishing: SK-8 artificial leather, Microlap polisher (Remet) and VelTex polisher (Buehler).

In addition, it has been established that mechanized polishing with the application of the proposed engineering auxiliaries in comparison with manual polishing enables an increase in accuracy of the

IOP Conf. Series: Materials Science and Engineering **151** (2016) 012043 doi:10.1088/1757-899X/151/1/012043

geometry of the processed plate surface in three times. A reserve of increase in shape accuracy and polishing quality is the use of polishers with equal thickness and structure of working surface.

Acknowledgments
This work was supported by Ministry of Education and Science of Russian Federation, Federal Project No. 9.1570.2014/K.

References

[1] Kang J and Hadfield M 2005 *Wear* **258** 2-12
[2] Matrenin S V and A I Slosman 2004 *Engineering Ceramics.* Guidebook (Tomsk: TPU).
[3] Artemov A S 2004 *Fiz. Tverd. Tela,* **46**(4) 89-97 (*in Russian*)
[4] Vinokurov V M 1967 *Study into Glass Polishing.* (Moscow: Mashinostroenie)
[5] Kim J D and Choi M S 1995 *J. Maters. Process. Technol.* **52** 368–385
[6] Khanov A M, Muratov K R and Gashev E A 2016 *Oriental journal of chemistry* **32**(1) 391-398
[7] Ardamatskii A L 1978 *Diamond Processing of Optical Items.* (Leningrad: Mashinostoenie) p 232
[8] Okatov M A (Ed) 2004 *Reference Book of Optician Engineer* (St. Petersburg: Politekhnika) p 679
[9] Aleinikov F K 1957 *Zh. Tekh. Fiz.* **27**(129) 2725
[10] Kukleva Z A 1960 *On some peculiarities of crystal polishing. Proc.*(Leningrad: GOI Procedings) **28** (157) p 98-104 (*in Russian*)
[11] Aleksandrov I E 1953 *On optimum polishing conditions of optical items.* (Moscow: Oborongiz) p 224
[12] Nekrasov V P 1989 Author's Certificate SU No. 1685686 B 24 B 1/00, 37/04. *Method of final polishing*
[13] Kumanin K G (Ed) 1962 *Shape generation of optical surfaces* (Moscow: Oborongiz) p 432
[14] Kachalov N N 1953 *Technology of grinding and polishing of glass* (Moscow, Leningrad: Publishing House of Academy of Sciences of USSR) p 382

IOP Conf. Series: Materials Science and Engineering **151** (2016) 012044 doi:10.1088/1757-899X/151/1/012044

An approach to organic field-effect transistor above-threshold drains current compact modeling that provides monotonic decrease of the output conductance with drain bias increasing

V O Turin[1], B A Rakhmatov[1], C H Kim[2] and B Iñiguez[3]

[1]Orel State University after Ivan Turgenev, Naugorskoe shosse 29, 302040 Orel, Russian Federation
[2]Gwangju Institute of Science and Technology, 61005 Gwangju, South Korea
[3]Rovira i Virgili University, 43840 Tarragona, Spain

Corresponding author e-mail address: voturin@ostu.ru

Abstract. For compact modeling of the organic field-effect transistor above-threshold drain current, we suggest using an approach that was recently proposed for the improved compact MOSFET model with the correct accounting of the nonzero output conductance in the saturation regime. This approach ensures a monotonic decrease of transistor output conductance from its maximum value in the linear regime to the minimum value in the saturation regime.

1. Introduction

Organic electronics is a relatively young area of science and technology and is aimed at the development of electronic devices based on organic materials [1]. Organic materials have some advantages compared with conventional monocrystal, polycrystalline and amorphous materials used in electronics. Organic materials can be flexible, cheap, biodegradable and easy to apply on large areas and provide almost room temperature fabrication process of electronic devices. The disadvantages of organic materials are rapid degradation and low radiation resistance of material parameters, the low mobility of the charge carriers and relatively large geometric dimensions of devices.

A key device for organic electronics, however, as for the conventional solid-state electronics, is a field-effect transistor [2-5]. Accordingly, for the effective development of organic electronics by using modern electronic CAD, based on the SPICE simulators, the correct compact model of the organic field effect transistor (OFET) is required. At the same time, the current development of the OFET compact model repeats the same steps as the conventional compact models of MOSFET and TFT already have passed.

2. Basics of modern above-threshold OFET drain current theory

In our work, we use as starting point an OFET compact model proposed in [2] and developed on the basis of the MOSFET Level 1 compact model. At low drain bias V_{DS} drain current of OFET is linearly dependent on V_{DS}:

$$I_{lin} = g_{ch} \cdot V_{DS} \,. \tag{1}$$

The output conductance g_{ch} of the OFET at low drain bias V_{DS} is determined by the following equation:

$$g_{ch} = \frac{K\,\mu_{FET}\,|V_{GS} - V_T|}{[1 + R_C\,K\,\mu_{FET}\,|V_{GS} - V_T|]}, \tag{2}$$

where V_{GS} - the gate voltage, V_T - threshold voltage, $K=W/L\times C_i$ - parameter, related to gate capacitor geometry and the characteristics of the material, W - the width of the gate, L - gate length, $C_i = \varepsilon_{ox}\,\varepsilon_0/d_{ox}$ - is the gate insulator capacitance per unit area, ε_{ox} - dielectric constant of the gate insulator, $\varepsilon_0 = 8.85\times10^{-12}$ F·m^{-1} - electric constant (vacuum permittivity), d_{ox} - the insulator thickness, R_C - contact resistance and μ_{FET} - field effect mobility:

$$\mu_{FET} = \frac{\mu_0}{V_{aa}^{\gamma}}|V_{GS} - V_T|^{\gamma}, \tag{3}$$

where μ_0 – is the conversion mobility set to 1 cm^2/V·s; V_{aa} - is the mobility enhancement voltage, γ - is the characteristic mobility exponent.

With the increase in the drain bias, the drain current saturates. The saturation voltage for OFET is $V_{sat} = \alpha_S\,|V_{GS} - V_T|$ with the saturation modulation parameter α_s. At the same time, the saturation current is $I_{sat} = g_{ch} \cdot V_{sat}$, where g_{ch} – OFET channel conductance (2). However, for realistic devices there is no perfect saturation due to variety of effects with the Channel Length Modulation effect as the most relevant. This imperfect saturation is usually taken into account through a saturation coefficient λ that enters equation for an asymptotic value of OFET output conductance for the case of large drain bias:

$$g_{asy} = \lambda\, I_{sat} \tag{4}$$

This saturation coefficient λ has V^{-1} dimension and is identical to the inverse of the Early voltage from the theory of bipolar transistors.

In [2] the asymptotic behavior of the drain current in the saturation regime, similar to those used in MOSFET Level 1 and RPI TFT models, is proposed:

$$I_{asy} = I_{sat} + g_{asy}|V_{DS}|\;. \tag{5}$$

In [4] for the asymptotic behavior of the drain current in saturation regime is proposed to use the equation from MOSFET BSIM3/4 compact models, that significantly improves the characteristics of the model, but does not eliminate it from some of the shortcomings analyzed in [6,7]:

$$I_{asy} = I_{sat} + g_{asy}\,(|V_{DS}| - V_{sat})\,. \tag{6}$$

For above-threshold OFET drain current in [2,4] is proposed to use equation, traditional for MOSFET Level 1 and BSIM3/4 models:

$$I = \frac{I_{lin}\, I_{asy}}{[\,|I_{lin}|^m + I_{sat}{}^m]^{\frac{1}{m}}}, \tag{7}$$

where m - the exponent that determines the smoothness of the transition from the linear transistor regime to the current saturation regime.

In [6, 7] the equation (7) were analyzed with I_{asy}, defined by both equations (5) and (6). It has been shown that the output conductance doesn't decrease monotonically with increasing drain bias V_{DS}, which is a drawback of MOSFET Level 1, RPI TFT and BSIM3/4 models. In addition, in the case when I_{asy} is given by equation (6) the output conductance of the OFET at zero drain bias (obtained by taking the derivative of I with respect to V_{DS} and supposing V_{DS}=0) is not equal to postulated value g_{ch}, given by equation (2).

3. New set of equations for the above-threshold OFET drain current

In this work, for OFET above-threshold drain current calculation, we suggest to use approach, that was recently proposed in [6,7] for the improved compact MOSFET model. This ensures the correct accounting of the nonzero output conductance in the saturation regime with a monotonic decrease of OFET output conductance from its maximum value in the linear regime to the minimum value in the saturation regime.

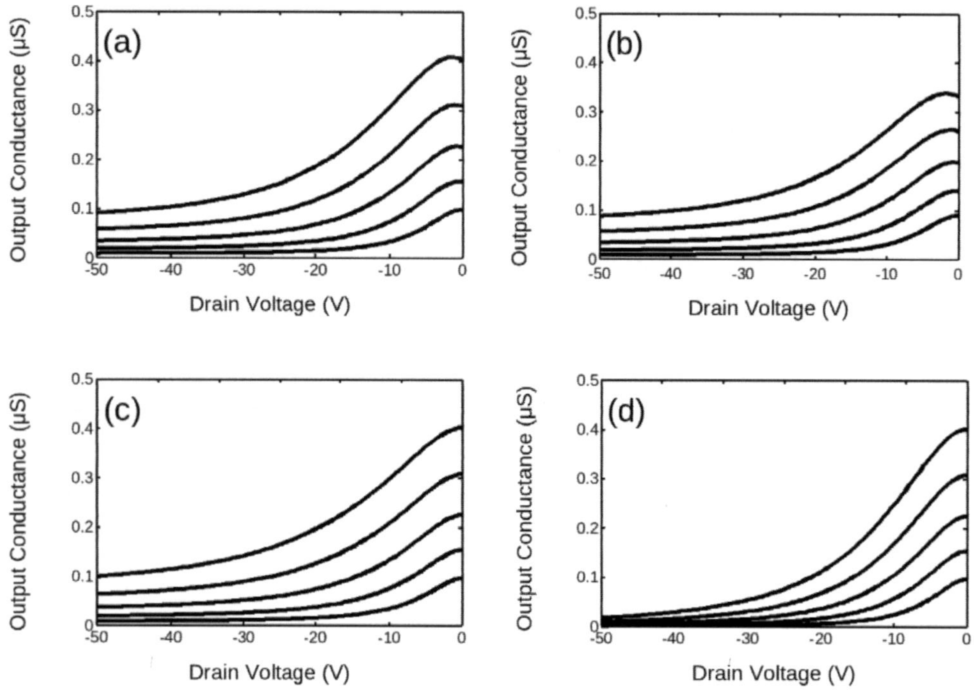

Figure 1. Output conductance-voltage characteristics for OFET calculated with different equations for drain current. In all the figures the lower curve for the gate voltage V_{GS} = -30 V, with further increments of 5 V to -50 V; (a): the equation for the drain current (7) with the asymptotic (5); (b) the equation for the drain current (7) with the asymptotic (6); (c) the drain current equations (8) and (9) with an asymptotic (6); (a), (b) and (c): $\lambda = 0.01$ V^{-1}; (d): $\lambda = 0$ V^{-1}.

At the same time, we preserve the equation for the asymptotic behavior of the drain current (6). Hence, the modified equation for the OFET drains current is:

$$I = \frac{I_{lin}\, I^*_{asy}}{\left[\, |I_{lin}|^m + {I^*_{asy}}^m \,\right]^{\frac{1}{m}}}, \tag{8}$$

where I^*_{asy} – is given by equation for the OFET drain current asymptotic (6) multiplied by a correction factor, which provides the correct asymptotic behavior of the equation (8):

$$I^*_{asy} = \frac{g_{ch}}{\left[g^m_{ch} - g^m_{asy}\right]^{\frac{1}{m}}} I_{asy}\,. \tag{9}$$

Note, that if we impose obvious condition $g_{asy}<g_{ch}$ on scope where we will use equations (8) and (9), we will ensure positive values of I^*_{asy} and $g^m_{ch} - g^m_{asy}$ that is necessary for correct use of equations (8) and (9) in numerical calculations, in respect that exponents m and $1/m$ is not necessarily even integer numbers.

Table 3. Parameters of the pentacene OFET and one's compact model [4].

γ	V_T (V)	$\mu_{FET}(V_{GS}$= -50 V) (cm^2/V·s)	α_S	R_C (kΩ)	m	λ (V^{-1})	L (um)	W (um)	C_i (nF/cm^2)
0.91	-12	0.13	0.46	24	1.8	1.2×10^{-3}	40	1000	3.3

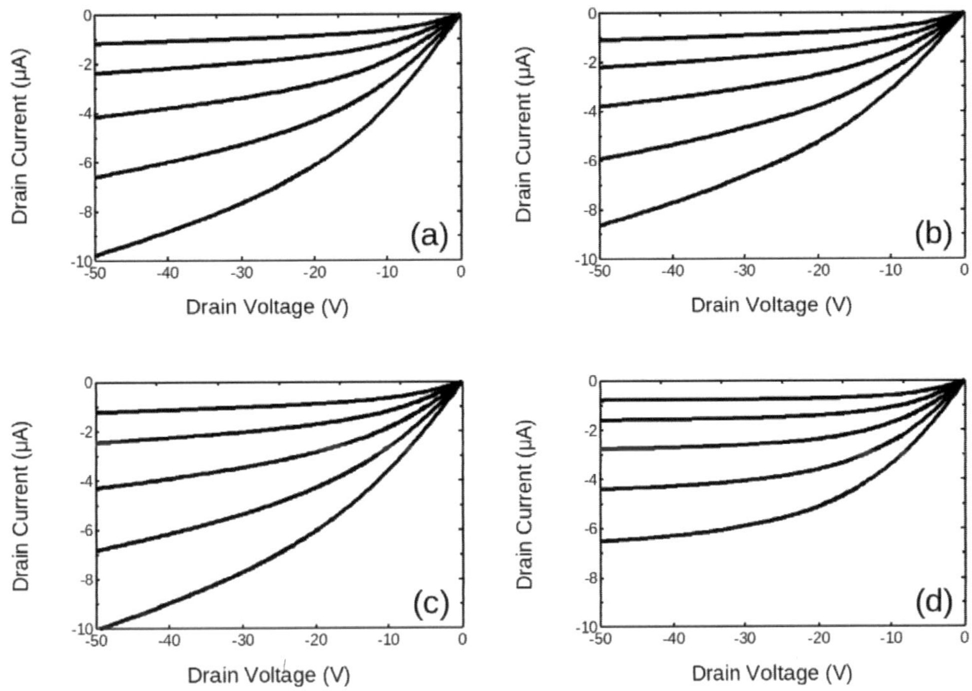

Figure 2. Current-voltage characteristics for OFET calculated with different equations for drain current. In all the figures the lower curve for the gate voltage V_{GS} = -30 V, with further increments of 5 V to -50 V; (a): the equation for the drain current (7) with the asymptotic (5); (b) the equation for the drain current (7) with the asymptotic (6); (c) the drain current equations (8) and (9) with an asymptotic (6); (a), (b) and (c): $\lambda = 0.01$ V^{-1}; (d): $\lambda = 0$ V^{-1}.

The numerical values of the compact model parameters are shown in Table 1 and are the same as extracted for pentacene OFET in [4]. By use of equation (3) for the field effect mobility with data for $\mu_{FET}(V_{GS}$= -50 V) from Table 1 we can calculate V_{aa} = 358 V, that is necessary to calculate field-effect mobility for different values of gate bias V_{GS}.

Calculations show that at $\lambda = 1.2\times10^{-3}$ V^{-1} new equations (8) - (9) do not give a significant difference from equation (7), that was used in [4] to plot OFET current-voltage and output conductance-voltage characteristics. However, in [5] sufficiently large value of $\lambda = 0.01$ V^{-1} was used, for which our calculations show a significant difference between the use of equations (8) - (9) instead of (7). New equations provide a monotonic decrease in the differential output conductance of OFET (Figure 1(c)). Equation (7) gives a significant deviation from the monotonic decrease of the OFET output conductance with the usage of both asymptotes of the OFET drain current (5) (see Figure 1(a)) and (6) (see Figure 1(b)). For comparison, Figure 1(d) shows the output conductance of OFET for the case of $\lambda = 0$ V^{-1}. On Figures 2(a), (b), (c) and (d) corresponded current-voltage characteristics are presented.

4. Conclusion

The development of the OFET compact model repeat the same steps as the conventional compact models of MOSFET and TFT already have passed. Hence, for above-threshold drain current in OFET compact model traditional equation for a smooth transition from the linear regime to saturation regime is usually used. However, in the case of accounting nonzero output conductance in saturation regime, this equation doesn't provide a monotonic decrease of the output conductance with increasing drain bias from its maximum value in the linear regime to the minimum value in the saturation regime. In this paper for OFET above-threshold drain current calculations, we offer to use a new approach that was proposed recently for the improved compact MOSFET model. This approach ensures monotonic decrease of the output conductance with an increase of the drain bias from the maximum value in a linear regime to a minimum value in the saturation regime. Our calculations show significant drawback in the usage of the traditional equation for parameters set extracted for typical OFET with a value of saturation coefficient $\lambda = 0.01\ V^{-1}$ or more. On the other hand, new equations for above-threshold drain current give a perfect monotonic decrease of output conductance from the maximum value in a linear regime to a minimum value in the saturation regime for all range of saturation coefficient λ. In further work, we plan to take into account source and drain resistances separately for a correct accounting one's effect on output characteristic both in linear and in the saturation regime follow approach developed in [8].

Acknowledgments
This work was partially supported by RFBR grant No. 12-02-97534 and by the Ministry of Education and Science of the Russian Federation grant No. 16.1117.2014/K (Goszadanie).

References
[1] Kuhto A V 2013 *Chemistry and Life – XXI century* **2** 3 (*In Russian*)
[2] Estrada M, Cerdeira A, Puigdollers J, Reséndiz L, Pallares J, Marsal L F, Voz C and Iñiguez B 2005 *Solid State Electron.* **49**(6) 1009
[3] Marinov O, Deen M J, Zschieschang U, Klauk H 2009 *IEEE T. Electron Devices* **56**(12) 2952
[4] Kim C H, Castro-Carranza A, Estrada M, Cerdeira A, Bonnassieux Y, Horowitz G and Iñiguez B 2013 *IEEE T. Electron Devices* **60**(3) 1136
[5] Kim C H, Bonnassieux Y and Horowitz G 2014 *IEEE T. Electron Devices* **61**(2) 278
[6] Turin V O, Zebrev G I, Makarov S V, Iñiguez B and Shur M S 2014 *Int. J. Numer. Model. El.* **27** 863.
[7] Turin V O, Sedov A V, Zebrev G I, Iñiguez B and Shur M S 2009 *Proc. SPIE* **7521** 75211H-1
[8] Turin V O, Zebrev G I, Makarov S V, Iñiguez B and Shur M S 2015 *Proc. Int. Conf. "Energy and Resource Saving – XXI"*, ed O V Pilipenko, A N Kachanov and Y S Stepanov (Orel: State University ESPC) p 150 (*In Russian*)

Thermal MEMS flow meter for gaseous working fluids on the basis of the hot-wire thermoanemometric sensor

T A Tsivinskaya, L G Avaeva, P V Grigoriev and S A Mileshin

Bauman Moscow State Technical University, 2-ya Baumanskaya st. 5/1, 105005 Moscow, Russian Federation

Corresponding author e-mail address: tsivinskaya_t@mail.ru

Abstract. This paper describes the main principles of constructing innovative MEMS flow meters for gaseous working fluids. MEMS flow meter contains hot-wire thermoanemometrisc sensor which response to the temperature change caused by convective working fluid heat transfer from a hot-wire. The advantages of using hot-wire thermoanemometric sensors were analyzed. The main emphasis of this work is on speed, construction simplicity and small size of the sensor. The new approach to the solution of a problem of sensor output signal relation with the working fluid temperature is presented. This approach is based on adding an extra temperature sensor and a special scheme for thermal compensation to the hot-wire sensor. The temperature scale between temperature sensors (thermistors) corresponds to the flow speed and the flow rate.

1. MEMS flow meters for gaseous working fluids analysis

The problem of precise gas flow calculation is important not only for modern industry but also for many people every day. The most well-known mechanical flow meters [1,2] are turbine and paddle-wheel meters. These flow meters need additional components when they are used in innovative embedded systems, such as "smart house". The disadvantages of mechanical flow meters are small lifetime caused by mechanical parts runout and sticking of fluid impurities to the sensor which leads to the corruption of measurement results. Moving parts of mechanical flow meters obstruct the gaseous or liquid flow creating an extra aerodynamic resistance which enlarges the measurement error.

Nowadays we can see an increasing need of modern sensors and transducers in all areas of human endeavour. The technology of silicon sensors constructing can be a solution of a problem of sensor integration into the digital signal processing system. The tendency of using silicon technology for precise sensors is also in modern digital flow meters design. The most common method for gas flow measuring [3-6] is a thermoanemometric method. This method uses convective working fluid heat transfer from hot-wire. The heat transfer is a function of gaseous working fluid speed. The advantages of the thermoanemometric sensor are high sensitivity, high speed, and construction simplicity. The use of silicon technology makes it possible to make the sensor so small that it does not affect the working fluid flow. Now let's see some innovative, constructive solutions which are used to create thermal MEMS flow meter for gaseous working fluids.

2. Constructive solutions for MEMS flow meter for gaseous working fluids creation

The main problem which occurs while gas flow is measuring with the thermoanemometric sensor is output signal relation with the working fluid temperature. Adding an extra temperature sensor and a special scheme for thermal compensation to the hot-wire thermoanemometric sensor should solve this problem [5].

In the proposed scheme temperature impact is compensated directly on silicon crystal by adding special thermistor into the Wheatstone bridge measuring electrical circuit. The additional thermistor should be placed along the gas flow directly in front of the heater. This opens up an opportunity to precisely control the heater supply voltage. Comparing the output signal with the plot shows the relation between the heater supply voltage and the thermistor temperature, which was measured while scheme testing on the test bench with a proper digital signal processing [6-8]. This allows us to detect the flow temperature change before touching the crystal and after the flow goes over the heater by measuring two thermistors resistance change. The bigger the difference between the first and the second thermistors temperature is, the higher is the flow speed and the flow rate.

Forming the measuring sensor directly on the silicon crystal provides us with higher flow meter sensitivity. However, some problems connected with crystal material characteristics occur (fig. 1). The first problem is silicon high thermal conductivity. If the heater is placed directly on the substrate, the most part of the thermal energy will expend on heating the substrate. In this case, thermistors would be heated from the silicone substrate to the temperature almost equal to the heater temperature. As a result, the temperature split between two thermistors while measuring the flow rate would be too small to get a stable output signal from the Wheatstone bridge. The second problem is silicon high sensitivity to stresses which occur while forming a conductive structure on the water surface or while placing the crystal into the flow meter body.

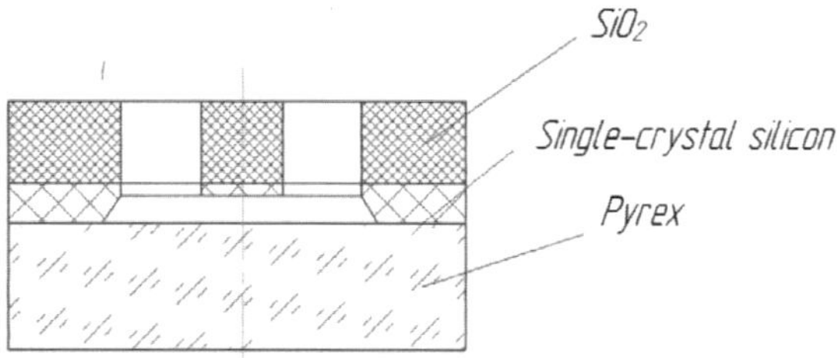

Figure 1. Flow meter sensor construction.

The proposed flow meter construction is shown in figure 1. The substrate is made of single-crystal silicon. As a solution to the problem of silicon high thermal conductivity, a 3×3 mm membrane is etched at the central part of the substrate. The thickness of the membrane is 30 μm. There is also a 1×3 mm bridge etched in the center of the membrane

A "thick" layer of SiO_2 is formed on the silicone substrate for additional heat insulation. The thickness of an insulation layer could be up to 1 mm. The measuring Wheatstone bridge scheme is formed directly on SiO_2 layer.

The problem of stresses which affect flow meter sensor output signal is multidimensional and should be solved at every stage of flow meter design. Silicon crystal is attached to the substrate made of Pyrex borosilicate glass with electrostatic welding authorship process [9], which provides an opportunity to combine different materials with minimal stresses in the contact zone. Pyrex grass not only has a coefficient of thermal expansion close to silicon but also reduces the number of dislocations in silicon crystal contact zone because of its physical and chemical properties. Reducing the number of dislocations reduces the number of stress regions in the material. Borosilicate glass also has amorph crystalline structure, so atoms have more freedom to move while energy balance re-establishing in the glass structure. This characteristic of Pyrex glass makes it possible to significantly reduce the internal stresses in the contact zone of glass and metal. The metal layer is the base of proposed flow meter construction. The silicon crystal on the Pyrex glass substrate and conversion board for flow meter sensor Wheatstone bridge scheme and signal data processor are placed directly on the metal layer. The sensor measuring scheme and the conversion board are connected after they are placed on the base layer. The Pyrex glass layer is an isolating intermediate zone which provides an opportunity to reduce residual stresses cases by the use of different materials while thermal annealing. The combination of base and intermediate isolating layer permits one to increase flow meter sensitivity and decrease the measurement error.

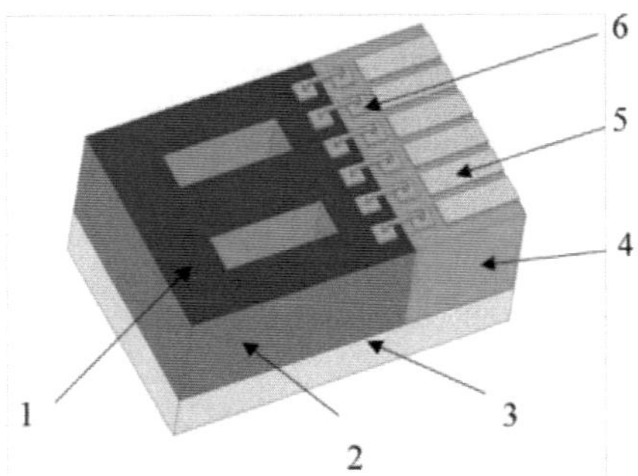

Figure 2. MEMS flow meter 3-D model. (1- flow meter sensor with Wheatstone bridge scheme, 2-Pyrex borosilicate glass, 3- kovar, 4- conversion board, 5- contact pads for signal data processor, 6 - conductor).

The flow meter base layer material is kovar. Kovar was chosen as a base layer material taking into consideration the fact that it was designed for combination with glass. Glass and kovar form strong chemical junctions in a thin contact layer after they are connected by presintering or electrostatic welding which helps to minimize dislocations appearance. In addition kovar thermal linear expansion coefficient is close to Pyrex glass coefficient value which improves thermal MEMS flow meter parameters.

Due to flow meter characteristics improvement, new materials for heater and thermistors forming are proposed. According to generalized analyses of different constructional solutions and rigid quality and reliability requirements general requirements for integrating heaters and temperature responsive thermistors were set up [6,10]. The requirements for integrating heaters are:

1. Should be formed on a thin film which coefficient of thermal resistance is not less than 0,3%/°C;
2. The heater should be thermally isolation from the substrate with non-conductive membrane with low thermal conductivity and air gap around it;
3. Non-conductive substrate the heater is placed on should have no internal stresses;
4. The heater material should be chemically stable to external agency or have special chemical protection;
5. The heater material should be temperature resistant and have low coefficient of thermal expansion;
6. Heater material coefficient of thermal expansion should be close to SiO_2 and the substrate values.

A thin pyrolytic silicon film produced by pyrolytic graphite sputter deposition meets all the requirements for integrating heater materials. Sputter deposition method doesn't change sputtering material structure and characteristics, so the thin pyrolytic silicon film has the same basic characteristics as pyrolytic graphite, including high coefficient of thermal resistance, corrosive environment, and erosion stability, high heat capacity. High heat capacity prevents inertial temperature rise and fall which leads to integrating heater stability.

MEMS flow meter measuring tool are four thermistors with 2.2 kOhm nominal resistance which form the Wheatstone bridge scheme with two measuring and two reference resistors. The two measuring resistors should be placed close to the heater so they could distinctly respond on the temperature change caused by the working fluid rate. The two reference resistors, which balance two legs of Wheatstone bridge scheme, should be placed apart from the flow and have high thermal stability. Nowadays measuring thermistors are made of platinum. However platinum is not only an expensive material but also the process of platinum thin film forming on the Ti underlay is quite complicated. It is proposed to form measuring thermistors from Ni thin film using sputter deposition. To form a thermistor with 2.2 kOhm nominal resistance a 0.3 μm Ni film is needed (Ni surface resistance is 0.2 ÷ 0.3 Ом/□)). The length of Ni thermistor should be 40 mm while 0.3 μm thermistor width. Due to the crystal length and width limits, the measuring Ni thermistor should be formed in a complex meander form. For the two reference resistors, thermal characteristics stability is more important than thermal sensitivity. It is proposed to form reference resistors from pyrolytic silicon thin film using sputter deposition method. Pyrolytic silicon film should be not more than 0.3 μm thick. The electrical scheme displacement on the crystal substrate is shown in figure 3.

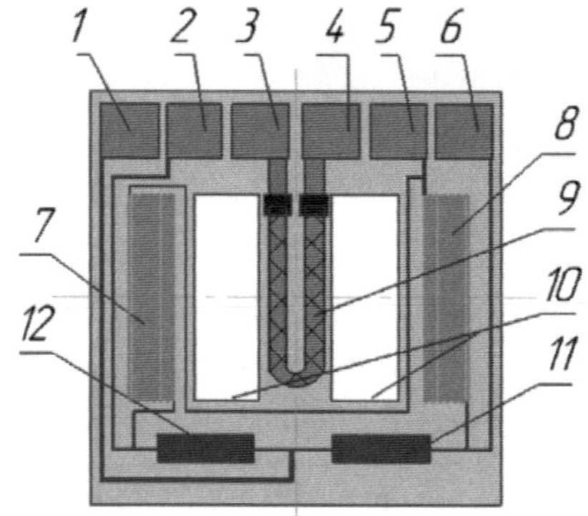

Figure 3. Termoamemometric flow meter sensor layout plot (1 – contact pad $-U_{in}$ for Wheatstone bridge; 2 – contact pad U_{out1}; 3 – contact pad $+U_{in}$ for the heater; 4 – contact pad $-U_{in}$ for the heater; 5 – contact pad $+U_{in}$ for Wheatstone bridge; 6 - contact pad U_{out2}; 7 – thermistor R_1; 8 – thermistor R_2; 9 – integrating heater; 10 – air gap; 11 – reference resistor R_3; 12 – reference resistor R_4).

Thermoamemometric flow meter for gaseous working fluids working principle is based on convective working fluid heat transfer from the heater. Thin silicon film with a high coefficient of thermal resistance is used as the heater. Heating is provided by DC current passing through the heater (silicon resistor). The heater temperature is regulated automatically by changing the heater supply voltage value according to the available data. Data acquisition is performed on a test bench to make a plot of the heater temperature and the heater supply voltage. The heater and the Wheatstone bridge scheme have different power supply lines to provide an opportunity of automatic heater temperature regulation

While gaseous working fluid flows through the resistor R_1 to the resistor R_2 (fig. 3, pos.7, 8) convective heat transfer is performed. As a result, the temperature of the thermistor R_1 becomes much smaller than the temperature of the thermistor R_2. The thermistors temperature change causes the thermistors resistance change. The difference between the resistances R_1 and R_2 disbalances the legs of Wheatstone bridge which is reflected in the potential difference between U_{out1} and U_{out2} scheme outputs (fig. 3, pos.2, 6). The potential difference increase suggests the increased working fluid flow. Volumetric flow rate of gaseous or liquid working fluids calculation is based on the relationship between flow meter sensor output signal and the temperature difference between thermistors R_2 and R_4.

3. Conclusion and recommendations

Proposed for consideration thermoanemometric flow meter sensor construction represents one unit, which elements, elements material and even the process of elements forming and manufacturing are combined to improve output signal stability and decrease the measurement error. Platinum integrating heater to silicon film heater change and sputter deposition process simplification would affect the cost of production.

The proposed flow meter sensor construction could be the base for further developments of constructions combining flow meter [2], pressure sensor [5], gas sensor [11-14] and data signal processor on single-crystal silicon crystal to get even more precise measuring results [6,8].

IOP Conf. Series: Materials Science and Engineering **151** (2016) 012045 doi:10.1088/1757-899X/151/1/012045

References

[1] Kolhare N R and Thorat P R 2013 *Int. J. Engineering Research & Technology* **2** 1-4.
[2] Vlasov A I, Zhuravleva L V, Sergeeva N A and Tsivinskaya T A 2015 *Current problems of energy saving and energy efficiency in technical systems (Tambov)* (Publ. Perhin R.V.) p 63-65
[3] Santhosh K V and Roy B K 2012 *Int. J. Control and Automation.* **5** 185-196
[4] Vlasov A I and Yudin A V 2011 *Communications in Computer and Information Science* **156** 180-192
[5] Borzov A B, Likhoedenko K P, Tsygankov V U, Vlasov A I, Tinyakov U N, Andreev K A and Tsivinskaya T A 2012 *Science and Education of the Bauman MSTU* **12**(11) 21
[6] Andreev K A, Vlasov A I and Shakhnov V A 2014 *Sensors and systems* **10** 54-57
[7] Vlasov A I, Steshenko V B, Nesterov V A and Myslovskiy E V 2012 *Science and Education of the Bauman MSTU* **5** 37.
[8] Nesterov V A, Vlasov A I and Rutkevich A V 2011 *Science and Education of the Bauman MSTU* **12** 39
[9] Andreev K A, Mileshin S A and Tsivinskaya T A 2013 *Sensors and systems* **165** 45-49
[10] Vlasov A I and Ivanov A M 2011 *Science and Education of the Bauman MSTU* **11** 34
[11] Veselov D S 2015 *Physics Procedia* **72** 500-503
[12] Veselov D S and Voronov Yu A 2015 *Physics Procedia* **71** 423-427
[13] Veselov D S and Voronov Yu A 2016 *J. Phys. Conf.* **747** 012022
[14] Veselov D S, Voronov Yu A and Vanyukhin K D 2015 *Physics Procedia* **72** 495-499

Attractor property for terminal manifold in control synthesis problem

A Diveev[1,2] and E Sofronova[2]

[1]Federal Research Centre "Computer Science and Control" Russian Academy of Sciences, Vavilova str. 44/2, 119333 Moscow, Russian Federation
[2] Peoples' Friendship University of Russia, Miklukho-Maklaya str. 6, 117198 Moscow, Russian Federation

Corresponding author e-mail address: sofronova_ea@mail.ru

Abstract. General statement of control synthesis problem is considered. The problem is to find control as a function depending on the state space vector. After substituting to the system of ordinary differential equations, the function should provide the property of attractor to a terminal manifold. Symbolic regression methods for numerical solution of the problem are proposed.A numerical criterion of the attractor is set. Several examples of problem solution by the network operator method are presented.

1. Introduction

Control synthesis the problem is to find a control function. If we substitute the control function in the system of ordinary differential equations, we obtain a system without control and with new properties [1]. A particular solution of the system will reach to terminal state and will provide the optimal value of quality criterion. The difficulty of finding control function is that the same control function must provide the optimal values of the quality criterion for all the initial conditions from some state space domain [2]. Control function should depend on the constraints, quality criterion, and terminal manifold. Searching control function for creating of the attractor from the terminal manifold, we call the general synthesis problem. Having solved synthesis problem the terminal manifold becomes attractor that provides new physical properties of the control object.

Symbolic regression allows us to create numerical algorithms for functions search on the set of their descriptions. Symbolic regression methods have appeared recently, and they differ in coding. Genetic programming codes formulas by a string of symbols [3-8]. Grammatical evolution codes mathematical expressions in Backus-Naur form [9-12]. Analytical programming presents formulas in the sequence of integer numbers [13-16]. For control problems, it is proposed to use the network operator method [17]. The method applies the principle of small variations of the basic solution [18] and code expressions in integer matrices. The network operator method searches for the best solution to a set of small variations of the basic solution. There are many other methods of symbolic regression, and all of them describe complex functions as a superposition of elementary functions. We study properties of superposition and define properties of complex functions. It is important to find out what basic functions are necessary to get certain properties of the complex function.

To solve the control synthesis problem we use a symbolic regression method. For this purpose, we replace the domain of initial conditions by the set of points. For terminal manifold of non-zero dimension, we set the requirement that the particular solutions of differential equations should reach several different points of the terminal manifold.

We used the network operator method to solve the general synthesis control problem for a nonlinear system and different terminal manifolds.

2. Theory of superposition

Consider a set of functions with no more than M arguments:

$$\begin{aligned} \mathrm{F} = \{ & f_{1,1}(z),\ldots,f_{1,n_1}(z), f_{2,1}(z_1,z_2),\ldots,f_{2,n_2}(z_1,z_2),\ldots \\ & \ldots,f_{M,1}(z_1,\ldots,z_M),\ldots,f_{M,n_M}(z_1,\ldots,z_M)\}. \end{aligned} \quad (1)$$

To construct a mathematical expression from elementary functions (1) we should poses together with the set (1) and a set of arguments or functions without arguments:

$$\mathrm{F}_0 = \{q_1,\ldots,q_{M_0}\} \quad (2)$$

Define a rule of description for superposition. We write a mathematical expression with elements of sets (1) and (2). Then we delete all brackets and insert operator icon "$\circ$" between elements. For example:

$$\begin{aligned} & f_{2,j_1}(f_{1,j_2}(\ldots f_{1,j_k}(q_\alpha), f_{1,j_{k+1}}(\ldots(f_{1,j_{k+l}}(q_\beta\underbrace{)\ldots)}_{k+l} = \\ & = f_{2,j_1}\circ f_{1,j_2}\circ\ldots f_{1,j_k}\circ q_\alpha\circ f_{1,j_{k+1}}\circ\ldots f_{1,j_{k+l}}\circ q_\beta. \end{aligned} \quad (3)$$

We call sets (1), (2) as a set of basic functions.

Definition 1. Sets F and F_0 have reach ability property if for any given limited nonzero number a, $|a|<\infty$, $a\neq 0$ and small positive value ε set F_0 includes a nonzero element $q\neq 0$ and we can make superposition from F as:

$$|a - A_1\circ\ldots\circ A_l\circ q| < \varepsilon, \quad (4)$$

where $A_i \in \mathrm{F}\cup\{q\}$, $i=1,\ldots,l$.

Definition 2. Sets (1), (2) have smooth property if for given real polynomial of power n $a\neq 0$ and small positive value ε set $\mathrm{F}_0=\{q,x\}$, $q\neq 0$, x is argument of polynomial, and we can make superposition from F as:

$$|x^n + a_{n-1}x^{n-1}\ldots + a_1 x + a_0 - A_2\circ\ldots\circ A_l\circ x| < \varepsilon. \quad (5)$$

Definition 3. Sets (1), (2) have piecewise continuous property if for any piecewise continuous function $g(x)\in KC$ given in interval $[x_0,x_n)$ with points of discontinuity $x_1,\ldots,x_{n-1}$ and it is presented in the form of polynomial between points discontinuity:

$$r_i(x) = \sum_{k_i=0}^{m_i} a_{k_i,i}x^{k_i}, \; x\in[x_i,x_{i+1}), \; i=0,\ldots,n-1, \quad (6)$$

and small positive value ε set $\mathrm{F}_0=\{q,x\}$, $q\neq 0$, x is an argument of polynomial, and we can make superposition from F as:

$$|r(x) - A_1\circ\ldots\circ A_l\circ x| < \varepsilon, \; \forall i\in\{0,\ldots,n-1\}, \quad (7)$$

where $A_i\in \mathrm{F}\cup\mathrm{F}_0$, $i=1,\ldots,l$,

$$r(x) = \begin{cases} r_0(x), \text{ if } x\in[x_0,x_1) \\ \vdots \\ r_{n-1}(x), \text{ if } x\in[x_{n-1},x_n) \end{cases} \quad (8)$$

We can easy show that the set of functions:

$$\begin{aligned} \mathrm{F} = \{ & f_{1,1}(z) = z, f_{1,2}(z) = z/2, f_{1,3}(z) = 2z, f_{1,4}(z) = -z, f_{1,5}(z) = \vartheta(z), \\ & f_{2,1}(z_1, z_2) = z_1 + z_2, f_{2,2}(z_1, z_2) = z_1 z_2 \}, \end{aligned} \tag{9}$$

where $\vartheta(z)$ is Heaviside function:

$$\vartheta(z) = \begin{cases} -1, \text{ if } z < 1 \\ 1 \text{ , otherwise} \end{cases}, \tag{10}$$

have reachability, smooth, and piecewise continuous properties.

3. Problem statement

Given mathematical model of control object as a system of ordinary differential equations:

$$\dot{\mathbf{x}} = \mathbf{f}(\mathbf{x}, \mathbf{u}), \tag{11}$$

where $\mathbf{x} \in \mathbb{R}^n$, $\mathbf{u} \in \mathrm{U} \subseteq \mathbb{R}^m, m \le n$, U is a compact limited set, $\mathbf{x} = [x_1 \; \dots \; x_n]^T$, $\mathbf{f}(\mathbf{x}, \mathbf{u}) = [f_1(\mathbf{x}, \mathbf{u}) \; \dots \; f_n(\mathbf{x}, \mathbf{u})]^T$, $\mathbf{u} = [u_1 \; \dots \; u_m]^T$.

Given domain of initial conditions:

$$\mathrm{X}_0 \subseteq \mathbb{R}^n, \tag{12}$$

where $\forall \mathbf{x} \in \mathrm{X}_0$, $\| \mathbf{x} \| \le \alpha < \infty$.

Given terminal manifold of *n-r* dimension:

$$\varphi_i(\mathbf{x}) = 0, \; i = 1, \dots, r, \tag{13}$$

where $r \le n$. If $r = n$ we have a manifold of zero dimension or a point in the space $\mathbb{R}^n$

Given quality criterion:

$$J = \int_0^{t_f} f_0(\mathbf{x}, \mathbf{u}) dt, \tag{14}$$

where $t_f = t$, if $t \le t^+$ and $\| \varphi(\mathbf{x}(t)) \| < \varepsilon$, $t^+ \le \beta < \infty$, $\varphi(\mathbf{x}) = [\varphi_1(\mathbf{x}) \; \dots \; \varphi_r(\mathbf{x})]^T$, ε is a small positive value.

It's necessary to find control function in the form:

$$\mathbf{u} = \mathbf{h}(\mathbf{x}), \tag{15}$$

where $\mathbf{h}(\mathbf{x}) : \mathbb{R}^n \to \mathbb{R}^m$, $\mathbf{h}(\mathbf{x}) \in \mathrm{U}$, $\forall \mathbf{x} \in \mathbb{R}^n$.

For any initial conditions $\forall \mathbf{x}(0) = \mathbf{x}^0 \in \mathrm{X}_0$ particular solution $\mathbf{x}(t, \mathbf{x}^0)$ of differential equations:

$$\dot{\mathbf{x}} = \mathbf{f}(\mathbf{x}, \mathbf{h}(\mathbf{x})) \tag{16}$$

for the time $t_f \le t^+$ reaches the terminal manifold (13) $\| \varphi(\mathbf{x}(t_f, \mathbf{x}^0)) \| < \varepsilon$ and provides the minimal value of quality criterion (14):

$$\int_0^{t_f} f_0(\mathbf{x}(t, \mathbf{x}^0), \mathbf{h}(\mathbf{x}(t, \mathbf{x}^0))) dt = \min_{\tilde{\mathbf{u}}(\cdot) \in \mathrm{U}} \int_0^{\tilde{t}_f} f_0(\tilde{\mathbf{x}}(t, \mathbf{x}^0), \tilde{\mathbf{u}}(t)) dt, \tag{17}$$

where $\tilde{\mathbf{u}}(\cdot)$ is any admissible control, $\tilde{\mathbf{u}}(\cdot)=\mathbf{u}(t)$, $t\in[0;\tilde{t}_f]$, which provides for a particular solution $\tilde{\mathbf{x}}(t,\mathbf{x}^0)$ of differential equations $\dot{\mathbf{x}}=\mathbf{f}(\mathbf{x},\tilde{\mathbf{u}}(t))$ from the same initial condition $\forall\mathbf{x}(0)=\mathbf{x}^0\in\mathrm{X}_0$ for a limited time $\tilde{t}_f\le t^+$ to reach the terminal manifold (13), $\|\varphi(\tilde{\mathbf{x}}(t_f,\mathbf{x}^0))\|<\varepsilon$.

Equations (11) - (17) correspond to the synthesis problem of control. To obtain general synthesis problem of control, we should find such control function to make terminal manifold (13) as an attractor. If $\exists t_f\le t^+$ and $\|\varphi(\mathbf{x}(t_f,\mathbf{x}^0))\|<\varepsilon$, then $\|\varphi(\mathbf{x}(t,\mathbf{x}^0))\|<\varepsilon$, $\forall t>t_f$, and if in (13) $r<n$, $\forall t'>t_f$, $\exists\delta>0$ and $\Delta t>0$ that:

$$\|\varphi(\mathbf{x}(t',\mathbf{x}^0))-\varphi(\mathbf{x}(t'+\Delta t,\mathbf{x}^0))\|>\delta. \quad (18)$$

The main difficulty solving general synthesis problem of control (11) - (18) is a domain (12) of the initial conditions. The problem is to check the possible solution of the problem for all initial conditions from the domain.

We replace the domain of initial conditions by the set of points of initial conditions:

$$\bar{\mathrm{X}}_0=\{\mathbf{x}^{0,1},\ldots,\mathbf{x}^{0,K}\}, \quad (19)$$

where $\mathbf{x}^{0,i}\in\mathrm{X}_0\subseteq\mathbb{R}^n$, $i=1,\ldots,K$.

We change quality criterion (14):

$$\bar{J}=\sum_{i=1}^{M}\left(\int_0^{t_f}f_0(\mathbf{x},\mathbf{u})dt\right)_i, \quad (20)$$

where low index at brackets $(A)_i$ indicates that the expression in parentheses A identified with the initial conditions $\mathbf{x}^{0,i}$, $1\le i\le K$.When calculating the criterion (20)should take into account different time to reach the terminal manifold (13). In the case of not getting on of the terminal manifold (13) during the time t^+ we set $t_f=t^+$.

We define the condition of achievement of the terminal manifold (14) for the numerical synthesis as an additional criterion.We introduce a penalty for failure to comply with conditions (18). As a result, we obtain the following quality criteria:

$$\bar{J}_1=\sum_{i=1}^{M}\left(\int_0^{t_f}f_0(\mathbf{x},\mathbf{u})dt\right)_i+D\alpha_1\to\min, \quad (21)$$

$$\bar{J}_2=\sum_{i=1}^{M}\left(\|\varphi(\mathbf{x}(t,\mathbf{x}^{0,i}))\|\right)_i+D\alpha_2\to\min, \quad (22)$$

where

$$t_f=\begin{cases}t, \text{ if } t\le t^+ \text{ and } \|\varphi(\mathbf{x}(t))\|<\varepsilon\\ t^+, \text{ otherwise}\end{cases}, \quad (23)$$

$$D=\begin{cases}L-|\mathrm{G}|, \text{ if } L>|\mathrm{G}|\\ 0, \text{ otherwise}\end{cases}, \quad (24)$$

where L is a predetermined number of different points on the terminal manifold, α_1, α_2 are positive penalty values, $\mathrm{G}=\{t_1,\ldots,t_d\}$, $d\le L$, $\forall t_p\in\mathrm{G}$, and $\forall t_q\in\mathrm{G}$ conditions are fulfilled $t_p\ge t_f$,

$t_q \geq t_f$, $t_p \neq t_q$, $\| \varphi(\mathbf{x}(t_p, \mathbf{x}^{0,i})) \| < \varepsilon$, $\| \varphi(\mathbf{x}(t_q, \mathbf{x}^{0,i})) \| < \varepsilon$, and $\| \varphi(\mathbf{x}(t_p, \mathbf{x}^{0,i})) - \varphi(\mathbf{x}(t_q, \mathbf{x}^{0,i})) \| > \delta$.
From here if $t_f < t^+$, then $t_1 = t_f$, and if $t_f = t^+$, then $G = \varnothing$, $D = L$.

4. Examples

Consider system of nonlinear equations with two controls:

$$\dot{x}_1 = x_2 + u_1,$$
$$\dot{x}_2 = -x_1 - x_1^3 + u_2,$$

where $-1 \leq u_i \leq 1$, $i = 1,2$.

Given initial conditions:

$$\overline{\mathrm{X}}_0 = \{[-1.1\ -1.1]^T, [-0.4\ -1.1]^T, [0.3\ -1.1]^T, [1\ -1.1]^T, [-1.1\ -0.4]^T, [-0.4\ -0.4]^T, [0.3\ -0.4]^T, [1\ -0.4]^T, [-1.1\ 0.3]^T, [-0.4\ 0.3]^T, [0.3\ 0.3]^T, [1\ 0.3]^T, [-1.1\ 1]^T, [-0.4\ 1]^T, [0.3\ 1]^T, [1\ 1]^T\}.$$

4.1. Example 1

Given a terminal manifold:

$$x_2 - 2x_1 = 0.$$

Defined functionals:

$$\tilde{J}_1 = D\alpha \to \min,$$
$$\tilde{J}_2 = |x_2(t^+) - 2x_1(t^+)| + D\alpha \to \min,$$

where $t^+ = 8$ s., $\alpha = 0.25$, D is defined from (24) with $\varepsilon = 0.01$, $\delta = 0.25$, $L = 4$.

We have obtained the following control function by the network operator method

$$u_i = \begin{cases} \operatorname{sgn}(u_i), \text{ if } |u_i| > 1 \\ \tilde{u}_i, \text{ otherwise} \end{cases}, \ i = 1,2, \tag{25}$$

where

$$\tilde{u}_1 = -q_1 x_1 + \operatorname{sgn}(-x_1)\ln(|q_1 x_1| + 1) - q_2 x_2 + \operatorname{sgn}(x_2)(\exp|x_2| - 1),$$
$$\tilde{u}_2 = \operatorname{sgn}(\tilde{u}_1)\ln(|\tilde{u}_1| + 1), \tag{26}$$

$q_1 = 2.425354$, $q_2 = 1.000305$.

Plots of particular solutions for extreme initial conditions $\mathbf{x}^{0,1} = [-1.1\ -1.1]^T$, $\mathbf{x}^{0,4} = [1\ -1.1]^T$, $\mathbf{x}^{0,13} = [-1.1\ 1]^T$, $\mathbf{x}^{0,16} = [1\ 1]^T$ are represented in Fig. 1

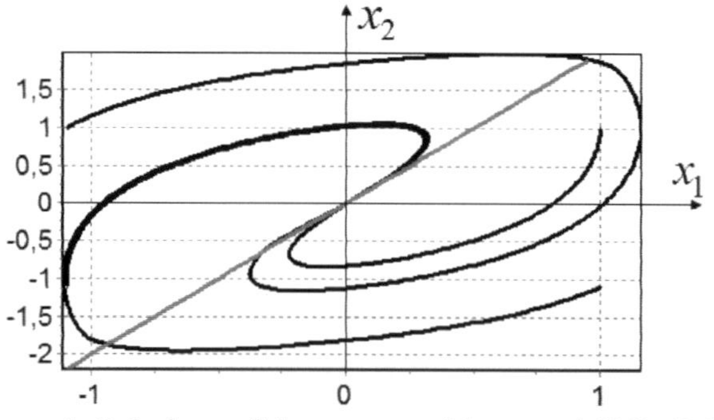

Figure 1. Solutions of the system with control (25), (26).

4.2. Example 2

Consider a terminal manifold:

$$x_1^2 + x_2^2 - 0.25 = 0.$$

We use functionals:

$$\bar{J}_1 = \int_0^{t^+} | x_1^2(t) + x_2^2(t) - 0.25 | dt + D\alpha \to \min,$$
$$\bar{J}_2 = | x_1^2(t^+) + x_2^2(t^+) - 0.25 | + D\alpha \to \min,$$

and values of parameters are taken from example 1.

We have obtained the following control for equation (25):

$$\tilde{u}_1 = \frac{\exp(q_2 - A) - \exp(q_2)}{1 + \exp(-A)} (\mu(2B^3) + 8B^9),$$
$$\tilde{u}_2 = \vartheta\left(\frac{\exp(q_2) - \exp(q_2 - A)}{1 + \exp(-A)} \right) + \tilde{u}_1 - \tilde{u}_1^3 + \mu(2B^3) + 8B^9, \quad (27)$$

where $q_1 = 2.221802$, $q_2 = 0.370178$,

$$A = \sqrt[3]{\mu(-q_2 x_2)} + \frac{1 - \exp(q_2 x_2)}{1 + \exp(q_2 x_2)} + \ln | x_2 |,$$

$$B = C \frac{\exp(q_2) - \exp(q_2 - A)}{1 + \exp(-A)}$$

$$C = \frac{1 - \exp(-F)}{1 + \exp(-F)} \operatorname{sgn}(-q_1 x_1 + (q_1 x_1)^3) \exp(| -q_1 x_1 + (q_1 x_1)^3 | - 1),$$

$$F = \frac{1 - \exp(q_1 x_1 - (q_1 x_1)^3)}{1 + \exp(q_1 x_1 - (q_1 x_1)^3)} \exp(| q_1 x_1 | - 1)$$

$$\mu(z) = \begin{cases} \operatorname{sgn}(z), & \text{if } |z| > 1 \\ z, & \text{otherwise} \end{cases},$$

$\vartheta(z)$ is a Heaviside function(10).

Plots in Fig. 2 show solutions of the system with control (25), (27) for extreme initial conditions.

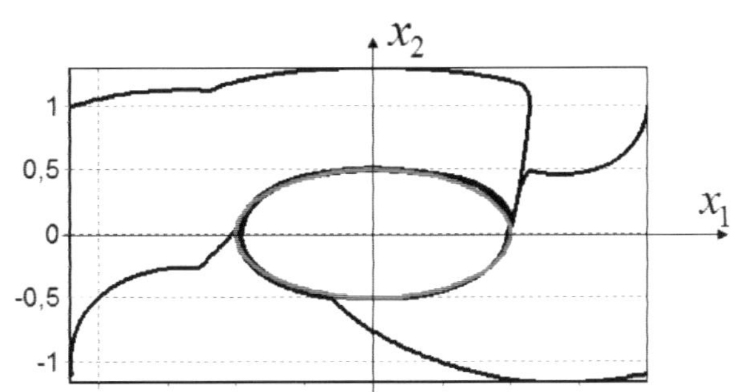

Figure 2. Solutions of system with control (25), (27).

4.3. Example 3

Consider terminal manifold:

$$-x_1 + x_1^3 + x_2 = 0$$

and functionals:

$$\bar{J}_1 = \int_0^{t^+} |-x_1(t) + x_1^3(t) + x_2(t)| dt + D\alpha \to \min,$$
$$\bar{J}_1 = \int_0^{t^+} |-x_1(t) + x_1^3(t) + x_2(t)| dt + D\alpha \to \min.$$

For equation (25) we have the following control:

$$\begin{aligned} \tilde{u}_1 &= \mu(G) + \sqrt[3]{C} - F + \operatorname{sgn}(C)(\exp(|C|) - 1), \\ \tilde{u}_2 &= \operatorname{sgn}(\tilde{u}_1)\ln(|\tilde{u}_1| + 1) - A + G - G^3 - B^3, \end{aligned} \quad (28)$$

where $A = \mu(\operatorname{sgn}(-x_1)\ln(|-q_1 x_1| + 1))$, $B = \sqrt[3]{\mu(-q_2 x_2) + q_2 x_2} + \operatorname{sgn}(x_2)\ln(|x_2| + 1)$, $G = \sqrt[3]{F} + \sqrt[3]{C} - B + A^3$, $F = \operatorname{sgn}(C)\ln(|C| + 1) + B^3$, $q_1 = 2.105469$, $q_2 = 1.405334$.

Fig. 3 shows particular solutions of the system with control (25), (28) for extreme initial conditions.

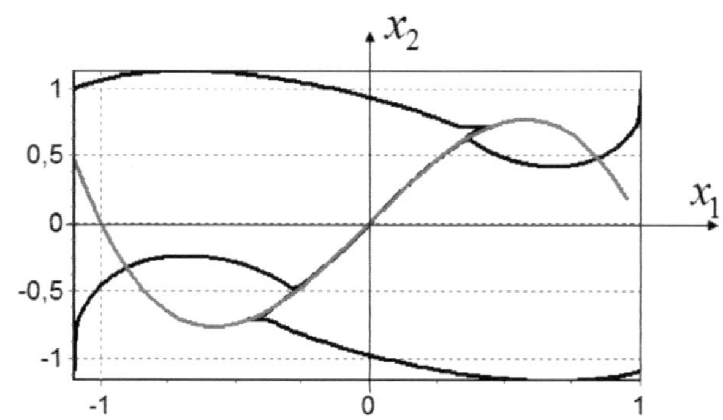

Figure 3. Solutions of the system with control (25), (28) for extreme initial conditions.

4.4. Example 4

Consider terminal manifold:

$$-x_1 + x_2 - x_2^3 = 0$$

and functionals:

$$\bar{J}_1 = \int_0^{t^+} |x_1(t) + x_2(t) - x_2^3(t)| dt + D\alpha \to \min,$$
$$\bar{J}_2 = |-x_1(t^+) + x_2(t^+) - x_2^3(t^+)| + D\alpha \to \min.$$

The network operator method found the following solution:

$$\begin{aligned} \tilde{u}_1 &= A + B + \mu(\operatorname{sgn}(x_2)(\exp|q_2 x_2| - 1)) + x_2 + \frac{1 - \exp(-x_2)}{1 + \exp(-x_2)}, \\ \tilde{u}_2 &= \frac{1 - \exp(\tilde{u}_1)}{1 + \exp(\tilde{u}_1)} + \tilde{u}_1 - \tilde{u}_1^3, \end{aligned} \quad (29)$$

where $q_1 = 2.556152$, $q_1 = 0.319153$, $A = \mu(1 + q_1 x_1 - (q_1 x_1)^3 + \operatorname{sgn}(x_1)\sqrt{|x_1|}) - q_1 x_1$,

$B = C - C^3 + (1-\exp(-x_2))(1+\exp(-x_2))^{-1}$, $C = \mathrm{sgn}(x_2)(\exp|q_2 x_2| - 1)$.

Plots in Fig. 4 show results of simulation of the system with control (25), (29) for extreme initial conditions.

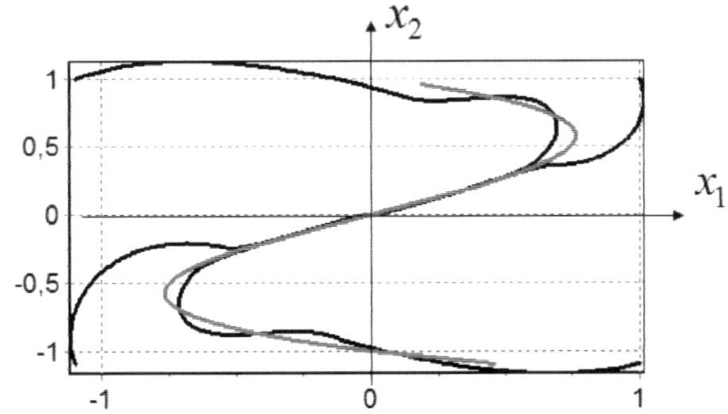

Figure 4. Solutions of the system with control (25), (29) for extreme initial conditions.

5. Conclusion

We formulated the general control synthesis problem. A solution of this problem makes terminal manifold an attractor. We offered to use symbolic regression to the numerical solution of the problem and specified criteria for numerical synthesis. Symbolic regression finds mathematical expressions as a superposition of elementary functions. Some tasks of synthesis for the nonlinear system have been solved by the network operator method. All examples show how terminal manifolds become attractors by solving general control synthesis problem.

References

[1] Lee E and Marcus L 1986 *Foundations of optimal control theory* (Krieger Pub Co) p 586
[2] Grimm G, Messina M, Tuna S and Teel A 2005 *IEEE T. Automat. Contr.* **50** 546-558
[3] Koza J 1992 *Genetic Programming: On the programming of computers by means of natural selection* (Cambridge, Massachusetts, London, MA: MIT Press) p 819
[4] Koza J, Keane M, Yu J, Bennett F, Mydlowec W and Stiffelman O 1999 *Proc. 38th Conf. on Decision and Control (Phoenix)*(IEEE) p 5292-5300
[5] Cpalka K, Lapa K and Przybyl A 2015 *Information technology and control* **44** 433-442
[6] Bourmistrova A and Khantsis S 2007 *Proc. Congress on Evolutionary Computation (Singapore)(IEEE)* p 1993–2000
[7] Imae J, Kikuchi Y, Ohtsuki N, Kobayashi T and Zhai G 2004 *Proc. 43rd Conf. on Decision and Control (Bahamas)*(IEEE) p 2734-2739
[8] Diveev A, Ibadulla S, Konyrbaev N and Shmalko E 2015 *IFAC-PapersOnLine* **48**(19) 106-111
[9] O'Neill M and Ryan C 2001 *IEEE T. Evolut. Comput.* **5** 349-358
[10] O'NeillM, RyanC, KeijzerM and CattolicoM 2003 *Genetic Programming and Evolvable Machines* **4**(1) 67-93
[11] Ortega A, De la Cruz M and Alfonseca M 2007 *IEEE T. Evolut. Comput.* **11** 77-90
[12] Diveev A, Kazaryan D, and Sofronova E 2013 *Proc. 21st Mediterranean Conf. on Control & Automation (Crete)* (IEEE) p 1148-1155
[13] Zelinka I, Davendra D, Senkerik R, Jasek R and Oplatkova Z 2011 *Evolutionary algorithm by edited Eisuke Kita* (InTech) *Chapter 9* p 149-176
[14] Zelinka I 2002 *Proc. 8th Int. Conf. on Soft Computing (Brno)* p 93-101
[15] Zelinka I, Oplatkova Z and Nolle L 2005 *IJSST* **6**(9) p 44-56
[16] Diveev A, Ibadulla S, Konyrbaev N and Shmalko E 2015 *IFAC-PapersOnLine* **48**(19) 75-80
[17] Diveev A and Sofronova E 2009 *Proc 7th Int. Conf. on Control and Automation (Christchurch)(IEEE)* p 701-708
[18] Diveev A 2015 *IFAC-PapersOnLine* **48**(25) p 28-33

Institute of Physics
Dirac House, Temple Back
Bristol BS1 6BE UK

ISSN: 1757-8981
ISBN 978-1-5108-3272-5
9 781510 832725

CURRAN ASSOCIATES INC.
proceedings.com

8th Joint European Magnetic Symposia (JEMS 2016)

Journal of Physics: Conference Series Volume 903

Glasgow, United Kingdom
21 – 26 August 2016

IOP
Institute of Physics